Lecture Notes in Mathematics 1848

Editors:
J.-M. Morel, Cachan
F. Takens, Groningen
B. Teissier, Paris

Subseries:
Fondazione C.I.M.E., Firenze
Adviser: Pietro Zecca

M. Abate   J. E. Fornaess   X. Huang
J.-P. Rosay   A. Tumanov

Real Methods in Complex and CR Geometry

Lectures given at the
C.I.M.E. Summer School
held in Martina Franca, Italy,
June 30 – July 6, 2002

Editors: D. Zaitsev
         G. Zampieri

Fondazione
C.I.M.E.

Editors and Authors

Marco Abate
Department of Mathematics
University of Pisa
via Buonarroti 2
56127 Pisa Italy

e-mail: abate@dm.unipi.it

John Erik Fornaess
Department of Mathematics
University of Michigan
East Hall, Ann Arbor
MI 48109, U.S.A.

e-mail: fornaess@umich.edu

Xiaojun Huang
Department of Mathematics
Rutgers University
New Brunswick
N.J. 08903, U.S.A.

e-mail: huangx@math.rutgers.edu

Jean-Pierre Rosay
Department of Mathematics
University of Wisconsin
Madison, WI 53706-1388, USA

e-mail: jrosay@math.wisc.edu

Alexander Tumanov
Department of Mathematics
University of Illinois
1409 W. Green Street
Urbana, IL 61801, U.S.A.

e-mail: tumanov@math.uiuc.edu

Dmitri Zaitsev
School of Mathematics
Trinity College
University of Dublin
Dublin 2, Ireland

e-mail: zaitsev@maths.tcd.ie

Giuseppe Zampieri
Department of Mathematics
University of Padova
via Belzoni 7
35131 Padova, Italy

e-mail: zampieri@math.unipd.it

Library of Congress Control Number: 2004094684

Mathematics Subject Classification (2000): 32V05, 32V40, 32A40, 32H50 32VB25, 32V35

ISSN 0075-8434
ISBN 3-540-22358-4 Springer Berlin Heidelberg New York
DOI: 10.1007/b98482

Springer is a part of Springer Science+Business Media
springeronline.com
© Springer-Verlag Berlin Heidelberg 2004
Printed in Germany

Typesetting: Camera-ready TeX output by the authors

Printed on acid-free paper
41/3142/du - 5 4 3 2 1 0

Preface

The C.I.M.E. Session "Real Methods in Complex and CR Geometry" was held in Martina Franca (Taranto), Italy, from June 30 to July 6, 2002. Lecture series were given by:

M. Abate: *Angular derivatives in several complex variables*
J. E. Fornaess: *Real methods in complex dynamics*
X. Huang: *On the Chern-Moser theory and rigidity problem for holomorphic maps*
J. P. Rosay: *Theory of analytic functionals and boundary values in the sense of hyperfunctions*
A. Tumanov: *Extremal analytic discs and the geometry of CR manifolds*

These proceedings contain the expanded versions of these five courses. In their lectures the authors present at a level accessible to graduate students the current state of the art in classical fields of the geometry of complex manifolds (Complex Geometry) and their real submanifolds (CR Geometry). One of the central questions relating both Complex and CR Geometry is the behavior of holomorphic functions in complex domains and holomorphic mappings between different complex domains at their boundaries. The existence problem for boundary limits of holomorphic functions (called boundary values) is addressed in the Julia-Wolff-Caratheodory theorem and the Lindelöf principle presented in the lectures of M. Abate. A very general theory of boundary values of (not necessarily holomorphic) functions is presented in the lectures of J.-P. Rosay. The boundary values of a holomorphic function always satisfy the tangential Cauchy-Riemann (CR) equations obtained by restricting the classical CR equations from the ambient complex manifold to a real submanifold. Conversely, given a function on the boundary satisfying the tangential CR equations (a CR function), it can often be extended to a holomorphic function in a suitable domain. Extension problems for CR mappings are addressed in the lectures of A. Tumanov via the powerful method of the extremal and stationary discs. Another powerful method coming from the formal theory and inspired by the work of Chern and Moser is presented in the lectures of X. Huang addressing the existence questions for CR maps. Finally, the dynamics of holomorphic maps in several complex variables is the topic of the lectures of J. E. Fornaess linking Complex Geometry and its methods with the theory of Dynamical Systems.

We hope that these lecture notes will be useful not only to experienced readers but also to the beginners aiming to learn basic ideas and methods in these fields.

We are thankful to the authors for their beautiful lectures, all participants from Italy and abroad for their attendance and contribution and last but not least CIME for providing a charming and stimulating atmosphere during the school.

Dmitri Zaitsev and Giuseppe Zampieri

CIME's activity is supported by:

Ministero degli Affari Esteri - Direzione Generale per la Promozione e la Cooperazione - Ufficio V;
Consiglio Nazionale delle Ricerche;
E.U. under the Training and Mobility of Researchers Programme.

Contents

Local Equivalence Problems for Real Submanifolds in Complex Spaces

Introduction to a General Theory of Boundary Values

Angular Derivatives
in Several Complex Variables

Marco Abate

Dipartimento di Matematica, Università di Pisa
Via Buonarroti 2, 56127 Pisa, Italy
abate@dm.unipi.it

1 Introduction

A well-known classical result in the theory of one complex variable, due to
Fatou [Fa], says that a bounded holomorphic function f defined in the unit
disk Δ admits non-tangential limit at almost every point $\sigma \in \partial\Delta$. As satisfying
as it is from several points of view, this theorem leaves open the question of
whether the function f admits non-tangential limit at a *specific* point $\sigma_0 \in \partial\Delta$.

Of course, one needs to make some assumptions on the behavior of f near
the point σ_0; the aim is to find the weakest possible assumptions. In 1920,
Julia [Ju1] identified the right hypothesis: assuming, without loss of generality,
that the image of the bounded holomorphic function is contained in the unit
disk then Julia's assumption is

$$\liminf_{\zeta \to \sigma_0} \frac{1 - |f(\zeta)|}{1 - |\zeta|} < +\infty. \tag{1}$$

In other words, $f(\zeta)$ must go to the boundary as fast as ζ (as we shall
show, it cannot go to the boundary any faster, but it might go slower). Then
Julia proved the following

Theorem 1.1 (Julia) *Let* $f \in \mathrm{Hol}(\Delta, \Delta)$ *be a bounded holomorphic function,
and take* $\sigma \in \partial\Delta$ *such that*

$$\liminf_{\zeta \to \sigma} \frac{1 - |f(\zeta)|}{1 - |\zeta|} = \beta < +\infty$$

for some $\beta \in \mathbb{R}$. *Then* $\beta > 0$ *and* f *has non-tangential limit* $\tau \in \partial\Delta$ *at* σ.

As we shall see, the proof is just a (clever) application of Schwarz-Pick
lemma. The real breakthrough in this theory is due to Wolff [Wo] in 1926
and Carathéodory [C1] in 1929: if f satisfies 1 at σ then the derivative f'
too admits finite non-tangential limit at σ — and this limit can be computed
explicitly. More precisely:

Theorem 1.2 (Wolff-Carathéodory) *Let $f \in \mathrm{Hol}(\Delta, \Delta)$ be a bounded holomorphic function, and take $\sigma \in \partial\Delta$ such that*

$$\liminf_{\zeta \to \sigma} \frac{1 - |f(\zeta)|}{1 - |\zeta|} = \beta < +\infty$$

for some $\beta > 0$. Then both the incremental ratio

$$\frac{f(\zeta) - \tau}{\zeta - \sigma}$$

and the derivative f' have non-tangential limit $\beta\tau\bar{\sigma}$ at σ, where $\tau \in \partial\Delta$ is the non-tangential limit of f at σ.

Theorems 1.1 and 1.2 are collectively known as the *Julia - Wolff - Carathéodory theorem*. The aim of this survey is to present a possible way to generalize this theorem to bounded holomorphic functions of several complex variables.

The main point to be kept in mind here is that, as first noticed by Korányi and Stein (see, e.g., [St]) and later theorized by Krantz [Kr1], the right kind of limit to consider in studying the boundary behavior of holomorphic functions of several complex variables depends on the geometry of the domain, and it is usually stronger than the non-tangential limit. To better stress this interdependence between analysis and geometry we decided to organize this survey as a sort of template that the reader may apply to the specific cases s/he is interested in.

More precisely, we shall single out a number of geometrical hypotheses (usually expressed in terms of the Kobayashi intrinsic distance of the domain) that when satisfied will imply a Julia-Wolff-Carathéodory theorem. This approach has the advantage to reveal the main ideas in the proofs, unhindered by the technical details needed to verify the hypotheses. In other words, the hard computations are swept under the carpet (i.e., buried in the references), leaving the interesting patterns *over* the carpet free to be examined.

Of course, the hypotheses can be satisfied: for instance, all of them hold for strongly pseudoconvex domains, convex domains with C^ω boundary, convex circular domains of finite type, and in the polydisk; but most of them hold in more general domains too. And one fringe benefit of the approach chosen for this survey is that as soon as somebody proves that the hypotheses hold for a specific domain, s/he gets a Julia-Wolff-Carathéodory theorem in that domain for free. Indeed, this approach has already uncovered new results: to the best of my knowledge, Theorem 4.2 in full generality and Proposition 4.8 have not been proved before.

So in Section 1 of this survey we shall present a proof of the Julia-Wolff-Carathéodory theorem suitable to be generalized to several complex variables. It will consist of three steps:

(a) A proof of Theorem 1.1 starting from the Schwarz-Pick lemma.

(b) A discussion of the Lindelöf principle, which says that if a (K-)bounded holomorphic function has limit restricted to a curve ending at a boundary point then it has the same limit restricted to any non-tangential curve ending at that boundary point.

(c) A proof of the Julia-Wolff-Carathéodory theorem obtained by showing that the incremental ratio and the derivative satisfy the hypotheses of the Lindelöf principle.

Then the next three sections will describe a way of performing the same three steps in a several variables context, providing the template mentioned above.

Finally, a few words on the literature. As mentioned before, Theorem 1.1 first appeared in [Ju1], and Theorem 1.2 in [Wo]. The proof we shall present here is essentially due to Rudin [Ru, Section 8.5]; other proofs and one-variable generalizations can be found in [A3], [Ah], [C1, 2], [J], [Kom], [LV], [Me], [N], [Po], [T] and references therein.

As far as I know, the first several variables generalizations of Theorem 1.1 were proved by Minialoff [Mi] for the unit ball $B^2 \subset \mathbb{C}^2$, and then by Hervé [He] in B^n. The general form we shall discuss originates in [A2]. For some other (finite and infinite dimensional) approaches see [Ba], [M], [W], [R], [Wl1] and references therein.

The one-variable Lindelöf principle has been proved by Lindelöf [Li1, 2]; see also [A3, Theorem 1.3.23], [Ru, Theorem 8.4.1], [Bu, 5.16, 5.56, 12.30, 12.31] and references therein. The first important several variables version of it is due to Čirka [Č]; his approach has been further pursued in [D1, 2], [DZ] and [K]. A different generalization is due to Cima and Krantz [CK] (see also [H1, 2]), and both inspired the presentation we shall give in Section 3 (whose ideas stem from [A2]).

A first tentative extension of the Julia-Wolff-Carathéodory theorem to bounded domains in $\mathbb{C}^2$ is due to Wachs [W]. Hervé [He] proved a preliminary Julia-Wolff-Carathéodory theorem for the unit ball of $\mathbb{C}^n$ using non-tangential limits and considering only incremental ratioes; the full statement for the unit ball is due to Rudin [Ru, Section 8.5]. The Julia-Wolff-Carathéodory theorem for strongly convex domains is in [A2]; for strongly pseudoconvex domains in [A4]; for the polydisk in [A5] (see also Jafari [Ja], even though his statement is not completely correct); for convex domains of finite type in [AT2]. Furthermore, Julia-Wolff-Carathéodory theorems in infinite-dimensional Banach and Hilbert spaces are discussed in [EHRS], [F], [MM], [SW], [Wl2, 3, 4], [Z] and references therein.

Finally, I would also like to mention the shorter survey [AT1], written, as well as the much more substantial paper [AT2], with the unvaluable help of Roberto Tauraso.

2 One Complex Variable

We already mentioned that Theorem 1.1 is a consequence of the classical Schwarz-Pick lemma. For the sake of completeness, let us recall here the relevant definitions and statements.

Definition 2.1 *The Poincaré metric on Δ is the complete Hermitian metric κ_Δ^2 of constant Gaussian curvature -4 given by*

$$\kappa_\Delta^2(\zeta) = \frac{1}{(1 - |\zeta|^2)^2} \, dz \, d\bar{z}.$$

The Poincaré distance ω on Δ is the integrated distance associated to κ_Δ.
It is easy to prove that

$$\omega(\zeta_1, \zeta_2) = \tfrac{1}{2} \log \frac{1 + \left| \frac{\zeta_1 - \zeta_2}{1 - \bar{\zeta}_2 \zeta_1} \right|}{1 - \left| \frac{\zeta_1 - \zeta_2}{1 - \bar{\zeta}_2 \zeta_1} \right|}.$$

For us the main property of the Poincaré distance is the classical Schwarz-Pick lemma:

Theorem 2.2 (Schwarz-Pick) *The Poincaré metric and distance are contracted by holomorphic self-maps of the unit disk. In other words, if $f \in \mathrm{Hol}(\Delta, \Delta)$ then*

$$\forall \zeta \in \Delta \qquad\qquad f^*(\kappa_\Delta^2)(\zeta) \le \kappa_\Delta^2(\zeta) \qquad\qquad (2)$$

and

$$\forall \zeta_1, \zeta_2 \in \Delta \qquad\qquad \omega\big(f(\zeta_1), f(\zeta_2)\big) \le \omega(\zeta_1, \zeta_2). \qquad\qquad (3)$$

Furthermore, equality in 2 for some $\zeta \in \Delta$ or in 3 for some $\zeta_1 \ne \zeta_2$ occurs iff f is a holomorphic automorphism of Δ.

A first easy application of this result is the fact that the $\liminf$ in 1 is always positive (or $+\infty$). But let us first give it a name.

Definition 2.3 *Let $f \in \mathrm{Hol}(\Delta, \Delta)$ be a holomorphic self-map of Δ, and $\sigma \in \partial\Delta$. Then the boundary dilation coefficient $\beta_f(\sigma)$ of f at σ is given by*

$$\beta_f(\sigma) = \liminf_{\zeta \to \sigma} \frac{1 - |f(\zeta)|}{1 - |\zeta|}.$$

If it is finite and equal to $\beta > 0$ we shall say that f is β-Julia at σ.

Then

Corollary 2.4 *For any $f \in \mathrm{Hol}(\Delta, \Delta)$ we have*

$$\frac{1 - |f(\zeta)|}{1 - |\zeta|} \geq \frac{1 - |f(0)|}{1 + |f(0)|} > 0 \tag{4}$$

for all $\zeta \in \Delta$; in particular,

$$\beta_f(\sigma) \geq \frac{1 - |f(0)|}{1 + |f(0)|} > 0$$

for all $\sigma \in \partial\Delta$.

Proof. The Schwarz-Pick lemma yields

$$\omega\big(0, f(\zeta)\big) \leq \omega\big(0, f(0)\big) + \omega\big(f(0), f(\zeta)\big) \leq \omega\big(0, f(0)\big) + \omega(0, \zeta),$$

that is

$$\frac{1 + |f(\zeta)|}{1 - |f(\zeta)|} \leq \frac{1 + |f(0)|}{1 - |f(0)|} \cdot \frac{1 + |\zeta|}{1 - |\zeta|} \tag{5}$$

for all $\zeta \in \Delta$. Let $a = (|f(0)| + |\zeta|)/(1 + |f(0)||\zeta|)$; then the right-hand side of 5 is equal to $(1 + a)/(1 - a)$. Hence $|f(\zeta)| \leq a$, that is

$$1 - |f(\zeta)| \geq (1 - |\zeta|)\frac{1 - |f(0)|}{1 + |f(0)||\zeta|} \geq (1 - |\zeta|)\frac{1 - |f(0)|}{1 + |f(0)|}$$

for all $\zeta \in \Delta$, as claimed. $\square$

The main step in the proof of Theorem 1.1 is known as *Julia's lemma*, and it is again a consequence of the Schwarz-Pick lemma:

Theorem 2.5 (Julia) *Let $f \in \mathrm{Hol}(\Delta, \Delta)$ and $\sigma \in \partial\Delta$ be such that*

$$\liminf_{\zeta \to \sigma} \frac{1 - |f(\zeta)|}{1 - |\zeta|} = \beta < +\infty.$$

Then there exists a unique $\tau \in \partial\Delta$ such that

$$\frac{|\tau - f(\zeta)|^2}{1 - |f(\zeta)|^2} \leq \beta \frac{|\sigma - \zeta|^2}{1 - |\zeta|^2}. \tag{6}$$

Proof. The Schwarz-Pick lemma yields

$$\left| \frac{f(\zeta) - f(\eta)}{1 - \overline{f(\eta)}f(\zeta)} \right| \leq \left| \frac{\zeta - \eta}{1 - \bar{\eta}\zeta} \right|$$

and thus

$$\frac{|1 - \overline{f(\eta)}f(\zeta)|^2}{1 - |f(\zeta)|^2} \leq \frac{1 - |f(\eta)|^2}{1 - |\eta|^2} \cdot \frac{|1 - \bar{\eta}\zeta|^2}{1 - |\zeta|^2} \tag{7}$$

for all $\eta, \zeta \in \Delta$. Now choose a sequence $\{\eta_k\} \subset \Delta$ converging to σ and such that

$$\lim_{k \to +\infty} \frac{1 - |f(\eta_k)|}{1 - |\eta_k|} = \beta;$$

in particular, $|f(\eta_k)| \to 1$, and so up to a subsequence we can assume that $f(\eta_k) \to \tau \in \partial\Delta$ as $k \to +\infty$. Then setting $\eta = \eta_k$ in 7 and taking the limit as $k \to +\infty$ we obtain 6.

We are left to prove the uniqueness of τ. To do so, we need a geometrical interpretation of 6.

Definition 2.6 *The horocycle $E(\sigma, R)$ of center σ and radius R is the set*

$$E(\sigma, R) = \left\{ \zeta \in \Delta \;\middle|\; \frac{|\sigma - \zeta|^2}{1 - |\zeta|^2} < R \right\}.$$

Geometrically, $E(\sigma, R)$ is an euclidean disk of euclidean radius $R/(1 + R)$ internally tangent to $\partial\Delta$ in σ; in particular,

$$|\sigma - \zeta| \le \frac{2R}{1 + R} < 2R \tag{8}$$

for all $\zeta \in E(\bar{\sigma}, R)$. A horocycle can also be seen as the limit of Poincaré disks with fixed euclidean radius and centers converging to σ (see, e.g., [Ju2] or [A3, Proposition 1.2.1]).

The formula 6 then says that

$$f\big(E(\sigma, R)\big) \subseteq E(\tau, \beta R)$$

for any $R > 0$. Assume, by contradiction, that 6 also holds for some $\tau_1 \ne \tau$, and choose $R > 0$ so small that $E(\tau, \beta R) \cap E(\tau_1, \beta R) =$. Then we get

$$\ne f\big(E(\sigma, R)\big) \subseteq E(\tau, \beta R) \cap E(\tau_1, \beta R) =,$$

contradiction. Therefore 6 can hold for at most one $\tau \in \partial\Delta$, and we are done. $\qquad\square$

In Section 4 we shall need a sort of converse of Julia's lemma:

Lemma 2.7 *Let $f \in \mathrm{Hol}(\Delta, \Delta)$, σ, $\tau \in \partial\Delta$ and $\beta > 0$ be such that*

$$f\big(E(\sigma, R)\big) \subseteq E(\tau, \beta R)$$

for all $R > 0$. Then $\beta_f(\sigma) \le \beta$.

Proof. For $t \in [0, 1)$ set $R_t = (1-t)/(1+t)$, so that $t\sigma \in \partial E(\sigma, R_t)$. Therefore $f(t\sigma) \in E(\tau, \bar{\beta} R_t)$; hence

$$\frac{1 - |f(t\sigma)|}{1 - t} \le \frac{|\tau - f(t\sigma)|}{1 - t} < 2\beta \frac{R_t}{1 - t} = \frac{2}{1 + t}\beta,$$

by 8, and thus

$$\beta_f(\sigma) = \liminf_{\zeta \to \sigma} \frac{1 - |f(\zeta)|}{1 - |\zeta|} \le \liminf_{t \to 1^-} \frac{1 - |f(t\sigma)|}{1 - t} \le \beta.$$

$\qquad\square$

To complete the proof of Theorem 1.1 we still need to give a precise definition of what we mean by non-tangential limit.

Definition 2.8 *Take $\sigma \in \partial\Delta$ and $M \geq 1$; the* Stolz region $K(\sigma, M)$ *of vertex σ and amplitude M is given by*

$$K(\sigma, M) = \left\{ \zeta \in \Delta \,\middle|\, \frac{|\sigma - \zeta|}{1 - |\zeta|} < M \right\}.$$

Geometrically, $K(\sigma, M)$ is an egg-shaped region, ending in an angle touching the boundary of Δ at σ. The amplitude of this angle tends to 0 as $M \to 1^+$, and tends to π as $M \to +\infty$. Therefore we can use Stolz regions to define the notion of non-tangential limit:

Definition 2.9 *A function $f \colon \Delta \to \mathbb{C}$ admits* non-tangential limit $L \in \mathbb{C}$ *at the point $\sigma \in \partial\Delta$ if $f(\zeta) \to L$ as ζ tends to σ inside $K(\sigma, M)$ for any $M > 1$.*

From the definitions it is apparent that horocycles and Stolz regions are strongly related. For instance, if ζ belongs to $K(\sigma, M)$ we have

$$\frac{|\sigma - \zeta|^2}{1 - |\zeta|^2} = \frac{|\sigma - \zeta|}{1 - |\zeta|} \cdot \frac{|\sigma - \zeta|}{1 + |\zeta|} < M|\sigma - \zeta|,$$

and thus $\zeta \in E(\sigma, M|\sigma - \zeta|)$.

We are then ready for the

Proof of Theorem 1.1: Assume that f is β-Julia at σ, fix $M > 1$ and choose any sequence $\{\zeta_k\} \subset K(\sigma, M)$ converging to σ. In particular, $\zeta_k \in E(\sigma, M|\sigma - \zeta_k|)$ for all $k \in \mathbb{N}$. Then Theorem 2.5 gives a unique $\tau \in \partial\Delta$ such that $f(\zeta_k) \in E(\tau, \beta M|\sigma - \zeta_k|)$. Therefore every limit point of the sequence $\{f(\zeta_k)\}$ must be contained in the intersection

$$\bigcap_{k \in \mathbb{N}} E(\tau, \beta M|\sigma - \zeta_k|) = \{\tau\},$$

that is $f(\zeta_k) \to \tau$, and we have proved that f has non-tangential limit τ at σ. $\qquad\square$

To prove Theorem 1.2 we need another ingredient, known as *Lindelöf principle*. The idea is that the existence of the limit along a given curve in Δ ending at $\sigma \in \partial\Delta$ forces the existence of the non-tangential limit at σ. To be more precise:

Definition 2.10 *Let $\sigma \in \partial\Delta$. A σ-curve in Δ is a continous curve $\gamma \colon [0, 1) \to \Delta$ such that $\gamma(t) \to \sigma$ as $t \to 1^-$. Furthermore, we shall say that a function $f \colon \Delta \to \mathbb{C}$ is* K-bounded *at σ if for every $M > 1$ there exists $C_M > 0$ such that $|f(\zeta)| \leq C_M$ for all $\zeta \in K(\sigma, M)$.*

Then Lindelöf [Li2] proved the following

Theorem 2.11 *Let* $f: \Delta \to \mathbb{C}$ *be a holomorphic function, and* $\sigma \in \partial\Delta$. *Assume there is a* σ-*curve* $\gamma: [0, 1) \to \Delta$ *such that* $f(\gamma(t)) \to L \in \mathbb{C}$ *as* $t \to 1^-$. *Assume moreover that*

(a) f *is bounded, or that*
(b) f *is* K-*bounded and* γ *is non-tangential, that is its image is contained in a* K-*region* $K(\sigma, M_0)$.

Then f *has non-tangential limit* L *at* σ.

Proof. A proof of case (a) can be found in [A3, Theorem 1.3.23] or in [Ru, Theorem 8.4.1]. Since each $K(\sigma, M)$ is biholomorphic to Δ and the biholomorphism extends continuously up to the boundary, case (b) is a consequence of (a). Furthermore, it should be remarked that in case (b) the existence of the limit along γ automatically implies that f is K-bounded ([Li1]; see [Bu, 5.16] and references therein).

However, we shall describe here an easy proof of case (b) when γ is radial, that is $\gamma(t) = t\sigma$, which is the case we shall mostly use.

First of all, without loss of generality we can assume that $\sigma = 1$, and then the Cayley transform allows us to transfer the stage to $H^+ = \{w \in \mathbb{C} \mid \operatorname{Im} w > 0\}$. The boundary point we are interested in becomes ∞, and the curve γ is now given by $\gamma(t) = i(1 + t)/(1 - t)$.

Furthermore if we denote by $K(\infty, M) \subset H^+$ the image under the Cayley transform of $K(1, M) \subset \Delta$, and by K_ε the truncated cone

$$K_\varepsilon = \{w \in H^+ \mid \operatorname{Im} w > \varepsilon \max\{1, |\operatorname{Re} w|\}\},$$

we have

$$K(\infty, M) \subset K_{1/(2M)} \qquad \text{and}$$

$$K_{1/(2M)} \cap \{w \in H^+ \mid \operatorname{Im} w > R\} \subset K(\infty, M'),$$

for every $R, M > 1$, where

$$M' = \sqrt{1 + 4M^2 \frac{R + 1}{R - 1}}.$$

The first inclusion is easy; the second one follows from the formula

$$\left| \frac{1 - \zeta}{1 - |\zeta|} \right|^2 = 1 + \frac{2}{|\zeta| + \operatorname{Re} \zeta} \left| \frac{\operatorname{Im} \zeta}{1 - |\zeta|} \right|^2, \tag{9}$$

true for all $\zeta \in \Delta$ with $\operatorname{Re} \zeta > 0$.

Therefore we are reduced to prove that if $f: H^+ \to \mathbb{C}$ is holomorphic and bounded on any K_ε, and $f \circ \gamma(t) \to L \in \mathbb{C}$ as $t \to 1^-$, then $f(w)$ has limit L as w tends to ∞ inside K_ε.

Choose $\varepsilon' < \varepsilon$ (so that $K_{\varepsilon'} \supset K_\varepsilon$), and define $f_n: K_{\varepsilon'} \to \mathbb{C}$ by $f_n(w) = f(nw)$. Then $\{f_n\}$ is a sequence of uniformly bounded holomorphic functions.

Furthermore, $f_n(ir) \to L$ as $n \to +\infty$ for any $r > 1$; by Vitali's theorem, the whole sequence $\{f_n\}$ is then converging uniformly on compact subsets to a holomorphic function $f_\infty \colon K_{\varepsilon'} \to \mathbb{C}$. But we have $f_\infty(ir) = L$ for all $r > 1$; therefore $f_\infty \equiv L$. In particular, for every $\delta > 0$ we can find $N \geq 1$ such that $n \geq N$ implies

$$|f_n(w) - L| < \delta \qquad \text{for all } w \in \bar{K}_\varepsilon \text{ such that } 1 \leq |w| \leq 2.$$

This implies that for every $\delta > 0$ there is $R > 1$ such that $w \in \bar{K}_\varepsilon$ and $|w| > R$ implies $|f(w) - L| < \delta$, that is the assertion. Indeed, it suffices to take $R = N$; if $|w| > N$ let $n \geq N$ be the integer part of $|w|$, and set $w' = w/n$. Then $w' \in \bar{K}_\varepsilon$ and $1 \leq |w'| \leq 2$, and thus

$$|f(w) - L| = |f_n(w') - L| < \delta,$$

as claimed. $\qquad\qquad\qquad\qquad\qquad\qquad\qquad\qquad\qquad\qquad\qquad\qquad\qquad$ $\square$

Example 1. It is very easy to provide examples of K-bounded functions which are not bounded: for instance $f(\zeta) = (1 + \zeta)^{-1}$ is K-bounded at 1 but it is not bounded in Δ. More generally, every rational function with a pole at $\tau \in \partial\Delta$ and no poles inside Δ is not bounded on Δ but it is K-bounded at every $\sigma \in \partial\Delta$ different from τ.

We are now ready to begin the proof of Theorem 1.2. Let then $f \in \mathrm{Hol}(\Delta, \Delta)$ be β-Julia at $\sigma \in \partial\Delta$, and let $\tau \in \partial\Delta$ be the non-tangential limit of f at σ provided by Theorem 1.1. We would like to show that f' has non-tangential limit $\beta\tau\bar\sigma$ at σ; but first we study the behavior of the incremental ratio $(f(\zeta) - \tau)/(\zeta - \sigma)$.

Proposition 2.12 *Let* $f \in \mathrm{Hol}(\Delta, \Delta)$ *be* β-*Julia at* $\sigma \in \partial\Delta$, *and let* $\tau \in \partial\Delta$ *be the non-tangential limit of* f *at* σ. *Then the incremental ratio*

$$\frac{f(\zeta) - \tau}{\zeta - \sigma}$$

is K-*bounded and has non-tangential limit* $\beta\tau\bar\sigma$ *at* σ.

Proof. We shall show that the incremental ratio is K-bounded and that it has radial limit $\beta\tau\bar\sigma$ at σ; the assertion will then follow from Theorem 2.11.(b).

Take $\zeta \in K(\sigma, M)$. We have already remarked that we then have $\zeta \in E(\sigma, M|\zeta - \sigma|)$, and thus $f(\zeta) \in E(\tau, \beta M|\zeta - \sigma|)$, by Julia's Lemma. Recalling 8 we get

$$|f(\zeta)' - \tau| < 2\beta M|\zeta - \sigma|,$$

and so the incremental ratio is bounded by $2\beta M$ in $K(\sigma, M)$.

Now given $t \in [0, 1)$ set $R_t = (1-t)/(1+t)$, so that $t\sigma \in \partial E(\sigma, R_t)$. Then $f(t\sigma) \in E(\tau, \beta R_t)$, and thus

$$1 - |f(t\sigma)| \leq |\tau - f(t\sigma)| \leq 2\beta R_t = 2\beta\frac{1-t}{1+t}.$$

Therefore

$$\frac{1-|f(t\sigma)|}{1-t} \leq \left|\frac{\tau - f(t\sigma)}{1-t}\right| \leq \frac{2}{1+t}\beta = \frac{2}{1+t}\liminf_{\zeta\to\sigma}\frac{1-|f(\zeta)|}{1-|\zeta|};$$

letting $t \to 1^-$ we see that

$$\lim_{t\to 1^-}\frac{1-|f(t\sigma)|}{1-t} = \lim_{t\to 1^-}\left|\frac{\tau - f(t\sigma)}{1-t}\right| = \beta, \tag{10}$$

and then

$$\lim_{t\to 1^-}\frac{|\tau - f(t\sigma)|}{1-|f(t\sigma)|} = 1. \tag{11}$$

Since $f(t\sigma) \to \tau$, we know that $\mathrm{Re}(\bar{\tau}f(t\sigma)) > 0$ for t close enough to 1; then 9 and 11 imply

$$\lim_{t\to 1^-}\frac{\tau - f(t\sigma)}{1-|f(t\sigma)|} = \tau,$$

and together with 10 we get

$$\lim_{t\to 1^-}\frac{f(t\sigma)-\tau}{t\sigma - \sigma} = \beta\tau\bar{\sigma},$$

as desired. □

By the way, the non-tangential limit of the incremental ratio is usually called the *angular derivative* of f at σ, because it represents the limit of the derivative of f inside an angular region with vertex at σ.

We can now complete the

Proof of Theorem 1.2: Again, the idea is to prove that f' is K-bounded and then show that $f'(t\sigma)$ tends to $\beta\tau\bar{\sigma}$ as $t \to 1^-$.

Take $\zeta \in K(\sigma, M)$, and choose $\delta_\zeta > 0$ so that $\zeta + \delta_\zeta\Delta \subset \Delta$. Therefore we can write

$$f'(\zeta) = \frac{1}{2\pi i}\int_{|\eta|=\delta_\zeta}\frac{f(\zeta+\eta)}{\eta^2}\,d\eta = \frac{1}{2\pi i}\int_{|\eta|=\delta_\zeta}\frac{f(\zeta+\eta)-\tau}{\zeta+\eta-\sigma}\cdot\frac{\zeta+\eta-\sigma}{\eta^2}\,d\eta \tag{12}$$

$$= \frac{1}{2\pi}\int_0^{2\pi}\frac{f(\zeta+\delta_\zeta e^{i\theta})-\tau}{\zeta+\delta_\zeta e^{i\theta}-\sigma}\left[1-\frac{\sigma-\zeta}{\delta_\zeta e^{i\theta}}\right]d\theta. \tag{13}$$

$$\tag{14}$$

Now, if $M_1 > M$ and

$$\delta_\zeta = \frac{1}{M}\frac{M_1-M}{M_1+1}|\sigma-\zeta|,$$

then it is easy to check that $\zeta + \delta_\zeta\Delta \subset K(\sigma, M_1)$; therefore 12 and the bound on the incremental ratio yield

$$|f'(\zeta)| \leq 2\beta M_1\left[1 + M\frac{M_1+1}{M_1-M}\right],$$

and so f' is K-bounded.

If $\zeta = t\sigma$, we can take $\delta_{t\sigma} = (1-t)(M-1)/(M+1)$ for any $M > 1$, and 12 becomes

$$f'(t\sigma) = \frac{1}{2\pi} \int_0^{2\pi} \frac{f(t\sigma + \delta_{t\sigma}e^{i\theta}) - \tau}{t\sigma + \delta_{t\sigma}e^{i\theta} - \sigma} \left[1 - \sigma\frac{M-1}{M+1}e^{-i\theta} \right] d\theta.$$

Since $t\sigma + \delta_{t\sigma}\Delta \subset K(\sigma, M)$, Proposition 2.12 yields

$$\lim_{t \to 1^-} \frac{f(t\sigma + \delta_{t\sigma}e^{i\theta}) - \tau}{t\sigma + \delta_{t\sigma}e^{i\theta} - \sigma} = \beta\tau\bar{\sigma}$$

for any $\theta \in [0, 2\pi]$; therefore we get $f'(t\sigma) \to \beta\tau\bar{\sigma}$ as well, by the dominated convergence theorem, and we are done.   □

It is easy to find examples of function $f \in \mathrm{Hol}(\Delta, \Delta)$ with $\beta_f(1) = +\infty$.

Example 2. Let $f \in \mathrm{Hol}(\Delta, \Delta)$ be given by $f(z) = \lambda z^k/k$ where $\lambda \in \mathbb{C}$ and $k \in \mathbb{N}$ are such that $k > |\lambda|$. Then $\beta_f(1) = +\infty$ for the simple reason that $|f(1)| = |\lambda|/k < 1$; on the other hand, $f'(1) = \lambda$.

Therefore if $\beta_f(\sigma) = +\infty$ both f and f' might still have finite non-tangential limit at σ, but we have no control on them. However, if we assume that $f(\zeta)$ is actually going to the boundary of Δ as $\zeta \to \sigma$ then the link between the angular derivative and the boundary dilation coefficient is much tighter. Indeed, the final result of this section is

Theorem 2.13 *Let $f \in \mathrm{Hol}(\Delta, \Delta)$ and $\sigma \in \partial\Delta$ be such that*

$$\limsup_{t \to 1^-} |f(t\sigma)| = 1. \tag{15}$$

Then

$$\beta_f(\sigma) = \limsup_{t \to 1^-} |f'(t\sigma)|. \tag{16}$$

In particular, f' has finite non-tangential limit at σ iff $\beta_f(\sigma) < +\infty$, and then f has non-tangential limit at σ too.

Proof. If the lim sup in 16 is infinite, then $f'(t\sigma)$ cannot converge as $t \to 1^-$, and thus $\beta_f(\sigma) = +\infty$ by Theorem 1.2.

So assume that the lim sup in 16 is finite; in particular, there is $M > 0$ such that $|f'(t\sigma)| \le M$ for all $t \in [0, 1)$. We claim that $\beta_f(\sigma)$ is finite too — and then the assertion will follow from Theorem 1.2 again.

For all $t_1, t_2 \in [0, 1)$ we have

$$|f(t_2\sigma) - f(t_1\sigma)| = \left| \int_{t_1}^{t_2} f'(t\sigma)\,dt \right| \le M|t_2 - t_1|. \tag{17}$$

Now, 15 implies that there is a sequence $\{t_k\} \subset [0, 1)$ converging to 1 and $\tau \in \partial\Delta$ such that $f(t_k) \to \tau$ as $k \to +\infty$. Therefore 17 yields

$$|\tau - f(t\sigma)| \leq M(1-t)$$

for all $t \in [0,1)$. Hence

$$\beta_f(\sigma) = \liminf_{\zeta \to \sigma} \frac{1-|f(\zeta)|}{1-|\zeta|} \leq \liminf_{t \to 1^-} \frac{1-|f(t\sigma)|}{1-t} \leq \liminf_{t \to 1^-} \frac{|\tau - f(t\sigma)|}{1-t} \leq M.$$

$\square$

So Julia's condition $\beta_f(\sigma) < +\infty$ is in some sense optimal.

3 Julia's Lemma

The aim of this section is to describe a generalization of Julia's lemma to several complex variables, and to apply it to get a several variables version of Theorem 1.1.

As we have seen, the one-variable Julia's lemma is a consequence of the Schwarz-Pick lemma or, more precisely, of the contracting properties of the Poincaré metric and distance. So it is only natural to look first for a generalization of the Poincaré metric.

Among several such generalizations, the most useful for us is the Kobayashi metric, introduced by Kobayashi [Kob1] in 1967.

Definition 3.1 *Let X be a complex manifold: the* Kobayashi (pseudo)metric *of X is the function $\kappa_X : TX \to \mathbb{R}^+$ defined by*

$$\kappa_X(z;v) = \inf\{|\xi| \mid \exists \varphi \in \mathrm{Hol}(\Delta, X) : \varphi(0) = z, d\varphi_0(\xi) = v\}$$

for all $z \in X$ and $v \in T_z X$. Roughly speaking, $\kappa_X(z;v)$ measures the (inverse of) the radius of the largest (not necessarily immersed) holomorphic disk in X passing through z tangent to v.

The Kobayashi pseudometric is an upper semicontinuous (and often continuous) complex Finsler pseudometric, that is it satisfies

$$\kappa_X(z; \lambda v) = |\lambda| \kappa_X(z; v) \tag{18}$$

for all $z \in X$, $v \in T_z X$ and $\lambda \in \mathbb{C}$. Therefore it can be used to compute the length of curves:

Definition 3.2 *If $\gamma : [a,b] \to X$ is a piecewise C^1-curve in a complex manifold X then its* Kobayashi (pseudo)length *is*

$$\ell_X(\gamma) = \int_a^b \kappa_X\big(\gamma(t); \dot{\gamma}(t)\big) \, dt.$$

The Kobayashi pseudolength of a curve does not depend on the parametrization, by 18; therefore we can define the *Kobayashi (pseudo)distance* $k_X: X \times X \to \mathbb{R}^+$ by setting

$$k_X(z, w) = \inf\{\ell_X(\gamma)\},$$

where the infimum is taken with respect to all the piecewise C^1-curves $\gamma: [a, b] \to X$ with $\gamma(a) = z$ and $\gamma(b) = w$. It is easy to check that k_X is a pseudodistance in the metric space sense. We remark that this is not Kobayashi original definition of k_X, but it is equivalent to it (as proved by Royden [Ro]).

The prefix "pseudo" used in the definitions is there to signal that the Kobayashi pseudometric (and distance) might vanish on nonzero vectors (respectively, on distinct points); for instance, it is easy to see that $\kappa_{\mathbb{C}^n} \equiv 0$ and $k_{\mathbb{C}^n} \equiv 0$.

Definition 3.3 *A complex manifold X is* (Kobayashi) *hyperbolic if k_X is a true distance, that is $k_X(z, w) > 0$ as soon as $z \neq w$; it is complete hyperbolic if k_X is a complete distance. A related notion has been introduced by Wu [Wu]: a complex manifold is taut if $\mathrm{Hol}(\Delta, X)$ is a normal family (and this implies that $\mathrm{Hol}(Y, X)$ is a normal family for any complex manifold Y).*

The main general properties of the Kobayashi metric and distance are collected in the following

Theorem 3.4 *Let X be a complex manifold. Then:*

(i) *If X is Kobayashi hyperbolic, then the metric space topology induced by k_X coincides with the manifold topology.*

(ii) *A complete hyperbolic manifold is taut, and a taut manifold is hyperbolic.*

(iii) *All the bounded domains of $\mathbb{C}^n$ are hyperbolic; all bounded convex or strongly pseudoconvex domains of $\mathbb{C}^n$ are complete hyperbolic.*

(iv) *A Riemann surface is Kobayashi hyperbolic iff it is hyperbolic, that is, iff it is covered by the unit disk (and then it is complete hyperbolic).*

(v) *The Kobayashi metric and distance of the unit ball $B^n \subset \mathbb{C}^n$ agree with the Bergmann metric and distance:*

$$\kappa_{B^n}(z; v) = \frac{1}{(1 - \|z\|^2)^2} \left[|(z, v)|^2 + (1 - \|z\|^2)\|v\|^2 \right]$$

for all $z \in B^n$ and $v \in \mathbb{C}^n$, where $(\cdot, \cdot)$ denotes the canonical hermitian product in $\mathbb{C}^n$, and

$$k_{B^n}(z, w) = \tfrac{1}{2} \log \frac{1 + \|\chi_z(w)\|}{1 - \|\chi_z(w)\|} \tag{19}$$

for all $z, w \in B^n$, where χ_z is a holomorphic automorphism of B^n sending z into the origin O. In particular, κ_Δ and k_Δ are the Poincaré metric and distance of the unit disk.

(vi) *The Kobayashi metric and distance are contracted by holomorphic maps: if $f: X \to Y$ is a holomorphic map between complex manifolds, then*

$$\kappa_Y\big(f(z); df_z(v)\big) \leq \kappa_X(z; v)$$

for all $z \in X$ and $v \in T_z X$, and

$$k_Y\big(f(z), f(w)\big) \leq k_X(z, w)$$

for all z, $w \in X$. In particular, biholomorphisms are isometries for the Kobayashi metric and distance.

For comments, proofs and much much more see, e.g., [A3, JP, Kob2] and references therein.

For us, the most important property of Kobayashi metric and distance is clearly the last one: the Kobayashi metric and distance have a built-in Schwarz-Pick lemma. So it is only natural to try and use them to get a several variables version of Julia's lemma. To do so, we need ways to express Julia's condition 1 and to define horocycles in terms of Kobayashi distance and metric.

A way to proceed is suggested by metric space theory (and its applications to real differential geometry of negatively curved manifolds; see, e.g., [BGS]). Let X be a locally compact complete metric space with distance d. We may define an embedding $\iota: X \to C^0(X)$ of X into the space $C^0(X)$ of continuous functions on X mapping $z \in X$ into the function $d_z = d(z, \cdot)$. Now identify two continuous functions on X differing only by a constant; let $\bar{X}$ be the image of the closure of $\iota(X)$ in $C^0(X)$ under the quotient map π, and set $\partial X = \bar{X} \setminus \pi\big(\iota(X)\big)$. It is easy to check that $\bar{X}$ and ∂X are compact in the quotient topology, and that $\pi \circ \iota: X \to \bar{X}$ is a homeomorphism with the image. The set ∂X is the *ideal boundary* of X.

Any element $h \in \partial X$ is a continuous function on X defined up to a constant. Therefore the sublevels of h are well-defined: they are the *horospheres* centered at the boundary point h. Now, a preimage $h_0 \in C^0(X)$ of $h \in \partial X$ is the limit of functions of the form d_{z_k} for some sequence $\{z_k\} \subset X$ without limit points in X. Since we are interested in $\pi(d_{z_k})$ only, we can force h_0 to vanish at a fixed point $z_0 \in X$. This amounts to defining the horospheres centered in h by

$$E(h, R) = \{z \in X \mid \lim_{k \to \infty} [d(z, z_k) - d(z_0, z_k)] < \tfrac{1}{2} \log R\} \qquad (20)$$

(see below for the reasons suggesting the appearance of $\frac{1}{2} \log$). Notice that, since d is a complete distance and $\{z_k\}$ is without limit points, $d(z, z_k) \to +\infty$ as $k \to +\infty$. On the other hand, $|d(z, z_k) - d(z_0, z_k)| \leq d(z, z_0)$ is always finite. So, in some sense, the limit in 20 computes one-half the logarithm of a (normalized) distance of z from the boundary point h, and the horospheres are a sort of distance balls centered in h.

In our case, this suggests the following approach:

Definition 3.5 *Let $D \subset \mathbb{C}^n$ be a complete hyperbolic domain in $\mathbb{C}^n$. The* (small) horosphere *of center $x \in \partial D$, radius $R > 0$ and pole $z_0 \in D$ is the set*

$$E_{z_0}^D(x, R) = \{z \in D \mid \limsup_{w \to x}[k_D(z, w) - k_D(z_0, w)] < \tfrac{1}{2} \log R\}. \qquad (21)$$

A few remarks are in order.

Remark 1. One clearly can introduce a similar notion of *large horosphere* replacing the lim sup by a lim inf in the previous definition. Large horospheres and small horospheres are actually different iff the geometrical boundary $\partial D \subset \mathbb{C}^n$ is smaller than the ideal boundary discussed above. It can be proved (see [A2] or [A3, Corollary 2.6.48]) that if D is a strongly convex C^3 domain then the lim sup in 21 actually is a limit, and thus the ideal boundary and the geometrical boundary coincide (as well as small and large horospheres).

Remark 2. The $\tfrac{1}{2}\log$ in the definition appears to recover the classical horocycles in the unit disk. Indeed, if we take $D = \Delta$ and $z_0 = 0$ it is easy to check that

$$\omega(\zeta, \eta) - \omega(0, \eta) = \tfrac{1}{2} \log \left(\frac{|1 - \bar{\eta}\zeta|^2}{1 - |\zeta|^2} \right) + \log \left(\frac{1 + \left| \frac{\eta - \zeta}{1 - \bar{\eta}\zeta} \right|}{1 + |\eta|} \right);$$

therefore for all $\sigma \in \partial \Delta$ we have

$$\lim_{\eta \to \sigma} [\omega(\zeta, \eta) - \omega(0, \eta)] = \tfrac{1}{2} \log \left(\frac{|\sigma - \zeta|^2}{1 - |\zeta|^2} \right),$$

and thus $E(\sigma, R) = E_0^\Delta(\sigma, R)$.

In a similar way one can explicitly compute the horospheres in another couple of cases:

Example 3. It is easy to check that the horospheres in the unit ball (with pole at the origin) are the classical horospheres (see, e.g., [Kor]) given by

$$E_O^{B^n}(x, R) = \left\{ z \in B^n \mid \frac{|1 - (z, x)|^2}{1 - \|z\|^2} < R \right\}$$

for all $x \in \partial B^n$ and $R > 0$. Geometrically, $E_O^{B^n}(x, R)$ is an ellipsoid internally tangent to ∂B^n in x, and it can be proved (arguing as in [A2, Propositions 1.11 and 1.13]) that horospheres in strongly pseudoconvex domains have a similar shape.

Example 4. On the other hand, the shape of horospheres in the unit polydisk $\Delta^n \subset \mathbb{C}^n$ is fairly different (see [A5]):

$$E_O^{\Delta^n}(x, R) = \left\{ z \in \Delta^n \mid \max_{|x_j|=1} \left\{ \frac{|x_j - z_j|^2}{1 - |z_j|^2} \right\} < R \right\} = E_1 \times \cdots \times E_n$$

for all $x \in \partial \Delta^n$ and $R > 0$, where $E_j = \Delta$ if $|x_j| < 1$ and $E_j = E(x_j, R)$ if $|x_j| = 1$.

Now we need a sensible replacement of Julia's condition 1. Here the key observation is that $1 - |\zeta|$ is exactly the (euclidean) distance of $\zeta \in \Delta$ from the boundary. Keeping with the interpretation of the lim sup in 20 as a (normalized) Kobayashi distance of $z \in D$ from $x \in \partial D$, one is then tempted to consider something like

$$\inf_{x \in \partial D} \limsup_{w \to x} [k_D(z, w) - k_D(z_0, w)] \tag{22}$$

as a sort of (normalized) Kobayashi distance of $z \in D$ from the boundary. If we compute in the unit disk we find that

$$\inf_{\sigma \in \partial \Delta} \limsup_{\eta \to \sigma} [\omega(\zeta, \eta) - \omega(0, \eta)] = \tfrac{1}{2} \log \frac{1 - |\zeta|}{1 + |\zeta|} = -\omega(0, \zeta).$$

So we actually find $\tfrac{1}{2} \log$ of the euclidean distance from the boundary (up to a harmless correction), confirming our ideas. But, even more importantly, we see that the natural lower bound $-k_D(z_0, z)$ of 22 measures exactly the same quantity.

Another piece of evidence supporting this idea comes from the boundary estimates of the Kobayashi distance. As it can be expected, it is very difficult to compute explicitly the Kobayashi distance and metric of a complex manifold; on the other hand, it is not as difficult (and very useful) to estimate them. For instance, we have the following (see, e.g., [A3, section 2.3.5] or [Kob2, section 4.5] for strongly pseudoconvex domains, [AT2] for convex C^2 domains, and [A3, Proposition 2.3.5] or [Kob2, Example 3.1.24] for convex circular domains):

Theorem 3.6 *Let $D \subset\subset \mathbb{C}^n$ be a bounded domain, and take $z_0 \in D$. Assume that*

(a) D is strongly pseudoconvex, or
(b) D is convex with C^2 boundary, or
(c) D is convex circular.

Then there exist c_1, $c_2 \in \mathbb{R}$ such that

$$c_1 - \tfrac{1}{2} \log d(z, \partial D) \leq k_D(z_0, z) \leq c_2 - \tfrac{1}{2} \log d(z, \partial D) \tag{23}$$

for all $z \in D$, where $d(\cdot, \partial D)$ denotes the euclidean distance from the boundary.

This is the first instance of the template phenomenon mentioned in the introduction. In the sequel, very often we shall not need to know the exact shape of the boundary of the domain under consideration; it will be enough to have estimates like the ones above on the boundary behavior of the Kobayashi distance. Let us then introduce the following template definition:

Definition 3.7 *We shall say that a domain $D \subset \mathbb{C}^n$ has the one-point boundary estimates if for one (and hence every) $z_0 \in D$ there are c_1, $c_2 \in \mathbb{R}$ such that*

$$c_1 - \tfrac{1}{2}\log d(z, \partial D) \leq k_D(z_0, z) \leq c_2 - \tfrac{1}{2}\log d(z, \partial D)$$

for all $z \in D$. In particular, D is complete hyperbolic.

So, again, if a domain has the one-point boundary estimates the Kobayashi distance from an interior point behaves exactly as half the logarithm of the euclidean distance from the boundary. We are then led to the following definition:

Definition 3.8 *Let $D \subset \mathbb{C}^n$ be a complete hyperbolic domain, $x \in \partial D$ and $\beta > 0$. We shall say that a holomorphic function $f \colon D \to \Delta$ is β-Julia at x (with respect to a pole $z_0 \in D$) if*

$$\liminf_{z \to x} \big[k_D(z_0, z) - \omega\big(0, f(z)\big) \big] = \tfrac{1}{2}\log \beta < +\infty.$$

The previous computations show that when $D = \Delta$ we recover the one-variable definition exactly. Furthermore, if the $\liminf$ is finite for one pole then it is finite for all poles (even though β possibly changes). Moreover, the $\liminf$ cannot ever be $-\infty$, because

$$k_D(z_0, z) - \omega\big(0, f(z)\big) \geq \omega\big(f(z_0), f(z)\big) - \omega\big(0, f(z)\big) \geq -\omega\big(0, f(z_0)\big),$$

and so $\beta > 0$ always. Finally, we explicitly remark that we might use a similar approach for holomorphic *maps* from D into another complete hyperbolic domain; but for the sake of simplicity in this survey we shall restrict ourselves to bounded holomorphic functions (see [A2, 4, 5] for more on the general case).

We have now enough tools to prove our several variables Julia's lemma:

Theorem 3.9 *Let $D \subset \mathbb{C}^n$ be a complete hyperbolic domain, and let $f \in \mathrm{Hol}(D, \Delta)$ be β-Julia at $x \in \partial D$ with respect to a pole $z_0 \in D$, that is assume that*

$$\liminf_{z \to x} \big[k_D(z_0, z) - \omega\big(0, f(z)\big) \big] = \tfrac{1}{2}\log \beta < +\infty.$$

Then there exists a unique $\tau \in \partial \Delta$ such that

$$f\big(E_{z_0}^D(x, R)\big) \subset E(\tau, \beta R)$$

for all $R > 0$.

Proof. Choose a sequence $\{w_k\} \subset D$ converging to x such that

$$\lim_{k \to +\infty} \big[k_D(z_0, w_k) - \omega\big(0, f(w_k)\big) \big] = \liminf_{z \to x} \big[k_D(z_0, z) - \omega\big(0, f(z)\big) \big].$$

We can also assume that $f(w_k) \to \tau \in \bar{\Delta}$. Being D complete hyperbolic, we know that $k_D(z_0, w_k) \to +\infty$; therefore we must have $\omega(0, f(w_k)) \to +\infty$, and so $\tau \in \partial \Delta$. Now take $z \in E_{z_0}^D(x, R)$. Then using the contracting property of the Kobayashi distance we get

$$
\begin{aligned}
\frac{1}{2} \log \frac{|\tau - f(z)|^2}{1 - |f(z)|^2} &= \lim_{\eta \to \tau} \left[\omega(f(z), \eta) - \omega(0, \eta) \right] \\
&= \lim_{k \to +\infty} \left[\omega(f(z), f(w_k)) - \omega(0, f(w_k)) \right] \\
&\leq \limsup_{k \to +\infty} \left[k_D(z, w_k) - \omega(0, f(w_k)) \right] \\
&= \limsup_{k \to +\infty} \left[k_D(z, w_k) - k_D(z_0, w_k) \right] + \\
&\quad + \lim_{k \to +\infty} \left[k_D(z_0, w_k) - \omega(0, f(w_k)) \right] \\
&= \limsup_{w \to x} \left[k_D(z, w) - k_D(z_0, w) \right] + \tfrac{1}{2} \log \beta < \tfrac{1}{2} \log(\beta R),
\end{aligned}
$$

and $f(z) \in E(\tau, \beta R)$. The uniqueness of τ follows as in the proof of Theorem 2.5. $\square$

The next step consists in introducing a several variables version of the classical non-tangential limit. Korányi [Kor] has been the first one to notice that in several complex variables the obvious notion of non-tangential limit (i.e., limit inside cone-shaped approach regions) is not the right one for studying the boundary behavior of holomorphic functions. Indeed, in B^n he introduced the *admissible approach region* $K(x, M)$ of vertex $x \in \partial B^n$ and amplitude $M > 1$ defined by

$$
K(x, M) = \left\{ z \in B^n \ \middle| \ \frac{|1 - (z, x)|}{1 - \|z\|} < M \right\}, \tag{24}
$$

and said that a function $f : B^n \to \mathbb{C}$ had *admissible limit* $L \in \mathbb{C}$ at $x \in \partial B^n$ if $f(z) \to L$ as $z \to x$ inside $K(x, M)$ for any $M > 1$. Admissible regions are a clear generalization of one-variable Stolz regions, but the shape is different: though they are cone-shaped in the normal direction to ∂B^n at x (more precisely, the intersection with the complex line $\mathbb{C}x$ is exactly a Stolz region), they are tangent to ∂B^n in complex tangential directions. Nevertheless, Korányi was able to prove a Fatou theorem in the ball: any bounded holomorphic function has admissible limit at almost every point of ∂B^n, which is a much stronger statement than asking only for the existence of the non-tangential limit.

Later, Stein [St] (see also [KS]) generalized Korányi results to any C^2 domain $D \subset\subset \mathbb{C}^n$ defining the admissible limit using the euclidean approach regions

$$
A(x, M) = \left\{ z \in D \mid |(z - x, \mathbf{n}_x)| < M \delta_x(z), \ \|z - x\|^2 < M \delta_x(z) \right\},
$$

where $x \in \partial D$, $M > 1$, $\mathbf{n}_x$ is the outer unit normal vector to ∂D at x, and

$$
\delta_x(z) = \min\{ d(z, \partial D), d(z, x + T_x \partial D) \};
$$

notice that $A(x, M) \subseteq K(x, M)$ if $D = B^n$. Furthermore, in the same period Čirka [Č] introduced another kind of approach regions, depending on the order of contact of complex submanifolds with the boundary of the domain.

Both Stein's and Čirka's approach regions are defined in euclidean terms, and so are not suited for our arguments casted in terms of invariant distances. Another possibility is provided by the approach regions introduced by Cima and Krantz [CK] (see also [Kr1, 2]):

$$\mathcal{A}(x, M) = \{z \in D \mid k_D(z, N_x) < M\},$$

where N_x is the set of points in D of the form $x - t\mathbf{n}_x$, with $t \in \mathbb{R}$, and $\mathbf{n}_x$ is the outer unit normal vector to ∂D at x. The approach regions $\mathcal{A}(x, M)$ in strongly pseudoconvex domains are comparable to Stein's and Čirka's approach regions — and thus yield the same notion of admissible limit. Unfortunately, the presence of the euclidean normal vector $\mathbf{n}_x$ is again unsuitable for our needs, and so we are forced to introduce a different kind of approach regions.

As we discussed before, the horospheres can be interpreted as sublevels of a sort of "distance" from the point x in the boundary, distance normalized using a fixed pole z_0. It turns out that a good way to define approach regions is taking the sublevels of the average between the Kobayashi distance from the pole z_0 and the "distance" from x. More precisely:

Definition 3.10 *Let $D \subset \mathbb{C}^n$ be a complete hyperbolic domain. The (small) K-region $K_{z_0}^D(x, M)$ of vertex $x \in \partial D$, amplitude $M > 1$ and pole $z_0 \in D$ is the set*

$$K_{z_0}^D(x, M) = \{z \in D \mid \limsup_{w \to x}[k_D(z, w) - k_D(z_0, w)] + k_D(z_0, z) < \log M\},$$

$$(25)$$

and we say that a function $f : D \to \mathbb{C}$ has K-limit $L \in \mathbb{C}$ at $x \in \partial D$ if $f(z) \to L$ as $z \to x$ inside $K_{z_0}^D(x, M)$ for any $M > 1$.

As usual, a few remarks and examples are in order.

Remark 3. Replacing the lim sup by a lim inf one obtains the definition of large K-regions, that we shall not use in this paper but that are important in the study of this kind of questions for holomorphic maps (instead of functions).

Remark 4. Changing the pole in K-regions amounts to a shifting in the amplitudes, and thus the notion of K-limit does not depend on the pole.

Example 5. It is easy to check that in the unit ball we recover Korányi's admissible regions exactly: $K_O^{B^n}(x, M) = K(x, M)$ for all $x \in \partial B^n$ and $M > 1$. More generally, it is not difficult to check (see [A2]) that in strongly pseudoconvex domains our K-regions are comparable with Stein's and Čirka's admissible regions, and so our K-limit is equivalent to their admissible limit.

Example 6. On the other hand, our K-regions are defined even in domains whose boundary is not smooth; for instance, in the polydisk we have ([A5])

$$K_O^{\Delta^n}(x, M) = \left\{z \in \Delta^n \,\middle|\, \frac{1 + \|z\|}{1 - \|z\|} \max_{|x_j|=1} \left\{\frac{|x_j - z_j|^2}{1 - |z_j|^2}\right\} < M^2\right\} \quad (26)$$

for all $x \in \partial \Delta^n$, where $\|z\| = \max\{|z_j|\}$; in particular, if $z \to x$ inside some $K_O^{\Delta^n}(x, M)$ then $z_j \to x_j$ non-tangentially if $|x_j| = 1$ while $z_j \to x_j$ without restrictions if $|x_j| < 1$.

Of course, one would like to compare K-regions with cone-shaped regions, that is K-limits with non-tangential limits. To do so, we again need to know something on the boundary behavior of the Kobayashi distance. More precisely, we need the following template definition:

Definition 3.11 We say that a domain $D \subset \mathbb{C}^n$ has the two-points upper boundary estimate at $x \in \partial D$ if there exist $\varepsilon > 0$ and $C > 0$ such that

$$k_D(z_1, z_2) \leq \tfrac{1}{2} \sum_{j=1}^{2} \log \left(1 + \frac{\|z_1 - z_2\|}{d(z_j, \partial D)} \right) + C \tag{27}$$

for all $z_1, z_2 \in B(x, \varepsilon) \cap D$, where $B(x, \varepsilon)$ is the euclidean ball of center x and radius ε.

Forstneric and Rosay [FR] have proved that C^2 domains have the two-points upper estimate, and a similar proof shows that this is true for convex circular domains too.

Assume then that $D \subset \mathbb{C}^n$ has the one-point boundary estimates and the two-points boundary estimate at $x \in \partial D$ (e.g., D is strongly pseudoconvex, or C^2 convex, or convex circular). Then if $z \in D$ is close enough to x we have

$$\limsup_{w \to x}[k_D(z, w) - k_D(z_0, w)] + k_D(z_0, z)$$
$$\leq \tfrac{1}{2} \log \left(1 + \frac{\|z - x\|}{d(z, \partial D)} \right) + \tfrac{1}{2} \log \frac{\|z - x\|}{d(z, \partial D)} + C - c_1 + c_2,$$

and thus cones with vertex at x are contained in K-regions. This means that the existence of a K-limit is stronger than the existence of a non-tangential limit, and that $x \in K(\bar{x}, M) \cap \partial D$, even though the latter intersection can be strictly larger than $\{x\}$ (this happens, for instance, in the polydisk).

Going back to our main concern, definition 25 allows us to immediately relate horospheres and K-regions: for instance it is clear that

$$z \in K_{z_0}^D(x, M) \quad \Longrightarrow \quad z \in E_{z_0}^D(x, M^2/R(z)), \tag{28}$$

where $R(z) > 0$ is such that $k_D(z_0, z) = \tfrac{1}{2} \log R(z)$. We are then able to prove a several variables generalization of Theorem 1.1:

Theorem 3.12 Let $D \subset \mathbb{C}^n$ be a complete hyperbolic domain, and let $f \in \text{Hol}(D, \Delta)$ be β-Julia at $x \in \partial D$ with respect to a pole $z_0 \in D$, that is assume that

$$\liminf_{z \to x} \left[k_D(z_0, z) - \omega(0, f(z)) \right] = \tfrac{1}{2} \log \beta < +\infty.$$

Assume moreover that $x \in K_{z_0}^D(\bar{x}, M)$ for some (and then for all large enough) $M > 1$. Then there exists a unique $\tau \in \partial \Delta$ such that f has K-limit τ at x.

Proof. It suffices to prove that if $y \in K_{z_0}^D(\bar{x}, M) \cap \partial D$ and $z \to y$ inside $K_{z_0}^D(x, M)$ then $f(z) \to \tau$, where $\tau \in \partial D$ is the point provided by Theorem 3.9. But indeed if $z \in K_{z_0}^D(x, M)$ then we just remarked that $z \in E_{z_0}^D(x, M^2/R(z))$; therefore Theorem 3.9 yields $f(z) \in E(\tau, \beta M^2/R(z))$. Since when z tends to the boundary of D we have $R(z) \to +\infty$, we get $f(z) \to \tau$ and we are done. $\square$

Actually, this proof yields slightly more than what is stated: it shows that $f(z) \to \tau$ as soon as z tends to any point in $K_{z_0}^D(\bar{x}, M) \cap \partial D$, even though this intersection might be strictly larger than $\{x\}$ (for instance in the polydisk).

4 Lindelöf Principles

The next step in our presentation consists in proving a Lindelöf principle in several complex variables. As first noticed by Čirka [Č], neither the non-tangential limit nor the K-limit (or admissible limit) are the right one to consider: the former is too weak, the latter too strong. But let us be more precise.

Definition 4.1 *Let $D \subset \mathbb{C}^n$ be a domain in $\mathbb{C}^n$, and $x \in \partial D$. An x-curve in D is again a continuous curve $\gamma : [0, 1) \to D$ such that $\gamma(t) \to x$ as $t \to 1^-$. Then, for us, a* Lindelöf principle *is a statement of the following form: "There are two classes $\mathcal{S}$ and $\mathcal{R}$ of x-curves in D such that: if $f : D \to \mathbb{C}$ is a bounded holomorphic function such that $f(\gamma^o(t)) \to L \in \mathbb{C}$ as $t \to 1^-$ for one curve $\gamma^o \in \mathcal{S}$ then $f(\gamma(t)) \to L$ as $t \to 1^-$ for all $\gamma \in \mathcal{R}$."*

In the classical Lindelöf principle, $\mathcal{S}$ is the set of all σ-curves in Δ, while $\mathcal{R}$ is the set of all non-tangential σ-curves. Remembering the previous section, one can be tempted to conjecture that in several variables one could take as $\mathcal{S}$ again the set of all x-curves, and as $\mathcal{R}$ the set of all x-curves contained in a K-region (or in an admissible region). But this is not true even in the ball, as remarked by Čirka:

Example 7. Take $f : B^2 \to \Delta$ given by

$$f(z, w) = \frac{w^2}{1 - z^2},$$

and $x = (1, 0)$. Then if $\gamma_0(t) = (t, 0)$ we have $f \circ \gamma_0 \equiv 0$, and indeed it is not difficult to prove that f has non-tangential limit 0 at x. On the other hand, for any $c \in \Delta$ we can consider the x-curve $\gamma_c : [0, 1) \to B^2$ given by $\gamma_c(t) = (t, c(1 - t)^{1/2})$; then

$$f(\gamma_c(t)) = \frac{c^2(1 - t)}{1 - t^2} = \frac{c^2}{1 + t} \to \frac{c^2}{2}.$$

So the existence of the limit along such a curve does not imply that f has the same radial limit. Conversely, the existence of the radial limit does not imply that f has the same limit along a curve γ_c even though such a curve is contained in a K-region: indeed,

$$\frac{\left|1-\left(\gamma_c(t),x\right)\right|}{1-\|\gamma_c(t)\|} \leq \frac{2\left|1-\left(\gamma_c(t),x\right)\right|}{1-\|\gamma_c(t)\|^2} = \frac{2}{1+t-|c|^2} \leq \frac{2}{1-|c|^2},$$

and so the image of γ_c is contained in $K\left(x,2/(1-|c|^2)\right)$. Finally, if $1 > \alpha > 1/2$ and $c \in \Delta$ we can consider the curve $\gamma_{c,\alpha}(t) = \left(t,c(1-t)^\alpha\right)$. This is not a non-tangential curve, because

$$\frac{\|\gamma_{c,\alpha}(t)-x\|}{1-\|\gamma_{c,\alpha}(t)\|} \geq \frac{\|\gamma_{c,\alpha}(t)-x\|}{1-\|\gamma_{c,\alpha}(t)\|^2} = \frac{1}{(1-t)^{1-\alpha}}\frac{\left((1-t)^{2(1-\alpha)}+|c|^2\right)^{1/2}}{1+t-|c|^2(1-t)^{2\alpha-1}} \to +\infty$$

as $t \to 1^-$. However,

$$f\left(\gamma_{c,\alpha}(t)\right) = \frac{c^2(1-t)^{2\alpha}}{1-t^2} = \frac{c^2(1-t)^{2\alpha-1}}{1+t} \to 0$$

as $t \to 1^-$, and so $f \circ \gamma(t)$ tends to zero for x-curves γ belonging to a family strictly larger than the one of non-tangential curves.

The conclusion of this example is that in general both classes $\mathcal{S}$ and $\mathcal{R}$ in a Lindelöf principle might not coincide with the classes of non-tangential curves, or of curves contained in a K-region, or of all x-curves. For instance, let us describe one of the Lindelöf principles proved by Čirka [Č]. Assume that the boundary ∂D is of class C^1 in a neighbourhood of a point $x \in \partial D$, and denote by $\mathbf{n}_x$ the outer unit normal vector to ∂D in x. Furthermore, let $H_x(\partial D) = T_x(\partial D) \cap iT_x(\partial D)$ be the holomorphic tangent space to ∂D at x, set $N_x = x + \mathbb{C}\mathbf{n}_x$, and let $\pi_x : \mathbb{C}^n \to N_x$ be the complex-linear projection parallel to $H_x(\partial D)$, so that $z - \pi_x(z) \in H_x(\partial D)$ for all $z \in \mathbb{C}^n$. Finally, set $H_z = z + H_x(\partial D)$. Then Čirka proved a Lindelöf principle taking: as $\mathcal{S}$ the set of x-curves γ such that the image of $\pi_x \circ \gamma$ is contained in D and such that

$$\lim_{t\to 1^-} \frac{\|\gamma(t)-\pi_x \circ \gamma(t)\|}{d\left(\pi_x \circ \gamma(t), H_{\pi_x\circ\gamma(t)} \cap \partial D\right)} = 0; \tag{29}$$

and as $\mathcal{R}$ the set of curves $\gamma \in \mathcal{S}$ such that $\pi_x \circ \gamma$ approaches x non-tangentially in $D \cap N_x$. Notice that $\mathcal{R}$ contains properly the set of all non-tangential curves. For instance, it is easy to check that if $D = B^2$ and $x = (1,0)$ then $\gamma_{c,\alpha} \in \mathcal{R}$ iff $\alpha > 1/2$.

Some years later another kind of Lindelöf principle (valid for normal holomorphic functions, not only for bounded ones) has been proved by Cima and Krantz [CK]. They supposed ∂D smooth at $x \in \partial D$, and used: as $\mathcal{S}$ the set of non-tangential x-curves; and as $\mathcal{R}$ the set of x-curves γ such that

$$\lim_{t \to 1^-} k_D\big(\gamma(t), \Gamma_x\big) = 0$$

for some cone Γ_x of vertex x inside D. Now, notice that, by continuity, $\gamma \in \mathcal{R}$ iff we can find a cone Γ_x and an x-curve γ_x whose image is contained in Γ_x such that

$$\lim_{t \to 1^-} k_D\big(\gamma(t), \gamma_x(t)\big) = 0. \tag{30}$$

If in Čirka's setting we take $\gamma_x = \pi_x \circ \gamma$ it is not difficult to see that 29 implies 30. Indeed, let $r(t) > 0$ be the largest r such that the image of $\Delta_r = r\Delta$ through the map $\psi(\zeta) = \gamma_x(t) + \zeta\big(\gamma(t) - \gamma_x(t)\big)$ is contained in D. Clearly,

$$r(t)\|\gamma(t) - \gamma_x(t)\| \ge d(\gamma_x(t), H_{\gamma_x(t)} \cap \partial D);$$

in particular, 29 implies $r(t) \to +\infty$. But then

$$k_D\big(\gamma(t), \gamma_x(t)\big) \le k_{\Delta_r(t)}(1,0) = \omega\left(0, \frac{1}{r(t)}\right) \to 0,$$

and so 30 holds.

It turns out that as soon as we have something like 30, to prove a Lindelöf principle it is just a matter of applying the one-variable Lindelöf principle and the contracting property of the Kobayashi distance. These considerations suggested in [A2] the introduction of a very general setting producing Lindelöf principles.

Definition 4.2 *Let $D \subset \mathbb{C}^n$ be a domain in C^n. A projection device at $x \in \partial D$ is given by the following data:*

(a) a neighbourhood U of x in $\mathbb{C}^n$;
(b) a holomorphically embedded disk $\varphi_x \colon \Delta \to U \cap D$ such that $\lim_{\zeta \to 1} \varphi_x(\zeta) = x$;
(c) a family $\mathcal{P}$ of x-curves in $U \cap D$;
(d) a device associating to every x-curve $\gamma \in \mathcal{P}$ a 1-curve $\tilde{\gamma}_x$ in Δ — or, equivalently, a x-curve $\gamma_x = \varphi_x \circ \tilde{\gamma}_x$ in $\varphi_x(\Delta) \subset U \cap D$.

When we have a projection device, we shall *always* use $\varphi_x(0)$ as pole for horospheres and K-regions.

It is very easy to produce examples of projection devices. For instance:

Example 8. The *trivial projection device.* Take $U = \mathbb{C}^n$, choose as φ_x any holomorphically embedded disk satisfying the hypotheses, as $\mathcal{P}$ the set of all x-curves, and to any $\gamma \in \mathcal{P}$ associate the radial curve $\tilde{\gamma}_x(t) = 1 - t$.

Example 9. The *euclidean projection device.* This is the device used by Čirka. Assume that ∂D is of class C^1 at x, let $\mathbf{n}_x$ be the outer unit normal at x, and set $N_x = x + \mathbb{C}\mathbf{n}_x$ as before. Choose U so that $(U \cap D) \cap N_x$ is simply connected with continuous boundary, and let $\varphi_x \colon \Delta \to (U \cap D) \cap N_x$ be a biholomorphism extending continuously up to the boundary with $\varphi_x(1) = x$. Let again $\pi_x: \mathbb{C}^n \to N_x$ be the orthogonal projection, and choose as $\mathcal{P}$ the set of x-curves γ in $U \cap D$ such that the image of $\pi_x \circ \gamma$ is still contained in $U \cap D$. Then for every $\gamma \in \mathcal{P}$ we set $\gamma_x = \pi_x \circ \gamma$.

Example 10. This is a slight variation of the previous one. If D is convex and of class C^1 in a neighbourhood of x, both the projection $\pi_x(D)$ and the intersection $D \cap N_x$ are convex domains in N_x; therefore there is a biholomorphism $\psi: \pi_x(D) \to D \cap N_x$ extending continuously to the boundary so that $\psi(x) = x$. Then we can take $U = \mathbb{C}^n$, φ_x as in Example 3.3, $\mathcal{P}$ as the set of all x-curves, and set $\gamma_x = \psi \circ \pi_x \circ \gamma$.

To describe the next projection device, that it will turn out to be the most useful, we need a new definition:

Definition 4.3 *A holomorphic map* $\varphi: \Delta \to X$ *in a complex manifold X is a complex geodesic if it is an isometry between the Poincaré distance ω and the Kobayashi distance k_X.*

Complex geodesics have been introduced by Vesentini [V1], and deeply studied by Lempert [Le] and Royden-Wong [RW]. In particular, they proved that if D is a bounded convex domain then for every $z_0, z \in D$ there exists a complex geodesic $\varphi: \Delta \to D$ passing through z_0 and z, that is such that $\varphi(0) = z_0$ and $z \in \varphi(\Delta)$. Moreover, there also exists a left-inverse of φ, that is a bounded holomorphic function $\tilde{p}: D \to \Delta$ such that $\tilde{p} \circ \varphi = \mathrm{id}_\Delta$ (see [A3, Chapter 2.6] or [Kob2, sections 4.6–4.8] for complete proofs). Furthermore, if there exists $z_0 \in D$ such that for every $z \in D$ we can find a complex geodesic φ continuous up to the boundary passing through z_0 and z (this happens, for instance, if D is strongly convex with C^3-boundary [Le], if it is convex of finite type [AT2], or if it is convex circular and $z_0 = O$ [V2]) then it is easy to prove (see, e.g., [A1]) that for any $x \in \partial D$ there is a complex geodesic φ continuous up to the boundary such that $\varphi(0) = z_0$ and $\varphi(1) = x$.

Example 11. The *canonical projection device*. Let $D \subset\subset \mathbb{C}^n$ be a bounded convex domain, and let $x \in \partial D$ be such that there is a complex geodesic $\varphi_x: \Delta \to D$ so that $\varphi_x(\zeta) \to x$ as $\zeta \to 1$. Then the canonical projection device is obtained taking $U = \mathbb{C}^n$, $\mathcal{P}$ as the set of all x-curves, and setting $\tilde{\gamma}_x = \tilde{p}_x \circ \gamma$, where $\tilde{p}_x: D \to \Delta$ is the left-inverse of φ_x. Notice that the canonical projection device is defined only using the Kobayashi distance; therefore it will be particularly well-suited for our aims.

Example 12. This is a far-reaching generalization of the previous example. Let D be any domain, $x \in \partial D$ any point, and $\varphi_x: \Delta \to D$ any holomorphically embedded disk satisfying the hypotheses. Choose a bounded holomorphic function $h: D \to \Delta$ such that $h(z) \to 1$ as $z \to x$ in D. Then we have a projection device just by choosing $U = \mathbb{C}^n$, $\mathcal{P}$ as the set of all x-curves, and setting $\tilde{\gamma}_x = h \circ \gamma$.

Example 13. All the previous examples can be localized: if there is a neigh-bourhood U of $x \in \partial D$ such that we can define a projection device at x for $U \cap D$, we clearly have a projection device at x for D. In particular, if D is locally biholomorphic to a convex domain in x (e.g., if D is strongly pseudo-convex in x), then we can localize the projection devices of Examples 3.3, 3.4 and 3.5.

We can now define the right kind of limit for Lindelöf principles.

Definition 4.4 *Let $D \subset \mathbb{C}^n$ be a domain equipped with a projection device at $x \in \partial D$. We shall say that a curve $\gamma \in \mathcal{P}$ is* special *if*

$$\lim_{t \to 1^-} k_{D \cap U}\big(\gamma(t), \gamma_x(t)\big) = 0; \tag{31}$$

and that it is restricted *if $\tilde{\gamma}_x$ is a non-tangential 1-curve. Let $\mathcal{S}$ denote the set of special x-curves, and $mathcal R \subset \mathcal{S}$ denote the set of special restricted x-curves. We shall say that a function $f \colon D \to \mathbb{C}$ has* restricted K-limit $L \in \mathbb{C}$ *at x if $f\big(\gamma(t)\big) \to L$ as $t \to 1^-$ for all $\gamma \in \mathcal{R}$.*

Remark 5. We could have defined the notion of special curve using k_D instead of $k_{D \cap U}$ in 31, and the following proofs would have worked anyway with a possibly larger set of curves. However, the chosen definition stresses the local nature of the projection device (as it should be, because it is a tool born to deal with local phenomena), allowing to replace D by $D \cap U$ everywhere. Furthermore, if D has the one-point boundary estimates, the two-two points upper boundary estimate at x and also the *two-points lower boundary esti-mate*, that is for any pair of distinct points $x_1 \neq x_2 \in \partial D$ there esist $\varepsilon > 0$ and $K \in \mathbb{R}$ such that

$$k_D(z_1, z_2) \geq -\tfrac{1}{2} \log d(z_1, \partial D) - \tfrac{1}{2} \log d(z_2, \partial D) + K$$

as soon as $z_1 \in B(x_1, \varepsilon) \cap D$ and $z_2 \in B(x_2, \varepsilon) \cap D$, then ([A3, Theorem 2.3.65])

$$\lim_{\substack{z, w \to x \\ z \neq w}} \frac{k_D(z, w)}{k_{D \cap U}(z, w)} = 1,$$

and so the two definitions of special curves coincide. Examples of domains having the two-points lower boundary estimate include strongly pseudoconvex domains ([FR], [A3, Corollary 2.3.55]).

The whole point of the definition of projection device is that the arguments used in [Č] and [CK] boil down to the following very general Lindelöf principle:

Theorem 4.5 *Let $D \subset \mathbb{C}^n$ be a domain equipped with a projection device at $x \in \partial D$. Let $f \colon D \to \mathbb{C}$ be a bounded holomorphic function such that $f\big(\gamma^\circ(t)\big) \to L \in \mathbb{C}$ as $t \to 1^-$ for one special curve $\gamma^\circ \in \mathcal{S}$. Then f has restricted K-limit L at x.*

Proof. We can assume that $f(D) \subset\subset \Delta$. If $\gamma \in \mathcal{S}$ we have

$$\omega\big(f\big(\gamma(t)\big), f\big(\gamma_x(t)\big)\big) \leq k_{D \cap U}\big(\gamma(t), \gamma_x(t)\big) \to 0$$

as $t \to 1^-$; therefore f has limit along γ iff it does along γ_x. In particular, it has limit L along γ_x^o; the classical Lindelöf principle applied to $f \circ \varphi_x$ shows then that f has limit L along γ_x for all restricted γ. But in turn this implies that f has limit L along all $\gamma \in \mathcal{R}$, and we are done. $\qquad\square$

The same proof, adapted as in [CK], works for normal functions too, not necessarily bounded.

Of course, the interest of such a result is directly proportional to how large the set $\mathcal{R}$ is. Let us see a few examples.

Example 14. The first one is a negative one: if $D = \Delta$ and $x = 1$ then the class $\mathcal{R}$ for the trivial projection device contains only 1-curves tangent in 1 to the radius, and so in this case Theorem 4.5 is even weaker than the classical Lindelöf principle. In other words, the trivial projection device probably is not that useful.

Example 15. The next one is much better: if $D \subset \mathbb{C}^n$ is of class C^1 we already remarked that all x-curves satisfying Čirka's condition 29 are special; therefore Theorem 4.5 recovers Čirka's result.

Example 16. If D is strongly convex, it is not difficult to check (see [A4]) that for the euclidean projection device a restricted x-curve is special iff

$$\lim_{t \to 1^-} \frac{\|\gamma(t) - \gamma_x(t)\|}{d\big(\gamma_x(t), \partial D\big)} = 0; \tag{32}$$

in particular, all non-tangential x-curves are special and restricted. A much harder computation (see [A3, Proposition 2.7.1]) shows that if D is strongly convex of class C^3 then for the canonical projection device a restricted x-curve γ is special again iff 32 holds (but this time γ_x is given by the canonical projection device, not by the euclidean one). Using this characterization it is possible to prove ([A3, Lemma 2.7.12]) that non-tangential x-curves are special and restricted for the canonical projection device too.

Example 17. In [A5] it is shown that in the polydisk Δ^n, if we use the canonical projection device associated to the complex geodesic $\varphi_x(\zeta) = \zeta x$, an x-curve γ is special iff

$$\lim_{t \to 1^-} \max_{|x_j|=1} \left\{ \frac{|\gamma_j(t) - (\gamma_x)_j(t)|}{1 - \|\gamma_x(t)\|} \right\} = 0.$$

Unfortunately, as already happened in one-variable, to prove the Julia-Wolff-Carathéodory theorem one needs a Lindelöf principle for not necessarily bounded holomorphic functions.

Definition 4.6 *We shall say that a function* $f\colon D \to \mathbb{C}$ *is K-bounded at* $x \in \partial D$ *if for every* $M > 1$ *there exists* $C_M > 1$ *such that* $|f(z)| < C_M$ *for all* $z \in K_{z_0}^D(x, M)$; *it is clear that this condition does not depend on the pole.*

Then we shall need a Lindelöf principle for K-bounded functions. As it can be expected, such a Lindelöf principle does not hold for any projection device: we need some connection between the projection device and K-regions. We shall express this connection in a template form.

Definition 4.7 *Let* $D \subset \mathbb{C}^n$ *be a domain. A projection device at* $x \in \partial D$ *is good if*

(i) for any $M > 1$ *there is* $\tilde{M} > 1$ *such that* $\varphi_x\big(K(1, M)\big) \subset K_{z_0}^{D \cap U}(x, \tilde{M})$;
(ii) if $\gamma \in \mathcal{R}$ *then there exists* $M_1 = M_1(\gamma)$ *such that*

$$\lim_{t \to 1^-} k_{K_{z_0}^{D \cap U}(x, M_1)}\big(\gamma(t), \gamma_x(t)\big) = 0.$$

Condition (i) actually is almost automatic. Indeed, it suffices that φ_x sends non-tangential 1-curves in non-tangential x-curves (this happens for instance if $\varphi_x'(1)$ exists and it is transversal to ∂D), and that non-tangential approach regions are contained in K-regions (as happens if D has the one-point boundary estimates and the two-point upper boundary estimate at x, as already noticed in the previous section).

Now, condition (i) implies that for any $\gamma \in \mathcal{R}$ there is $M > 1$ such that $\gamma(t),\ \gamma_x(t) \in K_{z_0}^{D \cap U}(x, M)$ for all t close enough to 1. Indeed, since γ is restricted, $\tilde{\gamma}_x(t)$ belongs to some Stolz region in Δ for t close enough to 1. Therefore, by condition (i), $\gamma_x(t)$ belongs to some K-region at x, and, being γ special, $\gamma(t)$ belongs to a slightly larger K-region for all t close enough to 1.

In particular, condition (ii) makes sense; but it is harder to verify. The usual approach is the following: for $t \in [0, 1)$ set

$$\psi_t(\zeta) = \gamma_x(t) + \zeta\big(\gamma(t) - \gamma_x(t)\big),$$

and

$$r(t, M_1) = \sup\{r > 0 \mid \psi_t(\Delta_r) \subset K_{z_0}^{D \cap U}(x, M_1)\}.$$

The contracting property of the Kobayashi distance yields

$$k_{K_{z_0}^{D \cap U}(x, M_1)}\big(\gamma(t), \gamma_x(t)\big) \leq \omega\left(0, \frac{1}{r(t, M_1)}\right);$$

therefore to prove (ii) it suffices to show that for any $\gamma \in \mathcal{R}$ there exists $M_1 > 1$ such that $r(t, M_1) \to +\infty$ as $t \to 1^-$. And to prove this latter assertion we need informations on the shape of $K_{z_0}^{D \cap U}(x, M)$ near the boundary, that is on the boundary behavior of the Kobayashi distance (and of the projection device).

Example 18. Both the euclidean and the canonical projection device are good on strongly convex domains (see [A2, 4]) — and thus their localized versions are good in strongly pseudoconvex domains. Furthermore, the canonical projection device is good in convex domains with C^ω boundary (or more generally in strictly linearly convex domains of finite type [AT2]) and in the polydisk [A5]; as far as I know, it is still open the question of whether the canonical projection device associated to the complex geodesic $\varphi_x(\zeta) = \zeta x$ is good in any convex circular domain.

Adapting the proof of Theorem 4.5 we get a Lindelöf principle for K-bounded functions:

Theorem 4.8 *Let $D \subset \mathbb{C}^n$ be a domain equipped with a good projection device at $x \in \partial D$. Let $f: D \to \mathbb{C}$ be a K-bounded holomorphic function such that $f(\gamma^o(t)) \to L \in \mathbb{C}$ as $t \to 1^-$ for one special restricted curve $\gamma^o \in \mathcal{R}$. Then f has restricted K-limit L at x.*

Proof. Take $\gamma \in \mathcal{R}$, and let $M_1 > 1$ be given by condition (ii). Then if f is bounded by C on $K_{z_0}^D(x, M_1)$ we have

$$k_{\Delta_C}\big(f(\gamma(t)), f(\gamma_x(t))\big) \leq k_{K_{z_0}^{D \cap U}(x, M_1)}\big(\gamma(t), \gamma_x(t)\big) \to 0$$

as $t \to 1^-$; therefore f has limit along γ iff it does along γ_x. In particular, it has limit L along γ_x^o; the classical Lindelöf principle for K-bounded functions in the disk Theorem 2.11.(b) applied to $f \circ \varphi_x$ (which is K-bounded thanks to condition (i) in the definition of good projection devices) shows then that f has limit L along γ_x for all restricted γ. But in turns this implies that f has limit L along all $\gamma \in \mathcal{R}$, and we are done. $\square$

As the proof makes clear, replacing K-regions by other kinds of approach regions (and changing conditions (i) and (ii) accordingly) one gets similar results. Let us describe a possible variation, which is useful for instance in convex domains of finite type.

Definition 4.9 *A projection device is* geometrical *if there is a holomorphic function $\tilde{p}_x: D \cap U \to \Delta$ such that $\tilde{p}_x \circ \varphi_x = \mathrm{id}_\Delta$ and $\tilde{\gamma}_x = \tilde{p}_x \circ \gamma$ for all $\gamma \in \mathcal{P}$.*

Example 19. The canonical projection device is a geometrical projection device.

Remark 6. In a geometrical projection device the map φ_x always is a complex geodesic of $U \cap D$: indeed

$$k_{U \cap D}\big(\varphi_x(\zeta_1), \varphi_x(\zeta_2)\big) \leq \omega(\zeta_1, \zeta_2) = \omega\big(\tilde{p}_x(\varphi_x(\zeta_1)), \tilde{p}_x(\varphi_x(\zeta_2))\big)$$
$$\leq k_{U \cap D}\big(\varphi_x(\zeta_1), \varphi_x(\zeta_2)\big)$$

for all $\zeta_1, \zeta_2 \in \Delta$.

Definition 4.10 *Let $D \subset \mathbb{C}^n$ be a domain equipped with a geometrical projection device at $x \in \partial D$. The T-region of vertex x, amplitude $M > 1$ and girth $\delta \in (0,1)$ is the set*

$$T(x, M, \delta) = \{z \in U \cap D \mid \tilde{p}_x(z) \in K(1, M), \, k_{U \cap D}(z, p_x(z)) < \omega(0, \delta)\},$$

where $p_x = \varphi_x \circ \tilde{p}_x$. A function $f: D \to \mathbb{C}$ is T-bounded at x if there is $0 < \delta_0 < 1$ such that f is bounded on each $T(x, M, \delta_0)$, with the bound depending on M as usual.

By construction, $\varphi_x(K(1, M)) \subset T(x, M, \delta)$ for all $0 < \delta < 1$. Furthermore, if $\gamma \in \mathcal{R}$ there always are $M > 1$ and $\delta \in (0, 1)$ such that $\gamma(t) \in T(x, M, \delta)$ for all t close enough to 1. Therefore we can introduce the following definition:

Definition 4.11 *A geometrical projection device at $x \in \partial D$ is T-good if for any $\gamma \in \mathcal{R}$ then there exist $M = M(\gamma) > 1$ and $\delta = \delta(\gamma) > 0$ such that*

$$\lim_{t \to 1^-} k_{T(x, M, \delta)}(\gamma(t), \gamma_x(t)) = 0.$$

Example 20. The canonical projection device is T-good in all convex domains of finite type ([AT2]).

Then arguing as before we get

Theorem 4.12 *Let $D \subset \mathbb{C}^n$ be a domain equipped with a T-good geometrical projection device at $x \in \partial D$. Let $f: D \to \mathbb{C}$ be a T-bounded holomorphic function such that $f(\gamma^o(t)) \to L \in \mathbb{C}$ as $t \to 1^-$ for one special restricted curve $\gamma^o \in \mathcal{R}$. Then f has restricted K-limit L at x.*

Of course, to compare such a result with Theorem 4.8 one would like to know whether K-bounded functions are T-bounded or not — and this boils down to compare K-regions and T-regions. It turns out that to make the comparison we need another property of the projection device:

Definition 4.13 *We shall say that a projection device at $x \in \partial D$ preserves horospheres if*

$$\varphi_x(E(1, R)) \subset E_{z_0}^{U \cap D}(x, R)$$

for all $R > 0$.

Proposition 4.14

Let $D \subset \mathbb{C}^n$ be a domain equipped with a geometrical projection device at $x \in \partial D$ preserving horospheres. Then

$$T(x, M, \delta) \subset K_{z_0}^{U \cap D}(x, M(1 + \delta)/(1 - \delta)), \tag{33}$$

for all $M > 1$ and $0 < \delta < 1$. Furthermore, a T-good geometrical projection device preserving horospheres is automatically good.

Proof. First of all, we claim that

$$\varphi_x\big(K(1, M)\big) \subset K_{z_0}^{U \cap D}(x, M)$$

for all $M > 1$. Indeed, if $\zeta \in K(1, M)$ we have $\zeta \in E\big(1, M^2/R(\zeta)\big)$, where $R(\zeta)$ satisfies

$$\tfrac{1}{2} \log R(\zeta) = \omega(0, \zeta) = k_{U \cap D}\big(z_0, \varphi_x(\zeta)\big)$$

as usual. But then $\varphi_x(\zeta) \in E_{z_0}^{U \cap D}\big(x, M^2/R(\zeta)\big)$, which immediately implies $\varphi_x(\zeta) \in K_{z_0}^{U \cap D}(x, M)$, as claimed.

Now take $z \in T(x, M, \delta)$. Then we have

$$k_{U \cap D}\big(z, w\big) - k_{U \cap D}(z_0, w) + k_{U \cap D}(z_0, z)$$
$$\leq 2k_{U \cap D}\big(z, p_x(z)\big) + k_{U \cap D}\big(p_x(z), w\big) - k_{U \cap D}(z_0, w) + k_{U \cap D}\big(z_0, p_x(z)\big)$$
$$< 2\omega(0, \delta) + k_{U \cap D}\big(p_x(z), w\big) - k_{U \cap D}(z_0, w) + k_{U \cap D}\big(z_0, p_x(z)\big)$$

for all $w \in U \cap D$. Now, $\tilde{p}_x(z) \in K(1, M)$; therefore $p_x(z) \in K_{z_0}^{U \cap D}(x, M)$ and

$$\limsup_{w \to x} \big[k_{U \cap D}(z, w) - k_{U \cap D}(z_0, w)\big] + k_{U \cap D}(z_0, z) < \log\left(\frac{1 + \delta}{1 - \delta}\right) + \log M,$$

that is 33. Then condition (ii) in the definition of good projection device follows immediately by the contraction property of the Kobayashi distance. $\square$

So if the projection device preserves horospheres we see that K-bounded functions are always T-bounded, and thus the hypotheses on the function in Theorem 4.12 are weaker than the hypotheses in Theorem 4.8.

Of course, one would like to know when a geometrical projection device preserves horospheres. It is easy to prove that φ_x sends horocycles into *large* horospheres for any geometrical projection device; so if large and small horospheres coincide (as it happens in strongly convex domains, for instance), we are done.

Another sufficient condition is the following:

Proposition 4.15 *Let $D \subset \mathbb{C}^n$ be a domain equipped with a geometrical projection device at $x \in \partial D$. Assume there is a neighbourhood $V \subseteq U$ of x and a family $\Psi: V \cap D \to \mathrm{Hol}(\Delta, U \cap D)$ of holomorphic disks in $U \cap D$ such that, writing $\psi_z(\zeta)$ for $\Psi(z)(\zeta)$, the following holds:*

(a) $\psi_z(0) = z_0 = \varphi_x(0)$ *for all $z \in V \cap D$;*
(b) *for all $z \in V \cap D$ there is $r_z \in [0, 1)$ such that $\psi_z(r_z) = z$;*
(c) ψ_z *converges to ϕ_x, uniformly on compact subsets, as $z \to x$ in $V \cap D$;*
(d) $k_{U \cap D}(z_0, z) - \omega(0, r_z)$ *tends to 0 as $z \to x$ in $V \cap D$.*

Then the projection device preserves horospheres.

Proof. Since we always have

$$\lim_{t\to1-}[\omega(\zeta,t)-\omega(0,t)] = \lim_{t\to1-}\big[k_{U\cap D}(\varphi_x(\zeta),\varphi_x(t))-k_{U\cap D}(z_0,\varphi_x(t))\big]$$
$$\leq \limsup_{w\to x}\big[k_{U\cap D}(\varphi_x(\zeta),w)-k_{U\cap D}(z_0,w)\big],$$

it suffices to prove the reverse inequality. In other words, it suffices to prove that for every $\varepsilon > 0$ there is $\delta > 0$ such that

$$k_{U\cap D}(\varphi_x(\zeta),w)-k_{U\cap D}(z_0,w) \leq \lim_{t\to1-}[\omega(\zeta,t)-\omega(0,t)]+\varepsilon$$

as soon as $w \in B(x,\delta)\cap D \subset V$.

First of all, we claim that for any $z \in U\cap D$ and $\psi \in \mathrm{Hol}(\Delta, U\cap D)$ the function

$$t \mapsto k_{U\cap D}(z,\psi(t))-\omega(0,t)$$

is not increasing. Indeed, if $t_1 \leq t_2$ we have

$$k_{U\cap D}(z,\psi(t_1))-\omega(0,t_1) = k_{U\cap D}(z,\psi(t_1))+\omega(t_1,t_2)-\omega(0,t_2)$$
$$\geq k_{U\cap D}(z,\psi(t_1))+k_{U\cap D}(\psi(t_1),\psi(t_2))-\omega(0,t_2)$$
$$\geq k_{U\cap D}(z,\psi(t_2))-\omega(0,t_2).$$

In particular, given $\zeta \in \Delta$ we can find $t_0 < 1$ so that

$$k_{U\cap D}(\varphi_x(\zeta),\varphi_x(t))-\omega(0,t) \leq \lim_{t\to1-}[\omega(\zeta,t)-\omega(0,t)]+\varepsilon/3$$

when $t \geq t_0$.

Now choose $\delta > 0$ such that when $w \in B(x,\delta)\cap D \subset V$ we have

$$\big|k_{U\cap D}(\varphi_x(\zeta),\psi_w(t_0))-k_{U\cap D}(\varphi_x(\zeta),\varphi_x(t_0))\big| \leq k_{U\cap D}(\varphi_x(t_0),\psi_w(t_0)) \leq \varepsilon/3$$

and

$$\omega(0,r_w) \geq k_{U\cap D}(z_0,w) \geq \max\{\omega(0,r_w)-\varepsilon/3,\omega(0,t_0)\}.$$

Then

$$k_{U\cap D}(\varphi_x(\zeta),w)-k_{U\cap D}(z_0,w) \leq k_{U\cap D}(\varphi_x(\zeta),\psi_w(r_w))-\omega(0,r_w)+\varepsilon/3$$
$$\leq k_{U\cap D}(\varphi_x(\zeta),\psi_w(t_0))-\omega(0,t_0)+\varepsilon/3$$
$$\leq k_{U\cap D}(\varphi_x(\zeta),\varphi_x(t_0))-\omega(0,t_0)+2\varepsilon/3$$
$$\leq \lim_{t\to1-}[\omega(\zeta,t)-\omega(0,t)]+\varepsilon,$$

and we are done.   □

So a geometrical projection device preserves horospheres if φ_x is embedded in a continuous family of "almost geodesic" disks sweeping a one-sided neighbourhood of x.

Example 21. In strongly convex domains ([Le]) and in strictly linearly convex domains of finite type ([AT2]) the conditions of the previous proposition can be satisfied using complex geodesics. In convex circular domains, it suffices to use linear disks (when φ_x is linear, as usual in this case). In particular, then, the canonical projection device is good in strictly linearly convex domains of finite type and in convex circular domains of finite type. But it might well be possible that the conditions in Proposition 4.15 are satisfied in other classes of domains too.

It follows that, at present, we can apply Theorem 4.12 to all convex domains of finite type, because we know that there the canonical projection device is T-good, whereas we can apply Theorem 4.8 only to strictly linearly convex (or convex circular) domains of finite type, because of the previous two Propositions — and to the polydisk, because we can prove directly that the canonical projection device is good there. So Theorems 4.8 and 4.12 are applicable to different classes of domains, and this is the reason we presented both. Anyway, it is very natural to conjecture that the canonical projection device is good in all convex domains of finite type and in all convex circular domains.

5 The Julia-Wolff-Carathéodory Theorem

We are almost ready to prove the several variables version of the Wolff-Carathéodory part of the Julia-Wolff-Carathéodory theorem. We only need to introduce another couple of concepts.

Definition 5.1 *A geometrical projection device at* $x \in \partial D$ *is* bounded *if* $d(z, \partial D)/|1 - \tilde{p}_x(z)|$ *is bounded in* $U \cap D$, *and the reciprocal quotient* $|1 - \tilde{p}_x(z)|/d(z, \partial D)$ *is* K-bounded in $U \cap D$.

Notice that this condition is local, because $d(z, \partial D) = d\big(z, \partial(D \cap U)\big)$ if $z \in D$ is close enough to x. Since the results we are seeking are local too, when using geometrical projection devices from now on *we shall assume* $U = \mathbb{C}^n$, so that $\tilde{p}_x$ is defined on the whole of D, effectively identifying D with $D \cap U$. For instance, results proved for strongly convex domains will apply immediately to strongly pseudoconvex domains.

Actually, geometric projection devices are very often bounded:

Lemma 5.2 *Let* $D \subset \mathbb{C}^n$ *be a domain having the one-point boundary estimates. Then every geometrical projection device in* D *is bounded.*

Proof. Let us first show that $d(z, \partial D)/|1 - \tilde{p}_x(z)|$ is bounded in D. Indeed we have

$$-\tfrac{1}{2} \log |1 - \tilde{p}_x(z)| \le -\tfrac{1}{2} \log(1 - |\tilde{p}_x(z)|) \le \omega\big(0, \tilde{p}_x(z)\big) \le k_D(z_0, z)$$
$$\le c_2 - \tfrac{1}{2} \log d(z, \partial D),$$

and thus $d(z, \partial D)/|1 - \tilde{p}_x(z)| \le \exp(2c_2)$ for all $z \in D$.

To prove K-boundedness of the reciprocal, we first of all notice that $\tilde{p}_x$ is 1-Julia at x. Indeed,

$$\liminf_{z \to x} \big[k_D(z_0, z) - \omega\big(0, \tilde{p}_x(z)\big)\big] \leq \liminf_{\zeta \to 1} \big[k_D\big(\varphi_x(0), \varphi_x(\zeta)\big) - \omega(0, \zeta)\big] = 0.$$

Therefore we can apply Theorem 3.9, with $\tau = 1$ because $\tilde{p}_x \circ \varphi_x = \mathrm{id}$. Take $z \in K_{z_0}^D(x, M)$, and define $R(z)$ by $k_D(z_0, z) = \frac{1}{2}\log R(z)$. Then 28 and Theorem 3.9 yield $\tilde{p}_x(z) \in E\big(1, M^2/R(z)\big)$, and so

$$\tfrac{1}{2}\log|1 - \tilde{p}_x(z)| \leq \tfrac{1}{2}\log(2M^2) - k_D(z_0, z) \leq \tfrac{1}{2}\log(2M^2) + \tfrac{1}{2}\log d(x, \partial D) - c_1,$$

that is $|1 - \tilde{p}_x(z)|/d(z, \partial D) \leq 2M^2 \exp(-2c_1)$ for all $z \in K_{z_0}^D(x, M)$, as claimed. $\qquad\square$

Now let D be a complete hyperbolic domain, and let $f: D \to \Delta$ be a bounded holomorphic function. If f is β-Julia at $x \in \partial D$, we know that it has K-limit $\tau \in \partial\Delta$ at x, and we want to discuss the boundary behavior of the partial derivatives of f.

If $v \in \mathbb{C}^n$, $v \neq O$, we shall write

$$\frac{\partial f}{\partial v} = \sum_{j=1}^{n} v_j \frac{\partial f}{\partial z_j}$$

for the partial derivative of f in the direction v. The idea is that the behavior of $\partial f/\partial v$ depends on the boundary behavior of the Kobayashi metric in the direction v. To be more specific, let us introduce the following definition:

Definition 5.3 *Let $D \subset \mathbb{C}^n$ be a domain in $\mathbb{C}^n$, and $x \in \partial D$. The* Kobayashi class $\mathbf{s}_x(v)$ *and the* Kobayashi type $s_x(v)$ *of a nonzero vector $v \in \mathbb{C}^n$ at x are defined by*

$$\begin{aligned}
\mathbf{s}_x(v) &= \{s \geq 0 \mid d(z, \partial D)^s \kappa_D(z; v) \text{ is } K\text{-bounded at } x\}, \quad \text{and} \\
s_x(v) &= \inf \mathbf{s}_x(v).
\end{aligned} \tag{34}$$

Example 22. If ∂D is of class C^2 near x, it is easy to prove (see, e.g., [AT2]) that $s_x(v) \leq 1$ for all $v \in \mathbb{C}^n$.

Example 23. If D is strongly pseudoconvex, Graham's estimates [G] show that $s_x(v) = 1$ if v is transversal to ∂D at x, that is if v is not orthogonal to $\mathbb{C}n_x$ (where $\mathbf{n}_x$ is the outer unit normal vector to ∂D at x, as before). On the other hand, $s_x(v) = 1/2$ if $v \neq O$ is complex-tangential to ∂D at x, and in both cases $s_x(v) \in \mathbf{s}_x(v)$.

Example 24. If D is convex of finite type $L \geq 2$ at x then (see [AT2]) we have $s_x(v) = 1 \in \mathbf{s}_x(v)$ if v is transversal to ∂D, and $1/L \leq s_x(v) \leq 1 - 1/L$ if $v \neq O$ is complex tangential to ∂D.

Example 25. If $D \subset \mathbb{C}^2$ is pseudoconvex of finite type $L \geq 2$, then Catlin's estimates [Ca] show that $1/L \leq s_x(v) \leq 1/2$ when v is complex tangential, and $s_x(v) = 1 \in \mathbf{s}_x(v)$ as always if v is transversal.

Example 26. The Kobayashi metric of the polydisk is given by (see, e.g., [JP, Example 3.5.6])

$$\kappa_{\Delta^n}(z;v) = \max_{j=1,\dots,n} \left\{ \frac{|v_j|}{1 - |z_j|^2} \right\};$$

therefore $s_x(v) \leq 1$ for all $x \in \partial\Delta^n$ and $v \in \mathbb{C}^n$, $v \neq O$. It is also clear that if $v_j = 0$ when $|x_j| = 1$, then $s_x(v) = 0 \in \mathbf{s}_x(v)$. On the other hand, if there is j such that $|x_j| = 1$ and $v_j \neq 0$ then $s_x(v) = 1 \in \mathbf{s}_x(v)$. Indeed, by 26 we see that if $z \in K_O^{\Delta^n}(x, M)$ we have

$$\frac{1 - \|z\|}{1 - |z_j|} \geq \frac{1}{2}\frac{1 - \|z\|}{1 + \|z\|}\frac{1 - |z_j|^2}{|x_j - z_j|^2} > \frac{1}{2M^2}$$

for all j such that $|x_j| = 1$. Therefore if $z \in K_O^{\Delta^n}(x, M)$ is close enough to x, setting

$$c = \min\{|v_j| \mid |x_j| = 1, |v_j| \neq 0\} > 0,$$

we have

$$(1 - \|z\|)\kappa_{\Delta^n}(z;v) \geq \frac{c}{2} \max_{\substack{|x_j|=1 \\ |v_j|\neq 0}} \left\{ \frac{1 - \|z\|}{1 - |z_j|} \right\} > \frac{c}{4M^2} > 0,$$

and so $s_x(v) = 1 \in \mathbf{s}_x(v)$ as claimed.

These results seem to suggest that the Kobayashi type might be the inverse of the D'Angelo type of ∂D at x along the direction v (that is, the highest order of contact of ∂D with a complex curve tangent to v at x), but we do not even try to prove such a statement here. Another open question is whether $s_x(v)$ always belongs to $\mathbf{s}_x(v)$ or not.

We can now state a very general Julia-Wolff-Carathéodory theorem:

Theorem 5.4 *Let $D \subset \mathbb{C}^n$ be a complete hyperbolic domain equipped with a bounded geometrical projection device at $x \in \partial D$. Let $f \in \mathrm{Hol}(D, \Delta)$ be β-Julia at x, that is such that*

$$\liminf_{z \to x}\left[k_D(z_0, z) - \omega\bigl(0, f(z)\bigr)\right] = \tfrac{1}{2}\log\beta < +\infty.$$

Then for every $v \in \mathbb{C}^n$ and $s \in \mathbf{s}_x(v)$ the function

$$d(z, \partial D)^{s-1}\frac{\partial f}{\partial v} \tag{35}$$

is K-bounded. Furthermore, it has K-limit zero at x if $s > s_x(v)$.

This statements holds, for instance, in domains locally biholomorphic to C^2 convex or to convex circular domains.

Remark 7. The previous statement is optimal with regard to K-boundedness, but it is only asymptotically optimal with regard to the existence of the limit, as the proof will make clear. The more interesting limit case, that is the behavior of 35 when $s = s_x(v) \in \mathbf{s}_x(v)$, requires deeper tools. As we shall discuss later, in several instances 35 will admit *restricted K-limit* but it will not admit K-limit at x when $s = s_x(v)$: see Example 4.6. So it will be necessary to use all the machinery we discussed in the previous section. Furthermore, the specific tools we shall use to deal with this case will depend more strongly on the actual shape of the domain; a fully satisfying template approach to the limit case is yet to be found.

Example 27. This example, due to Rudin [Ru, 8.5.8], shows that we cannot expect the function 35 to be K-bounded, let alone to have a restricted K-limit, if $s < s_x(v)$. Let $\psi \in \mathrm{Hol}(\Delta, \Delta)$ be given by

$$\psi(\zeta) = \exp\left(-\frac{\pi}{2} - i\log(1 - \zeta)\right).$$

As $\zeta \to 1$, the function $\psi(\zeta)$ spirals around the origin without limit; moreover,

$$\psi'(\zeta) = \frac{i}{1-\zeta}\psi(\zeta).$$

Let $f \in \mathrm{Hol}(B^2, \Delta)$ be given by

$$f(z_1, z_2) = z_1 + \frac{1}{2}z_2^2\psi(z_1).$$

Then f is 1-Julia at $x = (1, 0)$, and admits K-limit 1 at x. But

$$\frac{\partial f}{\partial z_1} = 1 + \frac{iz_2^2}{2(1 - z_1)}\psi(z_1), \qquad \frac{\partial f}{\partial z_2} = z_2\psi(z_1);$$

therefore $\partial f/\partial z_1$ has restricted K-limit 1 at x while $d(z, \partial B^2)^{s-1}\partial f/\partial z_1$ blows-up at x for all $s < 1$. Similarly, $d(z, \partial B^2)^{-1/2}\partial f/\partial z_2$ has restricted K-limit 0 at x but $d(z, \partial B^2)^{s-1}\partial f/\partial z_2$ is not K-bounded if $s < 1/2$. Notice furthermore that the K-limit of $\partial f/\partial z_1$ at x does not exist.

We can now start with the

Proof of Theorem 5.4: Take $s \in \mathbf{s}_x(v)$. We shall argue mimicking the proof of the one-dimensional Julia-Wolff-Carathéodory theorem. We shall first show that a sort of incremental ratio is K-bounded, and then we shall use a integral formula to prove K-boundedness of the partial derivative. After that, deriving the existence of the K-limit will be easy.

Let us first show that an incremental ratio is K-bounded. Take $z \in K_{z_0}^D(x, M)$ and set

$$\tfrac{1}{2} \log R(z) = \log M - k_D(z_0, z)$$

so that $z \in E_{z_0}^D\big(x, R(z)\big)$. Then $f(z) \in E\big(\tau, \beta R(z)\big)$, which implies as usual that

$$|\tau - f(z)| \leq 2\beta R(z).$$

On the other hand,

$$\tfrac{1}{2} \log R(z) \leq \log M - \omega\big(0, \tilde{p}_x(z)\big) \leq \tfrac{1}{2} \log\big[M^2|1 - \tilde{p}_x(z)|\big],$$

and so

$$\left| \frac{\tau - f(z)}{1 - \tilde{p}_x(z)} \right| \leq 2\beta M^2. \tag{36}$$

Now the integral formula. Take $z \in K_{z_0}^D(x, M)$. Since D is complete hyperbolic (and hence taut), we can find a holomorphic map $\psi \colon \Delta \to D$ with $\psi(0) = z$ and $\psi'(0) = v/\kappa_D(z; v)$. Then for any $r \in (0, 1)$ we can write

$$
\begin{aligned}
d(z, \partial D)^{s-1} \tfrac{\partial f}{\partial v}(z) &= d(z, \partial D)^{s-1} \kappa_D(z; v)(f \circ \psi)'(0) = \\
&= d(z, \partial D)^{s-1} \frac{\kappa_D(z; v)}{2\pi i} \int\limits_{|\zeta| = r} \frac{f\big(\psi(\zeta)\big)}{\zeta^2} \, d\zeta \\
&= \frac{1}{2\pi} \int_0^{2\pi} \frac{f\big(\psi(re^{i\theta})\big) - \tau}{\tilde{p}_x\big(\psi(re^{i\theta})\big) - 1} \, \frac{\tilde{p}_x\big(\psi(re^{i\theta})\big) - 1}{\tilde{p}_x(z) - 1} \left(\frac{\tilde{p}_x(z) - 1}{d(z, \partial D)} \right)^s \frac{d(z, \partial D)^s \kappa_D(z; v)}{re^{i\theta}} \, d\theta.
\end{aligned}
\tag{37}
$$

We must then prove that the four factors in the integrand are bounded as z varies in $K_{z_0}^D(x, M)$.

The last factor is K-bounded by the choice of s. The third factor is K-bounded because the projection device is bounded. To prove that the other two factors are K-bounded we shall need the following two lemmas:

Lemma 5.5 *Let $D \subset \subset \mathbb{C}^n$ be a domain equipped with a geometrical projection device at x. Then $\tilde{p}_x\big(K_{z_0}^D(x, M)\big) \subseteq K(1, M)$ for all $M > 1$.*

Proof. Take $z \in K_{z_0}^D(x, M)$. Then

$$
\begin{aligned}
\omega\big(\tilde{p}_x(z), \zeta\big) - \omega(0, \zeta) &+ \omega\big(0, \tilde{p}_x(\zeta)\big) \leq \\
&\leq k_D\big(z, \varphi_x(\zeta)\big) - \omega(0, \zeta) + k_D(z_0, z) \\
&= k_D\big(z, \varphi_x(\zeta)\big) - k_D\big(z_0, \varphi_x(\zeta)\big) + k_D(z_0, z) + k_D\big(z_0, \varphi_x(\zeta)\big) - \omega(0, \zeta) \\
&\leq k_D\big(z, \varphi_x(\zeta)\big) - k_D\big(z_0, \varphi_x(\zeta)\big) + k_D(z_0, z),
\end{aligned}
$$

and taking the lim sup as $\zeta \to 1$ we get the assertion. $\qquad\square$

Lemma 5.6 *Let $D \subset \subset \mathbb{C}^n$ be a complete hyperbolic domain, $z_0 \in D$, and $x \in \partial D$. If $M_1 > M > 1$ set*

$$r = \frac{M_1 - M}{M_1 + M}. \tag{38}$$

Then $\psi(\bar{\Delta}_r) \subset K_{z_0}^D(x, M_1)$ for all holomorphic maps $\psi \colon \Delta \to D$ such that $\psi(0) \in K_{z_0}^D(x, M)$.

Proof. Take $\zeta \in \bar{\Delta}_r$. Then

$$\limsup_{w \to x}\left[k_D\big(\psi(\zeta), w\big) - k_D(z_0, w)\right] + k_D\big(z_0, \psi(\zeta)\big) \leq$$
$$\leq 2k_D\big(\psi(\zeta), \psi(0)\big) + \limsup_{w \to x}\left[k_D\big(\psi(0), w\big) - k_D(z_0, w)\right] + k_D\big(z_0, \psi(0)\big)$$
$$< 2\omega(0, \zeta) + \log M \leq \log\left(M\tfrac{1+r}{1-r}\right) = \log M_1,$$

that is $\psi(\zeta) \in K_{z_0}^D(x, M_1)$. $\square$

Now we can deal with the first two factors in 37. Choose $M_1 > M$ and r as in 38. Then $\psi(re^{i\theta}) \in K_{z_0}^D(x, M_1)$ for all $\theta \in [0, 2\pi]$, and 36 yields

$$\left|\frac{f\big(\psi(re^{i\theta})\big) - \tau}{\tilde{p}_x\big(\psi(re^{i\theta})\big) - 1}\right| \leq 2\beta M_1^2,$$

that is the first factor is bounded too. Finally, Lemmas 5.5 and 5.6 imply that

$$\left|\frac{1 - \tilde{p}_x\big(\psi(\zeta)\big)}{1 - \tilde{p}_x(z)}\right| \leq M_1 \frac{1 - \big|\tilde{p}_x\big(\psi(\zeta)\big)\big|}{1 - |\tilde{p}_x(z)|} \leq \frac{M_1}{2} \frac{1 + |\tilde{p}_x(z)|}{1 - |\tilde{p}_x(z)|} \frac{1 - \big|\tilde{p}_x\big(\psi(\zeta)\big)\big|}{1 + \big|\tilde{p}_x\big(\psi(\zeta)\big)\big|}$$

for all $\zeta \in \bar{\Delta}_r$. Therefore

$$\tfrac{1}{2}\log\left|\frac{1 - \tilde{p}_x\big(\psi(\zeta)\big)}{1 - \tilde{p}_x(z)}\right| \leq \tfrac{1}{2}\log\frac{M_1}{2} + \omega\big(0, \tilde{p}_x(z)\big) - \omega\big(0, \tilde{p}_x(\psi(\zeta))\big)$$
$$\leq \tfrac{1}{2}\log\frac{M_1}{2} + \omega\big(\tilde{p}_x(\psi(0)), \tilde{p}_x(\psi(\zeta))\big) \leq \tfrac{1}{2}\log\frac{M_1}{2} + \omega(0, r),$$

and the second factor is bounded too.

We are left to show that 35 has K-limit 0 when $s > s_x(v)$. But indeed choose $s_x(v) < s_1 < s$; then we can write

$$d(z, \partial D)^{s-1}\frac{\partial f}{\partial v}(z) = d(z, \partial D)^{s-s_1}\left[d(z, \partial D)^{s_1 - 1}\frac{\partial f}{\partial v}(z)\right],$$

and thus it converges to 0 as $z \to x$ inside $K_{z_0}^D(x, M)$. $\square$

Let us now discuss what happens when $s = s_x(v) \in \mathbf{s}_x(v)$. As anticipated before, we shall need to apply Lindelöf principles and the material of the previous section. We shall still prove some general results, but the deeper theorems will work for specific classes of domains only.

We begin dealing with directions transversal to the boundary. If f being β-Julia at $x \in \partial D$ would imply $f \circ \varphi_x$ being β-Julia at 1, one could try to apply the classical Julia-Wolff-Carathéodory theorem to $f \circ \varphi_x$. It turns out that this approach is viable when the projection device preserves horospheres:

Lemma 5.7 *Let $D \subset \mathbb{C}^n$ be a complete hyperbolic domain equipped with a projection device at $x \in \partial D$ preserving horospheres. Then for any $f \in \mathrm{Hol}(D, \Delta)$ which is β-Julia at x, the composition $f \circ \varphi_x$ is β-Julia at 1. In particular, $(f \circ \varphi_x)'$ has non-tangential limit $\beta\tau$ at 1, where $\tau \in \partial\Delta$ is the K-limit of f at x.*

Proof. Since f is β-Julia and the projection device preserves horospheres we have

$$f \circ \varphi_x\big(E(1,R)\big) \subseteq E(\tau, \beta R)$$

for all $R > 0$. Then Lemma 2.7 yields

$$\liminf_{z \to x}\big[k_D(z_0, z) - \omega\big(0, f(z)\big)\big] =$$
$$= \tfrac{1}{2}\log\beta \geq \liminf_{\zeta \to 1}\big[\omega(0,\zeta) - \omega\big(0, f(\varphi_x(\zeta))\big)\big]$$
$$\geq \liminf_{\zeta \to 1}\big[k_D\big(z_0, \varphi_x(\zeta)\big) - \omega\big(0, f(\varphi_x(\zeta))\big)\big]$$
$$\geq \liminf_{z \to x}\big[k_D(z_0, z) - \omega\big(0, f(z)\big)\big],$$

and we are done. □

This is enough to deal with transversal directions, under the mild hypotheses that $s_x(v) \leq 1$ for all $v \in \mathbb{C}^n$ (which might be possibly true in all complete hyperbolic domains) and that the radial limit of φ_x' at 1 exists (this happens in convex domains of finite type or in convex circular domains, for instance). Indeed we have

Corollary 5.8 *Let $D \subset\subset \mathbb{C}^n$ be a complete hyperbolic domain equipped with a good bounded geometrical projection device at $x \in \partial D$ preserving horospheres. Assume moreover that $1 \in \mathbf{s}_x(v)$ for all $v \in \mathbb{C}^n$, $v \neq O$, and that the radial limit $\nu_x = \varphi_x'(1)$ of φ_x' at 1 exists. Finally, let $f \in \mathrm{Hol}(D, \Delta)$ be β-Julia at x, and denote by $\tau \in \partial\Delta$ its K-limit at x. Then:*

(i) $s_x(\nu_x) = 1$ and $\partial f/\partial \nu_x$ has restricted K-limit $\beta\tau \neq 0$ at x;
(ii) if moreover $s_x(v_T) < 1$ for all $v_T \neq O$ orthogonal to ν_x, then for all v_N not orthogonal to ν_x the function $\partial f/\partial v_N$ has non-zero restricted K-limit at x, and $s_x(v_N) = 1$.

Proof. The previous lemma implies that $(f \circ \varphi_x)'$ has radial limit $\beta\tau \neq 0$ at 1. Now write

$$\frac{\partial f}{\partial \nu_x}\big(\varphi_x(t)\big) = df_{\varphi_x(t)}(\nu_x) = (f \circ \varphi_x)'(t) + df_{\varphi_x(t)}\big(\nu_x - \varphi_x'(t)\big). \qquad (39)$$

Since $1 \in \mathbf{s}_x(v)$ for all $v \in \mathbb{C}^n$, Theorem 5.4 implies that the norm of df_z is K-bounded at x; therefore 39 yields

$$\lim_{t \to 1^-} \frac{\partial f}{\partial \nu_x}\big(\varphi_x(t)\big) = \beta\tau,$$

and (i) follows from Theorems 4.8 and 5.4, because $\beta\tau \neq 0$. If v_N is not orthogonal to ν_x, we can write $v_N = \lambda\nu_x + v_T$ with $\lambda \neq 0$ and v_T orthogonal to ν_x. Therefore

$$\frac{\partial f}{\partial v_N} = \lambda\frac{\partial f}{\partial \nu_x} + \frac{\partial f}{\partial v_T};$$

and (ii) follows from (i) and Theorem 5.4. □

We recall that for this statement to hold it is necessary to use restricted K-limits and not K-limits: see Example 4.6.

Remark 8. If D is strongly convex or convex of finite type then ([Le], [AT2]) ν_x is a complex multiple of $\mathbf{n}_x$. Therefore in these cases "orthogonal to ν_x" means "complex tangential to ∂D at x", and "not orthogonal to ν_x" means "transversal to ∂D at x". But the previous corollary does not a priori require any smoothness on ∂D.

Remark 9. As it stands, Corollary 5.8 applies for instance to projection devices that are the localization of the canonical projection device in strongly convex domains, or of the canonical projection device in strictly linearly convex domains of finite type. However, once this statement holds for a projection device, one might derive similar statements for not necessarily geometrical projection devices. For instance, in [A4] it is shown how to derive a similar result for the localization of the euclidean projection device in strongly pseudoconvex domains knowing Corollary 5.8 for the (geometrical, good and bounded) localization of the canonical projection device.

If D is a convex Reinhardt domain, we can use a completely different approach, not depending on the degree of smoothness of the boundary. We already noticed that in this case the canonical projection device is bounded and preserves horospheres; if moreover it is good (as, for instance, in the polydisk) we have the following

Theorem 5.9

Let $D \subset\subset \mathbb{C}^n$ be a convex Reinhardt domain, equipped with the canonical projection device at $x \in \partial D$ with $\varphi_x(\zeta) = \zeta x$. Assume that this projection device is good. Let $f \in \mathrm{Hol}(D, \Delta)$ be β-Julia at x, and let $v \in \mathbb{C}^n$ be such that $s_x(v) = 1 \in \mathbf{s}_x(v)$. Then $\partial f / \partial v$ has restricted K-limit at x.

Proof. By Theorems 4.8 and 5.4 it suffices to prove that $\partial f / \partial v(tx)$ converges as $t \to 1^-$.

Let M be the set of all n-uple of natural numbers $k = (k_1, \ldots, k_n) \in \mathbb{N}^n$ with $k_1, \ldots, k_n$ relatively prime and $|k| = k_1 + \cdots + k_n > 0$. For $z \in \mathbb{C}^n$ and $k \in \mathbb{N}^n$ we write $z^k = z_1^{k_1} \cdots z_n^{k_n}$. For each $k \in M$ we choose a point $y(k) \in \partial D$ such that

$$\max\{|x^k| \mid x \in \partial D\} = |y(k)^k|,$$

and we set

$$\Sigma_k = \{x \in \partial D \mid |x_j| = |y_j(k)| \text{ for } j = 1, \ldots, n\}.$$

For instance, if $D = \Delta^n$ we can take $y(k) = (1, \ldots, 1)$ and $\Sigma_k = (\partial \Delta)^n$ for all $k \in M$.

Let $\tau \in \partial\Delta$ be the K-limit of f at x. Since the function $(\tau + f)/(\tau - f)$ has positive real part, the generalized Herglotz representation formula proved in [KK] yields

$$\frac{\tau + f(z)}{\tau - f(z)} = \sum_{k \in M} \left[\int_{\Sigma_k} \frac{w^k + z^k}{w^k - z^k} \, d\mu_k(w) + C_k \right], \tag{40}$$

for suitable $C_k \in \mathbb{C}$ and positive Borel measures μ_k on Σ_k; the sum is absolutely converging.

Let $X_k = \{w \in \Sigma_k \mid w^k = x^k\}$, and set $\beta_k = \mu_k(X_k) \geq 0$ and $\mu_k^o = \mu_k - \mu_k|_{X_k}$, where $\mu_k|_{X_k}$ is the restriction of μ_k to X_k (i.e., $\mu_k|_{X_k}(E) = \mu_k(E \cap X_k)$ for every Borel subset E).

Using these notations 40 becomes

$$\frac{\tau + f(z)}{\tau - f(z)} = \sum_{k \in M} \left[\beta_k \frac{x^k + z^k}{x^k - z^k} + \int_{\Sigma_k} \frac{w^k + z^k}{w^k - z^k} \, d\mu_k^o(w) + C_k \right]. \tag{41}$$

In particular, if $z = tx = \varphi_x(t)$ we get

$$\frac{\tau + f(tx)}{\tau - f(tx)} = \sum_{k \in M} \left[\beta_k \frac{1 + t^{|k|}}{1 - t^{|k|}} + \int_{\Sigma_k} \frac{w^k + t^{|k|}x^k}{w^k - t^{|k|}x^k} \, d\mu_k^o(w) + C_k \right]. \tag{42}$$

Let us multiply both sides by $(1 - t)$, and then take the limit as $t \to 1^-$. The left-hand side, by Lemma 5.7 and Proposition 2.12, tends to $2\tau/\beta$. For the right-hand side, first of all we have

$$\frac{1 + t^{|k|}}{1 - t^{|k|}} (1 - t) = \frac{1 + t^{|k|}}{1 + \cdots + t^{|k|-1}} \longrightarrow \frac{2}{|k|}.$$

Next, if $|x^k| < |y(k)^k|$ it is clear that

$$(1 - t) \int_{\Sigma_k} \frac{w^k + t^{|k|}x^k}{w^k - t^{|k|}x^k} \, d\mu_k^o(w) \to 0. \tag{43}$$

therwise, since $\mu_k^o(X_k) = 0$, for every $\varepsilon > 0$ there exists an open neighborhood A_ε of X_k in Σ_k such that $\mu_k^o(A_\varepsilon) < \varepsilon$. Then

$$(1 - t) \left| \int_{\Sigma_k} \frac{w^k + t^{|k|}x^k}{w^k - t^{|k|}x^k} \, d\mu_k^o(w) \right|$$

$$= (1 - t) \left| \int_{A_\varepsilon} \frac{w^k + t^{|k|}x^k}{w^k - t^{|k|}x^k} \, d\mu_k^o(w) + \int_{\Sigma_k \setminus A_\varepsilon} \frac{w^k + t^{|k|}x^k}{w^k - t^{|k|}x^k} \, d\mu_k^o(w) \right|$$

$$\leq 2 \frac{1 - t}{1 - t^{|k|}} \varepsilon + (1 - t) \left| \int_{\Sigma_k \setminus A_\varepsilon} \frac{w^k + t^{|k|}x^k}{w^k - t^{|k|}x^k} \, d\mu_k^o(w) \right| \longrightarrow \frac{2}{|k|} \varepsilon.$$

Since this happens for all $\varepsilon > 0$, it follows that 43 holds in this case too. Summing up, we have found

$$\frac{\tau}{\beta} = \sum_{k \in M} \frac{\beta_k}{|k|};$$

$\hfill (44)$

in particular, the series on the right-hand side is converging.

Without loss of generality, we can assume that $v = \partial/\partial z_1$. Then differentiating 41 with respect to z_1 we get

$$\frac{\partial f}{\partial z_1}(z) = \bar{\tau}\big(\tau - f(z)\big)^2 \sum_{k \in M} k_1 \frac{z^k}{z_1} \left[\beta_k \frac{x^k}{(x^k - z^k)^2} + \int_{\Sigma_k} \frac{w^k}{(w^k - z^k)^2} \, d\mu_k^o(w) \right].$$

$\hfill (45)$

Then we have

$$\frac{\partial f}{\partial z_1}(tx) = \bar{\tau} \left(\frac{\tau - f(tx)}{1-t} \right)^2 \sum_{k \in M} k_1 t^{|k|-1} \, \bar{x}_1 \left[\beta_k \left(\frac{1-t}{1-t^{|k|}} \right)^2 \right.$$
$$\left. + x^k (1-t)^2 \int_{\Sigma_k} \frac{w^k}{(w^k - t^{|k|} x^k)^2} \, d\mu_k^o(w) \right].$$

The same argument used before shows that

$$(1-t)^2 \int_{\Sigma_k} \frac{w^k}{(w^k - t^{|k|} x^k)^2} \, d\mu_k^o(w) \to 0$$

as $t \to 1^-$. Therefore

$$\lim_{t \to 1^-} \frac{\partial f}{\partial z_1}(tx) = \beta^2 \tau \bar{x}_1 \sum_{k \in M} \beta_k \frac{k_1}{|k|^2},$$

$\hfill (46)$

where the series is converging because $\beta_k k_1 / |k|^2 \le \beta_k / |k|$, and we are done. $\hfill \square$

We end this survey by presenting two results for the case of complex tangential directions.

In the polydisk we have seen that either $s_x(v) = 1$ or $s_x(v) = 0$ for any $v \in \mathbb{C}^n$, and in the latter case $\kappa_D(z; v)$ is bounded for z close to x. The former case is dealt with in the previous theorem; but even in the latter case (which is the embodiment of "complex tangential" for the polydisk) we can prove that $\partial f / \partial v$ behaves as we expect:

Proposition 5.10 *Let $f \in \text{Hol}(\Delta^n, \Delta)$ be β-Julia at $x \in \partial \Delta^n$. Take $v \in \mathbb{C}^n$ such that $v_j = 0$ when $|x_j| = 1$, so that $s_x(v) = 0 \in \mathbf{s}_x(v)$. Then*

$$d(z, \partial D)^{-1} \frac{\partial f}{\partial v}(z)$$

has restricted K-limit 0 at x.

Proof. Since the canonical projection device is bounded, it suffices to prove that the holomorphic function

$$\left(1 - \tilde{p}_x(z)\right)^{-1} \frac{\partial f}{\partial v}(z) \tag{47}$$

has restricted K-limit 0; we recall that, in this case, $\tilde{p}_x \colon \Delta^n \to \Delta$ is given by (see [A5])

$$\tilde{p}_x(z) = \frac{1}{d_x}(z, \check{x}),$$

where d_x is the number of components of x of modulus 1, and

$$\check{x}_j = \begin{cases} x_j & \text{if } |x_j| = 1, \\ 0 & \text{if } |x_j| < 1. \end{cases}$$

Therefore it is enough to prove that 47 is K-bounded and that it tends to 0 when restricted to the radial curve $t \mapsto tx$.

We argue as in the proof of Theorem 5.4. For $j = 1, \ldots, n$ and $z \in \Delta^n$ set

$$w_j(z) = \frac{v_j}{(1 - |z_j|^2)\kappa_{\Delta^n}(z; v)}$$

and define $\psi_z \in \mathrm{Hol}(\Delta, \Delta^n)$ by

$$\psi_z(\zeta) = \left(\frac{\zeta w_1 + z_1}{1 + \bar{z}_1 w_1 \zeta}, \ldots, \frac{\zeta w_n + z_n}{1 + \bar{z}_n w_n \zeta} \right).$$

Then we have $\psi_z(0) = z$ and $\psi_z'(0) = v/\kappa_{\Delta^n}(z; v)$, so that we can write

$$\left(1 - \tilde{p}_x(z)\right)^{-1}\frac{\partial f}{\partial v}(z) = \left(1 - \tilde{p}_x(z)\right)^{-1}\kappa_{\Delta^n}(z; v)(f \circ \psi_z)'(0)$$
$$= \frac{1}{2\pi}\int_0^{2\pi} \frac{f\left(\psi_z(re^{i\theta})\right) - \tau}{\tilde{p}_x\left(\psi_z(re^{i\theta})\right) - 1} \frac{\tilde{p}_x\left(\psi_z(re^{i\theta})\right) - 1}{\tilde{p}_x(z) - 1} \frac{\kappa_{\Delta^n}(z; v)}{re^{i\theta}} \, d\theta, \tag{48}$$

for any $r \in (0, 1)$. The proof of Theorem 5.4 then shows that 47 is K-bounded as well as all the factors in the integrand.

Now let $z = \varphi_x(t) = tx$. Then Lemma 5.7, Theorem 4.8 and Proposition 2.12 imply that the first factor in the integrand converges to $\beta\tau$ as $t \to 1^-$, where $\tau \in \partial\Delta$ is the K-limit of f at x. The choice of v implies that

$$\lim_{t \to 1^-} \kappa_{\Delta^n}(tx; v) = \max_{|x_j| < 1} \left\{ \frac{|v_j|}{1 - |x_j|^2} \right\} = c < +\infty.$$

Finally, again the choice of v implies $\tilde{p}_x\left(\psi_{tx}(\zeta)\right) = t = \tilde{p}_x(tx)$ for all $t > 0$, and thus

$$\lim_{t \to 1^-} \left(1 - \tilde{p}_x(tx)\right)^{-1}\frac{\partial f}{\partial v}(tx) = \frac{c\beta\tau}{2r\pi}\int_0^{2\pi} \frac{d\theta}{e^{i\theta}} = 0,$$

and the assertion follows from Theorem 4.8. □

We remark that this result is more precise than [A5, Proposition 4.8].

We end with a final result in the case of complex tangential directions, whose proof has a fairly different flavor. We shall state the result for strongly convex domains only; but the same argument (due to Rudin [Ru, Proposition 8.5.7]) works in convex domains of finite type too (see [AT2]).

Proposition 5.11 *Let $D \subset\subset \mathbb{C}^n$ be a strongly convex domain, equipped with the canonical projection device at $x \in \partial D$. Let $v \in \mathbb{C}^n$ be complex tangential to ∂D at x, so that $s_x(v) = 1/2 \in \mathbf{s}_x(v)$. Let $f \in \mathrm{Hol}(D, \Delta)$ be β-Julia at x. Then*

$$d(z, \partial D)^{-1/2} \frac{\partial f}{\partial v}(z)$$

has restricted K-limit 0 at x.

Proof. As usual, it suffices to show that

$$\lim_{t \to 1} \frac{1}{(1-t)^{1/2}} \frac{\partial f}{\partial v}(\varphi_x(t)) = 0. \tag{49}$$

We need some preparation. Consider the map $\Phi \colon \Delta \times \mathbb{C} \to \mathbb{C}^n$ given by

$$\Phi(\zeta, \eta) = \varphi_x(\zeta) + \eta v.$$

Clearly, $\Phi^{-1}(D) \cap (\mathbb{C} \times \{0\}) = \Delta$ and $\Phi^{-1}(D) \cap (\{\zeta\} \times \mathbb{C})$ is convex for all $\zeta \in \Delta$. Furthermore, since D is strongly convex, v is complex tangential to ∂D at x and $t \mapsto \varphi_x(t)$ is transversal, there is an euclidean ball $B \subset \Phi^{-1}(D)$ of center $(t_0, 0)$ and radius $1 - t_0$ for a suitable $t_0 \in (0, 1)$.

Now define $\tilde{h} \colon B \to \Delta$ by

$$\tilde{h}(\zeta, \eta) = f(\Phi(\zeta, \eta)).$$

We remark that $\tilde{h}(\zeta, 0) = f(\varphi_x(\zeta))$ and $\partial \tilde{h}(\zeta, 0)/\partial \zeta = \partial f(\varphi_x(\zeta))/\partial v$. Hence we can write

$$\tilde{h}(\zeta, \eta) = f(\varphi_x(\zeta)) + \eta \frac{\partial f}{\partial v}(\varphi_x(\zeta)) + o(|\eta|).$$

Set

$$h(\zeta, \eta) = f(\varphi_x(\zeta)) + \tfrac{1}{2}\eta \frac{\partial f}{\partial v}(\varphi_x(\zeta)) = f(\varphi_x(\zeta)) + \eta(1 - \zeta)^{1/2} g(\zeta),$$

where $g(\zeta) = \tfrac{1}{2}(1 - \zeta)^{-1/2} \partial f(\varphi_x(\zeta))/\partial v$. Since h is the arithmetic mean of the first two partial sums of the power series expansion of $\tilde{h}$, it sends B into Δ. Furthermore, 49 is equivalent to $g(t) \to 0$ as $t \to 1$.

Choose $\varepsilon > 0$ and set $c = \beta^2/\varepsilon^2(1 - t_0)$. We wish to estimate

$$\limsup_{t \to 1} |g(t + ic(1 - t))|.$$

Set $\zeta_t = t + ic(1-t)$; it is easy to check that $(\zeta_t, 0) \in B$ if $(1-t) \leq 2(1 - t_0)/(1+c^2)$. Moreover

$$(1-t_0)^2 - |\zeta_t - t_0|^2 > (1-t_0)(1-t)$$

if $(1-t) < (1-t_0)/(1+c^2)$; hence if t is sufficiently close to 1 we can find $\eta_t \in \mathbb{C}$ such that

$$(1-t_0)^2 - |\zeta_t - t_0|^2 > |\eta_t|^2 > (1-t_0)(1-t) \tag{50}$$

and

$$\eta_t(1 - \zeta_t)^{1/2} g(\zeta_t) \in \mathbb{R}. \tag{51}$$

In particular, $(\zeta_t, \eta_t) \in B$ if $(1-t) < (1-t_0)/(1+c^2)$. By definition,

$$|1 - \zeta_t| = (1-t)\sqrt{1+c^2} \geq c\,(1-t);$$

hence 50 yields

$$|\eta_t(1 - \zeta_t)^{1/2} g(\zeta_t)| \geq (1-t_0)^{1/2} c^{1/2}(1-t)|g(\zeta_t)|. \tag{52}$$

Now, $\zeta_t \in K(1, 2\sqrt{1+c^2})$ if $(1-t) < (1-t_0)/(1+c^2)$; hence, by Lemma 5.7 and Proposition 2.12,

$$\frac{1 - f(\varphi_x(\zeta_t))}{1 - \zeta_t} = \beta + o(1)$$

as $t \to 1$, where without loss of generality we have assumed that the K-limit of f at x is 1. Therefore

$$f(\varphi_x(\zeta_t)) = 1 - (\beta + o(1))(1 - ic)(1-t). \tag{53}$$

Putting together 51, 52 and 53 we get

$$1 \geq \mathrm{Re}[h(\zeta, \eta)] \geq 1 - (\beta + o(1))(1-t) + (1-t_0)^{1/2} c^{1/2}(1-t)|g(\zeta_t)|,$$

that is

$$|g(\zeta_t)| \leq \frac{\beta + o(1)}{(1-t_0)^{1/2} c^{1/2}}.$$

Therefore

$$\limsup_{t \to 1} |g(t + ic(1-t))| \leq \frac{\beta}{(1-t_0)^{1/2} c^{1/2}} = \varepsilon.$$

Clearly the same estimate holds for $\zeta_t' = t - ic(1-t)$. Since $|g(\zeta)|$ is bounded in the angular region bounded by these two lines, it follows that

$$\limsup_{t \to 1} |g(t)| \leq \varepsilon.$$

Since ε is arbitrary, the assertion follows. $\square$

References

[A1]   M. Abate, *Common fixed points of commuting holomorphic maps*, Math. Ann., 283 1989 645-655

[A2]   M. Abate, *The Lindelöf principle and the angular derivative in strongly convex domains*, J. Analyse Math., 54 1990 189-228

[A3]   M. Abate, *Iteration theory of holomorphic maps on taut manifolds*, Mediterranean Press, Cosenza, 1989

[A4]   M. Abate, *Angular derivatives in strongly pseudoconvex domains*, Proc. Symp. Pure Math., 52, Part 2, 1991 23-40

[A5]   M. Abate, *The Julia-Wolff-Carathéodory theorem in polydisks*, J. Anal. Math., 74 1998 275-306

[AT1]   M. Abate, R. Tauraso, *The Julia-Wolff-Carathéodory theorem(s)*, Contemp. Math., 222 1999 161-172

[AT2]   M. Abate, R. Tauraso, *The Lindelöf principle and angular derivatives in convex domains of finite type*, To appear in J. Austr. Math. Soc., 73 2002 .-.

[Ah]   L. Ahlfors, *Conformal invariants*, McGraw-Hill, New York, 1973

[BGS]   W. Balmann, M. Gromov, V. Schroeder, *Manifolds of nonpositive curvature*, Birkhäuser, Basel, 1985

[Ba]   G. Bassanelli, *On horospheres and holomorphic endomorphisms of the Siegel disc*, Rend. Sem. Mat. Univ. Padova, 70 1983 147-165

[Bu]   R.B. Burckel, *An introduction to classical complex analysis*, Academic Press, New York, 1979

[C1]   C. Carathéodory, *Über die Winkelderivierten von beschränkten analytischen Funktionen*, Sitz. Preuss. Akad. Wiss., Berlin, 1929 39-54

[C2]   C. Carathéodory, *Theory of functions of a complex variable, II*, Chelsea, New York, 1960

[Ca]   D.W. Catlin, *Estimates of invariant metrics on pseudoconvex domains of dimension two*, Math. Z., 200 1989 429-466

[CK]   J.A. Cima, S.G. Krantz, *The Lindelöf principle and normal functions of several complex variables*, Duke Math. J., 50 1983 303-328

[Č]   E.M. Čirka, *The Lindelöf and Fatou theorems in $\mathbb{C}^n$*, Math. USSR, Sb., 21 1973 619-641

[D1]   P.V. Dovbush, *Existence of admissible limits of functions of several complex variables*, Sib. Math. J., 28 1987 83-92

[D2]   P.V. Dovbush, *On the Lindelöf theorem in $\mathbb{C}^n$*, Ukrain. Math. Zh., 40 1988 796-799

[DZ]   Ju.N. Drožžinov, B.I. Zav'jalov, *On a multi-dimensional analogue of a theorem of Lindelöf*, Sov. Math. Dokl., 25 1982 51-52

[EHRS] M. Elin, L.A. Harris, S. Reich, D. Shoikhet, *Evolution equations and geometric function theory in J^*-algebras*, J. Nonlinear Conv. Anal., 3 2002 81-121

[]   F K. Fan, *The angular derivative of an operator-valued analytic function*, Pacific J. Math., 121 1986 67-72

[Fa]   P. Fatou, *Séries trigonométriques et séries de Taylor*, Acta Math., 30 1906 335-400

[FR]   F. Forstneric, J.-P. Rosay, *Localization of the Kobayashi metric and the boundary continuity of proper holomorphic mappings*, Math. Ann., 279 1987 239-252

[G]   I. Graham, *Boundary behavior of the Carathéodory and Kobayashi metrics on strongly pseudoconvex domains in $\mathbb{C}^n$ with smooth boundary*, Trans. Am. Math. Soc., 207 1975 219-240

[H1]   K.T. Hahn, *Asymptotic behavior of normal mappings of several complex variables*, Canad. J. Math., 36 1984 718-746

[H2]   K.T. Hahn, *Nontangential limit theorems for normal mappings*, Pacific J. Math., 135 1988 57-64

[He]   M. Hervé, *Quelques propriétés des applications analytiques d'une boule à m dimensions dans elle-même*, J. Math. Pures Appl., 42 1963 117-147

[J]   H.L. Jackson, *Some remarks on angular derivatives and Julia's lemma*, Canad. Math. Bull., 9 1966 233-241

[Ja]   F. Jafari, *Angular derivatives in polydiscs*, Indian J. Math., 35 1993 197-212

[JP]   M. Jarnicki, P. Pflug, *Invariant distances and metrics in complex analysis*, Walter de Gruyter, Berlin, 1993

[Ju1]   G. Julia, *Extension nouvelle d'un lemme de Schwarz*, Acta Math., 42 1920 349-355

[Ju2]   G. Julia, *Principes géométriques d'analyse, I*, Gounod, Paris, 1930

[K]   Yu.V. Khurumov, *On Lindelöf's theorem in $\mathbb{C}^n$*, Soviet Math. Dokl., 28 1983 806-809

[Kob1] S. Kobayashi, *Invariant distances on complex manifolds and holomorphic mappings*, J. Math. Soc. Japan, 19 1967 460-480

[Kob2] S. Kobayashi, *Hyperbolic complex spaces*, Springer, Berlin, 1998

[Kom]   Y. Komatu, *On angular derivative*, Kōdai Math. Sem. Rep., 13 1961 167-179

[Kor]   A. Korányi, *Harmonic functions on hermitian hyperbolic spaces*, Trans. Am. Math. Soc., 135 1969 507-516

[KS]   A. Korányi, E.M. Stein, *Fatou's theorem for generalized halfplanes*, Ann. Sc. Norm. Sup. Pisa, 22 1968 107-112

[KK]   S. Kosbergenov, A.M. Kytmanov, *Generalizations of the Schwarz and Riesz-Herglotz formulas in Reinhardt domains*, (Russian) Izv. Vyssh. Uchebn. Zaved. Mat., n. 10 1984 60-63

[Kr1]   S.G. Krantz, *Invariant metrics and the boundary behavior of holomorphic functions on domains in $\mathbb{C}^n$*, J. Geom. Anal., 1 1991 71-97

[Kr2]   S.G. Krantz, *The boundary behavior of the Kobayashi metric*, Rocky Mountain J. Math., 22 1992 227-233

[LV]   E. Landau, G. Valiron, *A deduction from Schwarz's lemma*, J. London Math. Soc., 4 1929 162-163

[Le]   L. Lempert, *La métrique de Kobayashi et la représentation des domaines sur la boule*, Bull. Soc. Math. France, 109 1981 427-474

[Li1]   E. Lindelöf, *Mémoire sur certaines inégalités dans la théorie des fonctions monogénes, et sur quelques propriétés nouvelles de ces fonctions dans le voisinage d'un point singulier essentiel*, Acta Soc. Sci. Fennicae, 35 1909 3-5

[Li2]   E. Lindelöf, *Sur un principe générale de l'analyse et ses applications à la theorie de la représentation conforme*, Acta Soc. Sci. Fennicae, 46 1915 1-35

[MM]   M. Mackey, P. Mellon, *Angular derivatives on bounded symmetric domains*, Preprint, 2001

[M]   P. Mellon, *Holomorphic invariance on bounded symmetric domains*, J. Reine Angew. Math., 523 2000 199-223

[Me]   P.R. Mercer, *Another look at Julia's lemma*, Compl. Var. Th. Appl., 43 2000 129-138

[Mi]   A. Minialoff, *Sur une propriété des transformations dans l'espace de deux variables complexes*, C.R. Acad. Sci. Paris, 200 1935 711-713

[N]   R. Nevanlinna, *Remarques sur le lemme de Schwarz*, C.R. Acad. Sci. Paris, 188 1929 1027-1029

[Po]   C. Pommerenke, *Univalent functions*, Vandenhoeck & Ruprecht, Göttingen, 1975

[R]   A. Renaud, *Quelques propriétés des applications analytiques d'une boule de dimension infinie dans une autre*, Bull. Sci. Math., 97 1973 129-159

[Ro]   H.L. Royden, *Remarks on the Kobayashi metric*, Several Complex Variables, II, Lect. Notes Math. 189, Springer, Berlin, 1971, pp. 125-137

[RW]   H.L. Royden, P.-M. Wong, *Carathéodory and Kobayashi metric on convex domains*, Preprint, 1983

[Ru]   W. Rudin, *Function theory in the unit ball of* $\mathbb{C}^n$, Berlin, Springer, 1980

[SW]   A. Szałowska, K. Włodarczyk, *Angular derivatives of holomorphic maps in infinite dimensions*, J. Math. Anal. Appl., 204 1996 1-28

[St]   E.M. Stein, *Boundary behavior of holomorphic functions of several complex variables*, Princeton University Press, Princeton, 1972

[T]   M. Tsuji, *Potential theory in modern function theory*, Maruzen, Tokyo, 1959

[V1]   E. Vesentini, *Variations on a theme of Carathéodory*, Ann. Sc. Norm. Sup. Pisa, 6 1979 39-68

[V2]   E. Vesentini, *Complex geodesics*, Comp. Math., 44 1981 375-394

[W]   S. Wachs, *Sur quelques propriétés des transformations pseudo-conformes avec un point frontière invariant*, Bull. Soc. Math. Fr., 68 1940 177-198

[Wł1]   K. Włodarczyk, *Julia's lemma and Wolff's theorem for* J^**-algebras*, Proc. Amer. Math. Soc., 99 1987 472-476

[Wł2]   K. Włodarczyk, *The Julia-Carathéodory theorem for distance decreasing maps on infinite-dimensional hyperbolic spaces*, Atti Accad. Naz. Lincei, 4 1993 171-179

[Wł3]   K. Włodarczyk, *Angular limits and derivatives for holomorphic maps of infinite dimensional bounded homogeneous domains*, Atti Accad. Naz. Lincei, 5 1994 43-53

[Wł4]   K. Włodarczyk, *The existence of angular derivatives of holomorphic maps of Siegel domains in a generalization of* C^**-algebras*, Atti Accad. Naz. Lincei, 5 1994 309-328

[Wo]   J. Wolff, *Sur une généralisation d'un théorème de Schwarz*, C.R. Acad. Sci. Paris, 183 1926 500-502

[Wu]   H. Wu, *Normal families of holomorphic mappings*, Acta Math., 119 1967 193-233

[Z]   J.M. Zhu, *Angular derivatives of holomorphic maps in Hilbert spaces*, J. Math. (Wuhan), 19 1999 304-308

Real Methods in Complex Dynamics

John Erik Fornæss

Mathematics Department, The University of Michigan,
East Hall, Ann Arbor, Mi 48109 USA
fornaess@umich.edu

1 Lecture 1: Introduction to Complex Dynamics and Its Methods

1.1 Introduction

The main topic in this lecture series is complex dynamics in higher dimension, see Figure 1. Our focus will be on the methods used. In particular we are interested in real methods. We will interpret the term "real methods" quite broadly. First of all we will suppress to a large extent methods using complex analytic arguments and concepts. This includes algebraic geometry which has been used extensively recently in complex dynamics. On the other hand we will stress concepts and tools from other parts of dynamics such as the study of smooth diffeomorphisms. Topological dynamics and symbolic dynamics on the other hand will play a lesser role as they go beyond what one might call real methods. One topic which we will stress is random iteration.

```
┌─────────────────────────────────────────┐
│                                         │
│              Lecture 1                  │
│                                         │
│    Complex Dynamics in Dimension 1.     │
│                                         │
│                                         │
│              Lecture 2                  │
│                                         │
│     Fatou sets in higher dimension      │
│                                         │
│                                         │
│              Lecture 3                  │
│                                         │
│            Saddle points                │
│                                         │
│                                         │
│              Lecture 4                  │
│                                         │
│            Saddle sets                  │
│                                         │
└─────────────────────────────────────────┘
```

Fig. 1. The Lectures

Complex dynamics, Figure 2, can be divided in two parts, the Julia theory and the Fatou theory. With the Julia theory the focus is on the Julia set where the dynamics has chaotic features. In the Fatou theory the focus is on the orderly behaviour on the Fatou set. In addition there are sets which mix these two features. These are sets where the dynamics is chaotic in some directions and orderly in other directions, this is called saddle behaviour. In this lecture series we will focus on the orderly behaviour and on saddle behaviour. This is the part of the theory where we believe there are more basic open questions.

In the first lecture we start by making a few general remarks about dynamics. I have been told on occasion by people that they can read and understand papers on dynamics, but they don't see what is the motivation for the problems that are investigated. My hope is that those few elementary remarks will be useful. The motivations are to some extent the same these days as a century ago. We will touch on that at the end of the lecture. Basically the original motivations came from Newtons method and from Celestial mechanics. And in the meantime there has been very little progress. Researchers are still struggling with basic problems about Newtons method and the question of stability of the solar system is still as wide open as ever.

In the main part of this lecture, we will give a brief survey over complex dynamics in one dimension. This serves to introduce a few basic concepts, but the main purpose is to discuss methods used.

The second lecture introduces some topics about stable sets in complex dynamics. This is the only lecture where we focus on complex analytic methods. The basic concept is that of a Fatou set. Not much is known about the nature of Fatou sets. In one variable one has a complete description in the case of rational maps on the Riemann sphere. We discuss this already in the first lecture. For entire functions the situation is much less known. We focus on the case of holomorphic maps in higher dimension. There are some results known about what are possible types of Fatou components, but we will also point out many open questions. In addition to the Fatou set which is open, the iterates might be a normal family when restricted to certain low-dimensional submanifolds. This occurs for example at a fixed saddle point, i.e. a fixed point where some eigenvalues of the derivative have modulus strictly less than one and the other eigenvalues are strictly larger than one. Basic concepts in our discussion are saddle points and saddle sets. In such a setting it is natural to introduce random iteration. Basically the normal behaviour can be along a complex manifold which jumpes around under iteration.

In the third lecture the main focus is on a deep theorem by Bedford-Smillie-Lyubich ([BLS]) concerning the distribution of saddle points for complex Hénon maps. The proof uses real concepts and methods to a large extent. If F is a Hénon map, there is a certain probability measure which is invariant under F. This is the unique measure of maximal entropy. This measure is ergodic and the map is measure hyperbolic on the support J^* of the measure. This allows one to use Oseledets and Pesin theory from the theory of dynamics of real smooth diffeomorphisms to find compact Pesin boxes on which the map is uniformly hyperbolic. With these tools together with complex analytic methods, one can show that all saddle points are in J^*. The general theory of smooth dynamics allows one to conclude using a theorem of Katok that the saddle points are dense in J^*. This is the main result in this lecture and is the main tool needed for the last lecture.

Finally, in the last lecture we apply the results in Lecture 3 to discuss how one can prove uniform saddle behaviour for Hénon maps under suitable hypotheses. This extends work in ([F3]). In ([F3]) we showed that under these hypotheses the map is hyperbolic in J^*. One of the main open problems in the theory of Hénon maps is the equality of J and J^*. In this lecture we show how one can prove equality in some setting. It remains open if one can generalize this and prove that in general if F is hyperbolic on J^*, then F is hyperbolic on J (which would imply that $J = J^*$).

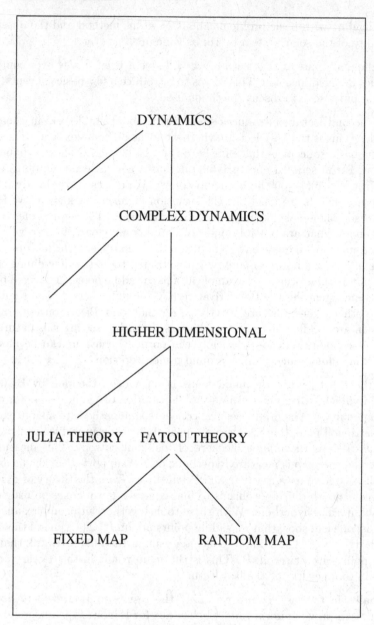

Fig. 2. Overview

1.2 General Remarks on Dynamics

Dynamics can be thought of as describing the interaction of 2 or more entities. There is a feeling that with 3 or more entities the interaction might have chaotic features. The reason is that the interaction between any two entities is constantly influenced by the third and this influence from the third unit keeps changing. It turns out that to actually prove that there are chaotic features in such a situation is very difficult and very little has actually been done.

One is particularly interested in how this interaction proceeds over time.

To make a scientific investigation, one can proceed by making a mathematical model of the interaction of the entities. This typically leads to a set X which describes at least approximately the relationship between the entities at some specific time, see Figure 3. One might then model the interaction of the entities by a map $F : X \to X$. So if $x \in X$ describes the situation at time T_1 then $y := F(x)$ describes the situation at time T_2. The time T_2 might be depending on x. This is the discrete version, there is a continuous version. In this case X might be a manifold and the map F is the integral of some vector field on X. So the dynamics is given by a flow. Also if one wants to model that the background situation varies over time, one might use variable maps F, perhaps varying randomly. Or one can replace $F(x)$ by $F(x) + \delta$, δ some small error.

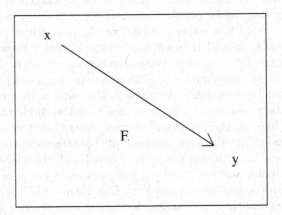

Fig. 3. Dynamical System

The computer is a tool in the investigation of dynamics. There are three main ways in which computers are used. Using the computer one can make approximate calculations, such as solving differential equations which are too complicated to be solved exactly. Basic problems are long-term stability of the solar system and weather prediction. An example of a "Theorem" "proved" in this way is the result of Laskar ([LM]). Laskar estimated that an error of Earths orbit doubles every 3.5 Million years, so an unmeasurable error such as a small fraction of the Planck length becomes as large as the radius of the Earths orbit after 500 Million Years, i.e. 10 percent of the age of the Earth.

Another way to use a computer is as an experimental tool to indicate conjectures that then can be proved rigorously. The Feigenbaum map is a famous example. Feigenbaum, see ([M]), discovered that by iterating a certain polynomial of the form $x^2 + c$ he got, after rescaling, functions that were close to the original one. This lead to the theory of renormalization in the field of complex dynamics in one variable. In fact rigorous proofs in this area all use complexification. An analogous theory has not been developed in higher dimension although computer pictures indicate that the phenomenon occurs also for Hénon maps in $\mathbb{R}^2$. The phenomenon is related to period doubling. In the complex domain one can ask the same in the case of behaviour under for example period tripling in the complex part of the Mandelbrot set. Although computer experiments indicate that there is renormalization also in this case, there is no rigorous proof yet.

A third way is to use computers to give rigorous proofs. An example is in ([FGa]). Here the authors investigate real Hénon maps. These Hénon maps have a fixed saddle point. This point then has an unstable manifold W^u and a stable manifold W^s. As one varies the parameters of the map, it appears from computer pictures that these two manifolds develop a tangency. The question then is whether this can be proved rigorously and whether the tangency has a generic nature, that is whether the tangency is quadratic and whether the stable and unstable manifolds move through the tangency linearly when the parameters change. This was proved using the computer in steps. The first step was to locate the tangency precisely. This was done using computer graphics and using gradually more digits of accuracy. The authors then observed that when the accuracy was about 10^{-10} the stable and unstable manifold came out as paralell lines on the computer pictures. After that it was necessary to convert to the use of Newton's method to locate the tangency. This process was continued in steps until the tangency was located with the precision of about 10^{-20}. At this point one could set up the equation for a generic tangency as the solution of a system of equations. One then could use the close by numerical values and calculate enough derivatives with enough precision that one could prove that the equations had a root somewhere very close. This last calculation was done using interval arithmetic in which only precise calcuation with integers was used to determine small intervals in which wanted numbers were located.

This has lead to theorems in complex dynamics in higher dimension, such as the existence of holomorphic endomorphisms with infinitely many attracting basins in $\mathbb{P}^2$ ([G]).

Next we turn to the important concept of phenomenology. In this context we simply mean the study of various topics using simplified mathematical models which (hopefully) capture the essential features. One can use dynamics as a systematic tool to study all sorts of diverse phenomena. Famous examples are the Lorentz attractor and the Hénon maps. It is common practice in science to study models of a situation. These models are always simplifications. In phenomenology one then makes even further simplifications to make it easier to analyze essential features. In the study of iteration of the map $F : X \to X$ one might reduce the dimension of the space X to a minimum and make maximal simplifications in the shape of the map F. This was the case for the Lorenz equations and the Hénon maps. These arose from simplifictions of equations describing the weather. Another more recent example is the controversial work by Wolfram ([W]) modeling the Universe by simple computer programs. In ([W]) a point $x \in X$ refers to a selection of colors on a given collection of squares and $F(x)$ is the next selection of colors.

The basic goals of a study of a dynamical system (F, X) can perhaps be thought of as the following two: Understanding the nature of a system (is the system predictable/unpredictable; is the description accurate/inaccurate) and learning how to manipulate it. So the question then is what information about the dynamical system would be useful towards these ends. In medical treatment one might play with parameters in the map $F = F(x, p)$ and change p around to improve the long term result $F^n(x)$, say administering a medicine at different time intervals, change the size of the doses etc. In weather prediction or calculation of orbits of satellites or planets, one calculates $F^n(x)$. For a satellite, one might in addition learn how to manipulate the answer $F^n(x)$ to a desired one. Basically one gives a rocket a small boost at a good time. This amounts to replacing $x_k = F^k(x)$ by a close by x'_k and then continue iterating to get $F^\ell(x'_k)$.

A big problem in dynamics arises when $F^n(x), F^n(y)$ become quite different for x close to y. In practice x is only known up to some error. This makes the system unpredictable except for small n. As the example of the weather or its simplifications to Lorenz and Hénon maps show, unpredictability does happen in important systems. This leads naturally to notions of disorderly or chaotic systems. It might happen that X has two pieces, a chaotic piece and an orderly piece. One might also have attractors. The Lorenz attractor and the Hénon attractor are compact sets $K \subset X$ to which $\{F^n(x)\}$ converge. On K the dynamics appers chaotic. Attractors have also been studied in complex dynamics. Let (X, d) be a compact metric space and f a continuous map from X to X. The sequence $(x_j)_{1 \le j \le n}$ is an ϵ-pseudo-orbit if $d(f(x_j), x_{j+1}) < \epsilon$ for $j = 1, \ldots, n - 1$. For $a, b \in X$, we write $a \succ b$ if for every $\epsilon > 0$ there is an ϵ-pseudo-orbit from a to b. We also write $a \succ a$. We write $a \sim b$ if $a \succ b$ and $b \succ a$, and denote by $[a]$ the equivalence class of a under this relation. Define an *attractor* to be a minimal equivalence class for $\sim$. Attractors for holomorphic maps on $\mathbb{P}^2$ were first investigated in ([FW1]).

1.3 An Introduction to Complex Dynamics and Its Methods

After these brief general remarks about dynamics we discuss complex dynamics from a historical perspective. Complex dynamics goes back to about 1871 with work by Schröder ([Sc]). Suppose that F is a holomorphic function near 0 in $\mathbb{C}$, $F(0) = 0$. The question is whether one can change coordinates so that F becomes a linear function. This leads to the Schröder functional equation for the coordinate change ϕ.

$$\phi \circ F = \lambda \phi$$

where $\phi'(0) = 1$ and $\lambda = F'(0)$. In the case when $0 < |\lambda| < 1$, this was resolved by the end of the century. The methods were purely complex, using power series, so they fall outside of our main topic.

The case when $F'(0) = 1$ is more difficult. One cannot solve the Schröder equation in this case. In this case there is generically an open region which is attracted to 0. The precise statement is usually called the flower theorem.

In the years up to 1920, Fatou and Julia developed the theory also in the global case. Let R be a rational function on $\overline{\mathbb{C}}$. The Fatou set F is the largest open set on which the sequence of iterates $\{R^n\}$ is a normal family. The Julia set J is the complement of the Fatou set. The main global method was normal families arguments. Again this falls within complex analytic methods.

In 1942 Siegel solved the Schröder equation in certain cases when $|\lambda| = 1$. Basically, λ had to be far enough from roots of unity. His methods were again to use power series and fall within complex analytic methods.

In 1965, Brolin introduced the use of the Green's function G. If P is a polynomial on $\mathbb{C}$, the Green's function

$$G(z) := \lim_{n \to \infty} \frac{\log^+ |P^n(z)|}{d^n},$$

d is the degree of P, measures the escape rate to infinity. This is a subharmonic function, harmonic in some places. Again this qualifies as a complex analytic tool.

The Mandelbrot set was introduced in 1978 by Brooks and Matelski and Mandelbrot, using the computer to draw it. Let $P_c(z) = z^2 + c$ denote the quadratic family of polynomials. The Mandelbrot set M measures how the dynamics of P_c depends on the parameter $c \in \mathbb{C}$. In particular, $c \in M$ if and only if the Julia set J_c of P_c is connected. This brings in other methods than just complex analysis into complex dynamics.

In the 1980s real methods and concepts from the general real theory of dynamics entered in a basic way into complex dynamics. We mention a few such results.

In 1983: Przytycki ([Pr]) found a formula for the **Lyapunov exponent** of a polynomial map in one complex dimension. The Lyapunov exponent measures the expansion of the mapping.

$$\lambda_0 = \log d + \sum_{c, f'(c)=0} \lim_{n \to \infty} \frac{1}{d^n} \log^+ |f^n(c)|.$$

In ([FLM]) Freire, Lopes and Mañe and in ([Ly]) Lyubich showed that rational maps have entropy $\log d$.

In 1983 Lyubich and Mañe showed that $\mu = dd^c G$ is the unique measure of maximal entropy in $\mathbb{C}$.

In 1983 ([MSS]), Mañe, Sad, Sullivan, showed that stable rational maps are dense in the space of rational maps. The question of density of hyperbolic rational maps is still open. A rational map $R : \overline{\mathbb{C}} \to \overline{\mathbb{C}}$ is said to be hyperbolic if there are constants $c > 0, \lambda > 1$ so that $|(R^n)'(z)| > c\lambda^n$ for all $n \geq 1, z \in J$. One can show that a rational map is hyperbolic if and only if the critical orbit does not cluster on the Julia set. The similar question for holomorphic maps in higher dimension is still open.

There are Julia sets with positive area for rational maps, but it is an open question whether this is possible for the quadratic family.

In 1985, Sullivan completed the classification of Fatou domains with his non-wandering Theorem, using quasiconformal maps, a complex method. Sullivans theorem implies that all Fatou components are preperiodic. The periodic components are of 4 types.

(1) Periodic attracting basins.

(2) Periodic parabolic basins. Here there is a periodic point p, $F^n(p) = p$ and $(F^n)'(p)$ is a root of unity. The periodic point in this case is on the boundary of the basin of attraction.

(3) Siegel Discs.

(4) Herman rings.

In the last two cases the map is conjugate to rotation on a disc or an annulus. Keeping in mind the possible generalization of this to higher dimension it is more reasonable to call these domains by the common name Siegel domains.

Mary Rees ([R]), 1986 found a family of positive measure in the set of rational maps which are **ergodic** with respect to Lebegues measure. This was based on work in the real case by Benedicks-Carleson and Jacobsson. The real case was an investigation into the real quadratic maps $1 - ax^2$ with a close to 2. This was also a precursor to the important work by Benedicks and Carleson proving that for some parameters the real Hénon maps exhibit strange attractors. This lead to the work ([FGa]) mentioned above.

Shishikura introduced the use of quasiconformal surgery in 1987. There is no analogue yet of this in higher dimension. Quasiconformal maps can be used to study small perturbations $z^2 + \epsilon$ of z^2 for example. In this case the Julia set is a quasicircle. But it is not known how to make an analogous study of $(z^2 + \epsilon w, w^2 + \delta z)$ say.

1989: Eremenko-Lyubich showed that μ is the unique measure of maximal entropy for rational maps on $\overline{\mathbb{C}}$.

One serious objection to the theory of iteration of a map is that in most realistic situation the surroundings are always changing. One way to investigate situations more realistically is to use random iteration. This was first done in complex dynamics in ([FS1]). A rational map $R(z) : \overline{\mathbb{C}} \to \overline{\mathbb{C}}$ can be extended to a holomorphic map $R(z,c) : \overline{\mathbb{C}} \times \Delta(0,\epsilon) \to \overline{\mathbb{C}}$ where c is a parameter. Under random iteration one considers iterates of functions $R_n(z), z \to R(z,c_n)$ for a sequence $\{c_n\} \subset \Delta(\epsilon)$ and the composites $R_{(n)} = R_n \circ \cdots \circ R_1$. Random iteration is a natural topic also when one investigates dynamics in higher dimension. A simple example is polynomial skew products, $R(z,w) = (P(z), Q(z,w))$ where P and Q are polynomials. One can view the dynamics in the w direction as a random iteration as the z variable changes at each iteration. Random iteration in higher dimension was studied in ([FW2]).

We end this lecture by discussing the original motivations for investigating complex dynamics. And we revisit this from a more modern point of view.

There were two initial motivations for studying complex dynamics. One was the study of Newtons method to find roots of polynomials. If $P(z)$ is a polynomial, let
$$R(z) = z - \frac{P(z)}{P'(z)}.$$
Then $R(z)$ is a rational function on $\overline{\mathbb{C}}$. If z_0 is a root of P, $P(z_0) = 0$, then $R(z_0) = z_0$ and $|R'(z_0)| < 1$. Hence if z is in some neighborhood of z_0, then the iterates $R^n(z) \to z_0$. This was basically only successful for polynomials of degree 2. In the quadratic case one can after a change of coordinates assume that $P(z) = z^2 - 1$. In that case the roots are ± 1 and the corresponding basins are the open left and right half planes. The rational function $R = \frac{z^2+1}{2z}$ which maps the imaginary axis $\cup \infty$ to itself chaotically. One searched in vain for a smiliar simple description for higher degrees.

This problem was central in reviving complex dynamics around 1980 when Hubbard rediscovered that the degree 3 case was extremely difficult. This problem was a strong motivation in ([FS3]) for investigating rational maps in higher dimension. The point is that although in one complex dimension, Newtons method gives rise to a rational function, this function is still holomorphic, i.e. has no singularities as a map on $\overline{\mathbb{C}}$, when one uses Newtons method in higher dimension, one gets quotients of polynomials as in one variable, but in this case, the zero sets of the numerator and the denominator can no longer be cancelled, and one gets points of indeterminacy. These are singularities when one extends the map to projective space. Much of the work on this topic in higher dimension is concerned with how to manage these singularities, such as blowing up points etc.

The other motivation came from celestial mechanics. Poincare studied the 3-body problem, Figure 4, via a Poincare map describing how a massless third body hit an imaginary plane after each rotation. This gave rise to a planar endomorphism which is not given by an explicit formula. Hence the idea was to learn about the situation by iteration of simple explicit maps such as holomorphic polynomials.

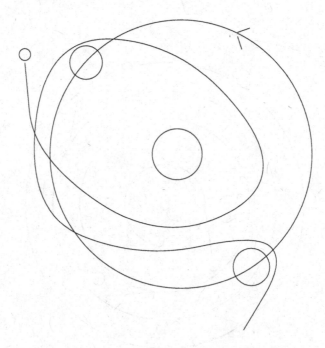

Fig. 4. 3 Body Problem

More recently, KAM theory has been used to find invariant circles which can trap the orbit of the third body. These are real versions of the Siegel discs arising in the complex case. In particular, Hénon used computer graphics to investigate this assuming that the Poincare map could be modelled by an area-preserving real planar Hénon map.

In higher dimension one has Siegel domains that can trap orbits in the complex but in the reals one only has invariant tori which do not trap points because of their high codimension. From a more modern viewpoint, one has also to take into account that there are random disturbances from other celestial bodies which are likely to kick a body from one orbit to another, so that if the orbit is inside a KAM circle or a Siegel domain, it is likely to migrate off over time, Figure 5.

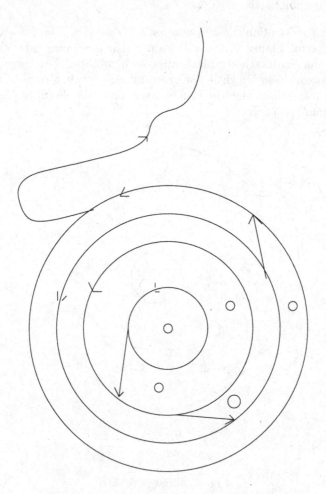

Fig. 5. Random Migration off KAM circles

In ([FS4], [FS6]) the authors investigated the dynamics of holomorphic symplectomorphisms, i.e. biholomorphic maps which preserve the symplectic form $\sum dz_j \wedge dw_j$. This generalizes the situation of the gravational many body problem which are given by real symplectomorphims. The motivation from the real situation prompted Michel Herman to ask about the long-term behaviour of orbits in this complex setting. In fact Herman conjectured that there is a dense G_δ set of holomorphic symplectomorphisms of $\mathbb{C}^2$ for which the set of points with bounded orbit has empty interior. The main result in ([FS4]) was the verification of this conjecture. This was generalized to higher dimension in ([FS6]).

In the real gravitational setting it is known that in the case of 5 bodies it is possible that the system explodes in finite time. i.e. at least one of the bodies clusters at infinity in finite time. It was therefore natural to ask the prevalence of this phenomenon in the complex setting. This was done in ([FGr]). There it was shown that for a dense set of holomorphic Hamiltonian flows, the set of points which explode is also dense.

Another development after the beginning of complex dynamics is progress in physics. Instead of just stability questions of the solar system one has many body problems on a wide ranging scale from subatomic to superclusters of galaxies, from questions of origin and growth of structures to their stability. In quantum mechanics one can study iteration of unitary maps on complex Hilbert space. This leads naturally to the question of how to extend the theory of complex dynamics to infinite dimension. In quantum mechanics there are various notions of quantum chaos that have been explored. Since there seems to be chaotic phenomena in classical mechanics, and since one expects quantum mechanics to approach classical mechanics in the limit of higher energy, one expects some signs of this chaos in quantum mechanics. Complex dynamics in the setting of infinite dimension has been explored in ([F2]). The results in that paper were generalized by Weickert in ([We2]). Weickert discussed spectral properties for a unitary quantum map on Hilbert space, $U(\alpha)$ depending on a real parameter α. Under a Diophantine condition on α, $U(\alpha)$ has a pure point spectrum. However, the corresponding classical dynamics is shown to be unbounded for all irrational α. A sufficient condition for purely continuous spectrum is also provided in ([We2]). In ([FW3]) the authors investigated the quantization of Hénon maps and investigated the similarity of the dynamics of the Hénon maps and their quantization. In ([FS9]) the authors investigated more general unitary maps on Hilbert space from the viewpoint of complex dynamics. In the papers ([F2]) and ([We2]) the main topic was to give rigorous proofs of the phenomenon of localization. Localization is considered a situation where chaotic behaviour of the classical system is suppressed in the quantized version. This still leaves open what is the better notion of quantum chaos, or how to define chaotic behaviour in the setting of complex dynamics in infinite dimension.

In ([BF]) the authors investigate the growth of structures. Let M be a topological space and $f : M \to M$ a continuous map. Let ρ be a continuous pseudometric on M and let μ be a probability measure on M. A distinguished point p in M is said to *absorb* all points $q \in M$ whose orbit get closer than a given number, called *radius*, $\epsilon_p > 0$, i.e., $\rho(f^n(p), f^n(q)) \le \epsilon_p$ for some integer $n \ge 0$. The basin of attraction, $B(p)$ of p consists of those q which gets absorbed by p. Hence a point p absorbs all points which at least once gets close to p. The authors investigate this phenomenon in the context of real and complex dynamics, Figure 6.

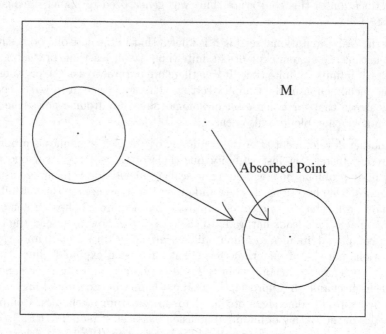

Fig. 6. Basin

2 Lecture 2: Basic Complex Dynamics in Higher Dimension

2.1 Local Dynamics

We recall the solution of the Schröder equation in the attracting case in one complex dimension.

Theorem 2.1 *(Koenig 1884, [Ko]) Let $F : \mathbb{C}_0 \to \mathbb{C}_0$ be a germ of a holomorphic function fixing the origin. Suppose that $0 < |F'(0)| < 1$. Then there exists a coordinate change ϕ so that $\phi \circ F = F'(0)\phi$, i.e., in a new coordinate system F is linear.*

This works just as well if $|F'(0)| > 1$.

In the case $F(z) = \alpha z^k + \cdots, \alpha \neq 0, \ k \geq 2$, the analogous Böttcher equation $\phi \circ F = \phi^k$ is solvable with a coordinate change ϕ ([Bo]).

Theorem 2.2 *(Siegel 1942 ([Si1])) Let $F : \mathbb{C}_0 \to \mathbb{C}_0$ be a germ of a holomorphic function fixing the origin. Suppose $F'(0) = a = e^{2\pi i\omega}$ has modulus one. Suppose also that there exist $\lambda > 0, \mu > 1$ so that $|\omega - \frac{m}{n}| > \lambda n^{-\mu}$ (diophantine condition) for all integers $m, n \geq 1$. Then there is a change of coordinates ϕ so that $\phi \circ F = F'(0)\phi$, i.e. in a new coordinate system F becomes linear.*

In 1952, Siegel ([Si2]) proved a similar result for vector-fields near equilibrium points. We are interested here in the higher dimensional case of these results. We recall first from the proof of the theorem of Siegel the problem of resonance. So write $F(z) = a_1 z + \sum_{j=2}^{\infty} a_j z^j$. We can rewrite the Schröder equation as $F \circ \phi(z) = \phi(a_1 z)$. We look for a change of coordinates of the form $\phi(z) = z + \sum_{r=2}^{\infty} c_r z^r$ and get:

$$a_1(z + \sum_{r=2}^{\infty} c_r z^r) + \sum_{j \geq 2} a_j(z + \sum_{r=2}^{\infty} c_r z^r)^j = a_1 z + \sum_{r \geq 2} c_r(a_1 z)^r$$

which reduces to

$$\sum_{k \geq 2} c_k(a_1^k - a_1)z^k = \sum_{j \geq 2} a_j(z + \sum_{r \geq 2} c_r z^r)^j.$$

We get inductive formulas for the coefficients c_k.

$$c_2 = \frac{a_2}{a_1^2 - a_1}$$

$$c_k = \frac{[\cdots]}{a_1^k - a_1}.$$

In general we cannot expect the top to vanish, which leads to the resonance condition $a_1^k \neq a_1$ for all $k \geq 2$. For convergence we need that the $|a_1^k - a_1| = |a_1^{k-1} - 1|$ are not too small. The rest are delicate power series estimates.

The theorem generalizes to $\mathbb{C}^n$ ([Ste], [He]).

Theorem 2.3 *Let $F : \mathbb{C}_0^k \to \mathbb{C}_0^k$ be a germ of a holomorphic mapping fixing the origin. Assume that F' is diagonalizable at the origin: $F(z_1, z_2, \ldots, z_k) = (\lambda_1 z_1, \ldots, \lambda_k z_k) + \hat{F}, \hat{F} = \mathcal{O}(\|z\|^2), |\lambda_j| \leq 1, j \leq k$. Suppose that there exist constants $\lambda > 0, \mu > 0$ so that*

$$|\lambda_1^{i_1} \cdots \lambda_k^{i_k} - \lambda_j| \geq \lambda(i_1 + \cdots + i_k)^{-\mu} \ \forall \ 0 \leq i_1, \cdots i_k, i_1 + \cdots i_k \geq 2.$$

Then there is a change of coordinates so that the map equals the linear part of F.

The proof in this case is an analogous power series estimate, and in this case the resonance condition is that

$$\lambda_1^{i_1} \cdots \lambda_k^{i_k} - \lambda_j \neq 0$$

and these expressions need to be not too small. One can find the proof in 1 complex dimension in ([CG]). Our proof is an adaption to several variables.

Poincaré ([Po]) proved this theorem in the attracting case, $0 < |\lambda_j| < 1$ and nonresonant.

Proof: We consider the functional equation $G = \psi^{-1} \circ F \circ \psi$, $\psi'(0) = \text{Id}$, where we would like to have

$$G = \lambda z = (\lambda_1 z_1, \ldots, \lambda_k z_k).$$

Write $\psi = z + \hat{\psi}$.

$$\begin{aligned} \psi(\lambda z) &= F \circ \psi \\ &= [\lambda \cdot + \hat{F}] \circ \psi \\ &= \lambda \psi + \hat{F}(\psi) \end{aligned}$$

We will approach ψ stepwise in an inductive procedure, so we consider the simplified equation:

$$\begin{aligned} \psi(\lambda z) &= \lambda \psi + \hat{F}(z) \\ \lambda z + \hat{\psi}(\lambda z) &= \lambda z + \lambda \hat{\psi} + \hat{F} \\ \hat{\psi}(\lambda z) - \lambda \hat{\psi} &= \hat{F}(z) \end{aligned}$$

We write

$$\hat{F} = \sum_{|I| \geq 2} b_I z^I, \quad b_I = (b_I^1, \ldots, b_I^k)$$

$$\hat{\psi}(z) = \sum_{|I| \geq 2} c_I z^I$$

$$\hat{\psi}(\lambda z) - \lambda \hat{\psi} = \sum c_I (\lambda^I - \lambda) z^I$$

$$c_I(\lambda^I - \lambda) = b_I$$

$$c_I^j(\lambda_{i_1} \cdots \lambda_{i_k} - \lambda_j) = b_I^j$$

$$c_I^j = \frac{b_I^j}{\lambda^I - \lambda_j}$$

The hypothesis of the Theorem implies that

$$\frac{1}{|\lambda^I - \lambda_j|} \le C|I|^\mu.$$

Assume next that we have the following estimate on $\hat{F}$ defined on a slightly larger polydisc than $\Delta^k(0, r)$:

$$\max \left| \frac{\partial \hat{F}^j}{\partial z_\ell} \right| =: |D\hat{f}| \le \delta$$

on $\Delta^k(0, r)$.

We estimate next $\hat{F}$ on $\Delta^k(0, r(1 - \eta)), 0 < \eta < \frac{1}{5}$.

$$\hat{F}(z) = \left(\hat{F}^1, \ldots, \hat{F}^k \right)$$
$$= \left(\sum b_I^1 z^I, \ldots, \sum b_I^k z^I \right)$$

$$\frac{\partial \hat{F}^j}{\partial z_\ell} = \sum_{i_\ell \ge 1} b_I^j i_\ell z_1^{i_1} \cdots z_\ell^{i_\ell - 1} \cdots z_k^{i_k}$$

$$b_I^j i_\ell = \frac{1}{(2\pi i)^k} \int_{|\zeta_1| = \cdots = |\zeta_k| = r} \frac{\frac{\partial \hat{F}^j}{\partial z_\ell}(\zeta)}{\zeta_1^{i_1 + 1} \cdots \zeta_\ell^{i_\ell} \cdots \zeta_k^{i_k + 1}} d\zeta_1 \cdots d\zeta_k$$

$$\left| b_I^j i_\ell \right| \le \delta \frac{r^k}{r^{|I| + k - 1}}$$

$$\left| b_I^j i_\ell \right| \le \frac{\delta}{r^{|I| - 1}}$$

Next we estimate $D\hat{\psi}$ on $\Delta(0, (1 - \eta)r)$.

$$\frac{\partial \hat{\psi}^j}{\partial z_\ell} = \sum c_I^j i_\ell z_1^{i_1} \cdots z_\ell^{i_\ell - 1} \cdots z_k^{i_k}$$

$$= \sum \frac{b_I^j i_\ell}{\lambda^I - \lambda_j} z^{\hat{I}_\ell}$$

$$\left| \frac{\partial \hat{\psi}^j}{\partial z_\ell} \right| \le \sum \frac{\delta}{r^{|I| - 1}} \frac{[r(1 - \eta)]^{\hat{I}_\ell}}{|\lambda^I - \lambda_j|}$$

$$\le \delta C \sum_{|I| \ge 2} (1 - \eta)^{|I| - 1} |I|^\mu$$

$$\le \frac{5\delta C}{4} \sum_{|I| \ge 2} (1 - \eta)^{|I|} |I|^\mu$$

There are at most $|I|^k$ terms for each $i_1 + \cdots + i_k = |I|$.

$$\left|\frac{\partial\hat{\psi}^j}{\partial z_\ell}\right| \le \frac{5\delta C}{4}\sum_{n\ge2} n^k(1-\eta)^n n^\mu$$

$$= \frac{5\delta C}{4}\sum_{n\ge2}(1-\eta)^n n^{k+\mu}$$

$$\le \frac{5\delta C}{4}\sum_{n\ge2}(1-\eta)^n n^{k+[\mu]+1}$$

Next recall that $\sum_{n\ge0}(1-\eta)^n = \frac{1}{\eta}$. Hence after differentiation:

$$\frac{(k+[\mu]+1)!}{\eta^{k+[\mu]+2}} = \sum_{n\ge k+[\mu]+1} n(n-1)\cdots(n-(k+[\mu]))(1-\eta)^{n-(k+[\mu]+1)}$$

$$= \sum_{n\ge0}(n+k+[\mu]+1)\cdots(n+1)(1-\eta)^n$$

$$\ge \sum_{n\ge0} n^{k+[\mu]+1}(1-\eta)^n$$

$$\ge \sum_{n\ge2} n^{k+[\mu]+1}(1-\eta)^n$$

$$\left|\frac{\partial\hat{\psi}^j}{\partial z_\ell}\right| \le \frac{5\delta C}{4}\sum_{n\ge2}(1-\eta)^n n^{k+[\mu]+1}$$

$$\le \frac{5\delta C(k+[\mu]+1)!}{4\eta^{k+[\mu]+2}}$$

Assume next that δ and η are such that

$$\frac{5\delta C(k+[\mu]+1)!}{4\eta^{k+[\mu]+2}} \le \frac{\eta}{k}.$$

Then $|D\hat{\psi}| \le \frac{\eta}{k}$. Let $G = \psi^{-1}\circ F\circ\psi$. By these estimates we have that ψ maps $\Delta^k(0, r(1-4\eta))$ to $\Delta^k(0, r(1-3\eta))$ and if $\delta < \frac{\eta}{k}$ then F map $\Delta^k(0, r(1-3\eta))$ to $\Delta^k(0, r(1-2\eta))$. Moreover, ψ must map $\Delta^k(0, r(1-\eta))$ 1-1 to a domain containing $\Delta^k(0, r(1-2\eta))$ so ψ^{-1} is well defined on $\Delta^k(0, r(1-2\eta))$. Hence G maps $\Delta^k(0, r(1-4\eta))$ into $\Delta^k(0, r(1-\eta))$. We write $G = \lambda z + \hat{G}$. Next we estimate $\hat{G}$.

$$G = \psi^{-1}\circ F\circ\psi$$
$$\psi\circ G = F\circ\psi$$

$$G + \hat{\psi} \circ G = \lambda\psi + \hat{F} \circ \psi$$
$$\lambda z + \hat{G} + \hat{\psi} \circ G = \lambda z + \lambda\hat{\psi} + \hat{F} \circ \psi$$
$$\hat{G} = -\hat{\psi} \circ G + [\lambda\hat{\psi}] + \hat{F} \circ \psi$$

Recall: $\hat{F}(z) = \hat{\psi}(\lambda z) - \lambda\hat{\psi}$

$$\hat{G} = -\hat{\psi} \circ G + [\hat{\psi}(\lambda z) - \hat{F}(z)] + \hat{F} \circ \psi$$
$$\hat{G} = \hat{\psi}(\lambda z) - \hat{\psi}(\lambda z + \hat{G}) + \hat{F}(z + \hat{\psi}) - \hat{F}(z)$$

Set $M = \max_j |\hat{G}^j|$ on $\Delta^k(0, r(1 - 4\eta))$

$$M \leq \max_j \left|\hat{\psi}^j(\lambda z) - \hat{\psi}^j(\lambda z + \hat{G})\right| + \max_j \left|\hat{F}^j(z + \hat{\psi}) - \hat{F}^j(z)\right|$$
$$\leq k\frac{\eta}{k}M + k\delta\max\left|\hat{\psi}^\ell\right|$$
$$\leq \eta M + k\delta kr\frac{5\delta C(k + [\mu] + 1)!}{4\eta^{k+[\mu]+2}}$$
$$M \leq \frac{5\delta^2 k^2 rC(k + [\mu] + 1)!}{4\eta^{k+[\mu]+2}(1 - \eta)}$$

Next we do Cauchy estimates on $\Delta(r(1 - 5\eta))$ to obtain:

$$\left|D\hat{G}\right| \leq \frac{5\delta^2 k^2 C(k + [\mu] + 1)!}{4\eta^{k+[\mu]+3}(1 - \eta)} =: \tilde{\delta}.$$

We want to use an inductive argument where we pass from F to G. For the induction to work we need good estimates on η and $\delta, \tilde{\delta}$. We summarize the hypotheses we have used:

$$\left|D\hat{F}\right| \leq \delta$$
$$\frac{5\delta C(k + [\mu] + 1)!}{4\eta^{k+[\mu]+2}} \leq \frac{\eta}{k} \text{ i.e.}$$
$$C\delta \leq \frac{4\eta^{k+[\mu]+3}}{5k(k + [\mu] + 1)!}$$
$$\delta < \frac{\eta}{k}$$

We will determine a constant $t > 0$ so that these estimates hold inductively with sequences δ_n, η_n. So assume that

$$C\delta_n < \frac{4t\eta_n^{k+[\mu]+3}}{5k(k + [\mu] + 1)!}.$$

Then with δ_n as δ and $\tilde{\delta}$ as δ_{n+1} and with the choice $\eta_{n+1} = \frac{\eta_n}{2}$ we get

$$
\begin{aligned}
C\delta_{n+1} &= C\frac{5\delta_n^2 k^2 \acute{C}(k + [\mu] + 1)!}{4\eta_n^{k+[\mu]+3}(1 - \eta_n)} \\
&= [C\delta_n]^2 \frac{5k^2(k + [\mu] + 1)!}{4\eta_n^{k+[\mu]+3}(1 - \eta_n)} \\
&\leq \left[\frac{4t\eta_n^{k+[\mu]+3}}{5k(k + [\mu] + 1)!}\right]^2 \frac{5k^2(k + [\mu] + 1)!}{4\eta_n^{k+[\mu]+3}(1 - \eta_n)} \\
&= \frac{t^2}{1 - \eta_n} \frac{4\eta_n^{k+[\mu]+3}}{5(k + [\mu] + 1)!} \\
&\leq \frac{2^{k+[\mu]+3}tk}{1 - \eta_n} \frac{4t\eta_{n+1}^{k+[\mu]+3}}{5k(k + [\mu] + 1)!} \\
&\leq t\frac{5k2^{k+[\mu]+3}}{4} \frac{4t\eta_{n+1}^{k+[\mu]+3}}{5k(k + [\mu] + 1)!}
\end{aligned}
$$

It follows that if $t < \frac{4}{5k2^{k+[\mu]+3}}$ then the inductive estimate on $C\delta_n$ holds. It is also evident from this estimate that $\delta_{n+1} < \frac{\eta_{n+1}}{k}$ shrinking t again if necessary.

We next set up the inductive procedure. So we define a sequence of maps F_n defined on $\Delta^k(0, r_n)$ and ψ_n defined on $\Delta^k(0, r_n(1-\eta_n))$. By the inductive procedure above, $r_{n+1} = r_n(1 - 5\eta_n)$. Since the η_n decrease geometrically, the r_n decreases to a strictly positive limit r_∞. We have the functional equations

$$
\begin{aligned}
F_{n+1} &= \psi_n^{-1} \circ F_n \circ \psi_n \\
&= (\psi_0 \circ \cdots \circ \psi_n)^{-1} \circ F_0 \circ (\psi_0 \circ \cdots \circ \psi_n).
\end{aligned}
$$

Passing to the limit we get a map ψ_∞ and a map F_∞ so that

$$
\begin{aligned}
\lambda z &= F_\infty \\
&= \psi_\infty^{-1} \circ F \circ \psi_\infty.
\end{aligned}
$$

$\square$

We remark that the diophantine condition in the above theorems can be relaxed to the Brjuno condition. In the one-variable case, let $\{\frac{p_n}{q_n}\}$ denote the sequence of rationals approximating ω from the continued fraction expansion of ω. The Brjuno condition ([B]) is that $\sum_{n=1}^{\infty} \frac{\log q_{n+1}}{q_{n+1}} < \infty$. This suffices for the existence of linearizing coordinates. Similarly in higher dimension ([He]).

Yoccoz ([Y]) proved that in the case of quadratic maps in $\mathbb{C}$, this condition is also necessary.

The analogue of Yoccoz's result is open in higher dimension.

In one variable one can show that if a subsequence of the expressions a_1^k converges to 1 fast enough, then one can by choosing appropriate higher order terms find a function which does not satisfy the Schröder equation. The reason is that there is a sequence of periodic points converging to zero. Such points are called Cremer points.

Cremer ([Cr]) showed that convergence of the formal solution ϕ breaks down when

$$\limsup_{m \to \infty} \left(-\frac{1}{m} \log \left(\inf_{2 \le |I| \le m, 1 \le j \le k} |\lambda^I - \lambda_j| \right) \right) = \infty.$$

Exercise 1. If $\lambda_1, \cdots, \lambda_k$ are nonzero complex numbers for which $\lambda_1^{i_1} \cdots \lambda_k^{i_k} - \lambda_j \ne 0$ always but some subsequence

$$|\lambda_1^{i_1} \cdots \lambda_k^{i_k} - \lambda_j|^{\frac{1}{i_1 + \cdots i_k}} \to 0$$

can we find higher order terms so that the map becomes non linearizable.

In the case there are resonances, one might still be able to find coordinates in which the map is simplified.

Theorem 2.4 *(Sternberg ([Ste]), Rosay-Rudin ([RR])) Let $F : \mathbb{C}_0^k \to \mathbb{C}_0^k$ be a germ of a holomorphic mapping with the origin as an attracting fixed point. We can assume that $F'(0) = A$ is lower triangular with diagonal entries: $1 > |\lambda_1| \ge \cdots \ge |\lambda_k| > 0$. Then there exists a triangular polynomial automorphism G of $\mathbb{C}^k$, $g_1 = \lambda_1 z_1, g_j = \lambda_j z_j + h_j(z_1, \ldots, z_{j-1}), j \ge 2$, and for any $m \ge 2$ a polynomial selfmap $T_m : \mathbb{C}^k \to \mathbb{C}^k$, $T_m(0) = 0, T_m'(0) = Id$ such that $T_m \circ F - G \circ T_m = \mathcal{O}(|z|^m)$.*

Remark 1. Rosay and Rudin ([RR]) didn't need for their use to know whether T_m could be chosen independent of m. However, this is true as was proved in ([JV]) in their generalization of the ([RR]) paper to be discussed below.

Exercise 2. Is there a similar result for the case of Siegel discs allowing finitely many resonances and otherwise diophantine estimates. How about if some eigenvalues are larger than 1?

Proof of the Theorem: We find T_m and lower triangular G_m inductively so that $G_m \circ T_m - T_m \circ F = \mathcal{O}(|z|^m)$. For $m = 2$, set $G_2 = A, T_2 = Id$. Suppose next that for $m \ge 2$

$$G_m \circ T_m - T_m \circ F = \mathcal{O}(|z|^m)$$
$$= H + \mathcal{O}(|z|^{m+1}),$$

H homogeneous of degree m.

Suppose at first that m is such that there is no resonance of the form

$$\lambda_j = \lambda_1^{i_1} \cdots \lambda_k^{i_k}, i_1 + \cdots i_k = m.$$

[Resonance can only happen as long as $|\lambda_1|^m \geq |\lambda_n|$, so only finitely times.]

Let D be the diagonal part of A. For any homogeneous H of degree m, consider the expression $D \circ H - H \circ D$. This is composed of monomials of the form $\lambda_j z^I - (\lambda z)^I$ and the only way the expression can be zero is for there to be a resonance. Hence the linear map $H \to D \circ H - H \circ D$ is $1-1$ and hence bijective. Next after a linear change of coordinates, all the off diagonal coefficients in A can be as small as you wish, hence the operator $H \to A \circ H - H \circ A$ is also surjective. Hence we can write $H = A \circ H' - H' \circ A$.
We get

$$G_m \circ T_m - T_m \circ F = H + \mathcal{O}(|z|^{m+1})$$
$$= A \circ H' - H' \circ A + \mathcal{O}(|z|^{m+1})$$
$$= G_m \circ H' - H' \circ F + \mathcal{O}(|z|^{m+1})$$
$$G_m \circ (T_m - H') - (T_m - H') \circ F = \mathcal{O}(|z|^{m+1})$$

Set $T_{m+1} = T_m - H'$ and $G_{m+1} = G_m$.

The next case is when there is a resonance for m. Then the argument is the same except that H can in addition contain nonzero terms which are not in the range of the commutator map. But then these terms are lower triangular and can be absorbed by G_m.   □

Remark 2. If some eigenvalues λ_j are zero, the local dynamics is less well understood, but see ([FJ]).

Next suppose that $F : \mathbb{C}_0^n \to \mathbb{C}_0^n$ is a holomorphic germ of a diffeomorphism. We say that 0 is a saddle point if the tangentspace at 0 has complementary subspaces E^s, E^u each of dimension at least 1 so that $F'(E^{s/u}) = E^{s/u}$ and all eigenvalues of $F'_{|E^s}$ have modulus strictly smaller than 1 and all eigenvalues of $F'_{|E^u}$ have modulus strictly larger than 1. We call $E^{s/u}$ the stable and unstable subspace respectively. We define the local (un)stable set W^s_{loc} (W^u_{loc}) to be the set of those points z for which $F^n(z) \to 0$ ($F^{-n}(z) \to 0$) while all points in the sequence stay close to 0 (Figure 7)

Theorem 2.5 *(Stable manifold Theorem) The set W^s_{loc} (W^u_{loc}) is a closed complex submanifold of a neighborhood of the origin and the tangentspace at 0 is E^s (E^u).*

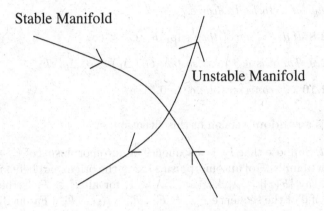

Stable Manifold

Unstable Manifold

Fig. 7. Stable and Unstable Manifolds

Proof: We prove first that the local stable set is a complex manifold with tangentspace E^s. First we can make an affine change of coordinates so that $E^s = \{z_1 = \cdots = z_r = 0\}$ and $E^u = \{z_{r+1} = \cdots = z_n = 0\}$. For any point $p = (0, \ldots, 0, z^0_{r+1}, \ldots, z^n)$ close to the origin consider the polydisc

$$\Delta_p = \{(z_1, \ldots, z_r, z^0_{r+1}, \ldots, z^0_n); |z_j| < \delta, j \leq r\}.$$

By the expansion of F along E^u it follows that the images $F^k(\Delta_p)$ contain graphs over Δ_p and they expand so that in the limit there is exactly one point in Δ_p which stays forever in these graphs. This point depends holomorphically on q and this parametrizes the stable manifold. The tangency is clear.

Next we consider the local unstable manifolds. But these are simply the limit graphs of the $F^k(\Delta_p)$.   □

The theorem is still valid if some of the eigenvalues of F' vanish at the origin.

2.2 Global Dynamics

Theorem 2.6 *([RR]) Let F be a global automorphism of $\mathbb{C}^k$ fixing 0 and assume that 0 is an attracting fixed point. Then the basin of attraction of 0 is biholomorphic to $\mathbb{C}^k$ and the map F is conjugate to G (as above) there.*

Proof: The proof goes by showing that the sequence of $G^{-n} \circ T_m \circ F^n$ converges on the basin of attraction to a biholomorphic map onto $\mathbb{C}^k$.

We list some of the key lemmas to give an indication of how the proof goes. See ([RR]) for the details.

Lemma 2.7 *If the g_j are polynomials, degree $g_j \leq d$, then $G = (\lambda_1 z_1, \lambda_2 z_2 + g_1(z_1), \ldots, \lambda_k z_k + g_k(z_1, \ldots, z_{k-1}))$ is an automorphism of $\mathbb{C}^k$. The inverse is lower triangular with finite degree.*

Lemma 2.8 *If $\deg G < \infty$, then $\sup_k \deg G^k < \infty$.*

Lemma 2.9 *There is a $\beta < \infty$ so that $G^k(\Delta_k) \subset \beta^k \Delta_k, \forall k$.*

Lemma 2.10 *G^k converges u.c.c. to 0.*

There is a **random** version of this situation:

Question 1. Suppose that F_n is a sequence of automorphisms of $\mathbb{C}^k$ contained in a compact family $\mathcal{F}$ of automorphisms fixing the origin and being uniformly attracting, say $|F(z)| \leq \lambda |z|, |z| < \epsilon$, $\lambda < 1$, for all $F \in \mathcal{F}$. Is the basin of attraction of 0 of the sequence $F_{(n)} = F_n \circ F_{n-1} \circ \cdots \circ F_1$ a Fatou-Bieberbach domain (biholomorphic to $\mathbb{C}^k$)?

The following question gives a good example how random dynamics is relevant to dynamics in higher dimension.

Question 2. (Bedford) Suppose that F is an automorphism of $\mathbb{C}^3$ and that K is an invariant compact subset (i.e. $F(K) \subset K$) on which F is hyperbolic, more precisely, the tangentspace of $\mathbb{C}^3$ at points in K has two continuous invariant complex subbundles, one of dimension 2, E^s, the stable bundle and one, E^u, the unstable bundle of dimension 1. Suppose that F is uniformly contracting along E^s and uniformly expanding along E^u. Then the stable manifold of any point in K is a two dimensional complex manifold. Is it biholomorphic to $\mathbb{C}^2$?

There is a partial result by ([JV]): The answer is yes for almost every $p \in K$ for any invariant probability measure on K. The proof in ([JV]) goes by using a theorem of Oseledec to show that for a full measure of orbits, the eigenvalues vary sufficiently slowly along the orbit that one can basically reduce it to a modification of the Rosay Rudin case. The theorem is valid more generally. The set K only needs to be measure hyperbolic and the result holds in any dimension on any complex manifold. We will discuss the theorem of Oseledec in the next lecture.

Let $\{F_n(z, w) = (z^2 + a_n z, b_n w)\}$ be a sequence of polynomial automorphisms of $\mathbb{C}^2$, let $c_n = \max\{|a_n|, |b_n|\}$. Then $a_n, b_n \neq 0$ so $c_n > 0$. Let $\Omega := \{z; F_{(n)}(z) = F_n \circ \cdots F_1(z) \to 0\}$, Figure 8.

Theorem 2.11 *([F4]) Suppose that $c_n \leq c^{2^n}$ for some $c < 1$. Then there is a nonconstant bounded plurisubharmonic function on Ω. In particular, Ω is not biholomorphic to $\mathbb{C}^2$.*

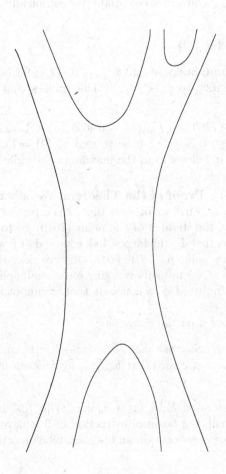

Fig. 8. Ω

Proof: We can write $F_n = (f_1^n, f_2^n)$. Set $\psi_n = \max\{|f_1^n|, |f_2^n|, c^{2^n}\}$. Then $\phi_n := \frac{\log \psi_n}{2^n}$ is plurisubharmonic on $\mathbb{C}^2$.

We have immediately that $\phi_{n+1} \leq 2\phi_n^2$. Hence we get that

$$\psi_{n+1} = \frac{\log \phi_{n+1}}{2^{n+1}}$$

$$\leq \frac{\log 2}{2^{n+1}} + \frac{\log \phi_n}{2^n}$$
$$= \frac{\log 2}{2^{n+1}} + \psi_n.$$

It follows that ψ_n converges essentially monotonically to a plurisubharmonic function ψ on $\mathbb{C}^2$.

Lemma 2.12 $\Omega = \{\psi < 0\}$.

Proof of the Lemma: Suppose that $\psi(z) < 0$. Then for large n, $\psi_n < -d < 0$. Hence $\log \phi_n < -d2^n$, so $\phi_n < e^{-2^n d}$. This implies that $F_{(n)} \to 0$. Hence $z \in \Omega$.

Suppose that $z \in \Omega$. Then $F_{(n)}(z) \to 0$ so $\phi_n(z) < 1$ for all large n. Hence $\psi_n(z) < 0$ so $\psi(z) \leq 0$. Next we observe that $\psi_n(0) = \log c$ so also $\psi(0) = \log c < 0$. But then it follows from the maximum principle that $\psi(z) < 0$ on Ω. □

Continuation of the Proof of the Theorem: We only need to show that ψ is nonconstant on Ω. First we observe that Ω is a proper subset of $\mathbb{C}^2$. We see for example that the iterates of the point $(10, 0)$ go to infinity, hence is not in Ω. This means that B, the largest ball centered at 0 which is contained in Ω has a boundary point $p \in \partial B \cap \partial \Omega$. Observe next that by the above lemma $\psi(p) \geq 0$. Next we use subaveraging on a small sphere centered at p and upper semicontinuity of ψ to conclude that ψ cannot be constant equal to $\log c$ on Ω. □

One can also prove a partial converse.

Theorem 2.13 *([F4]) Suppose that $F_n = (z^2 + b_n z, b_n w)$, $0 < |b_n| < c < 1$ and that there is some $t < 2$ such that $|b_{n+1}| \geq |b_n|^t$. Then Ω is biholomorphic to $\mathbb{C}^2$.*

Han Peters investigated $F_n = (z^2 + a_n w, a_n z)$, $|a_n| \nearrow 1$. He showed that in the case a_n are real, the basin of attraction of 0 is never open. However, the basin can in some cases contain an open set biholomorphic to $\mathbb{C}^2$.

2.3 Fatou Components

One of the more delicate Fatou components in one variable are the attracting basins of indifferent fixed points. The higher dimensional analogue was first studied by Ueda. Ueda ([Ue]) investigated the case of automorphisms F fixing the origin with the property that $F'(0)$ has one eigenvalue equal to 1 and one eigenvalue of modulus strictly less than one. He showed that the basin of attraction of the origin contains a connected component biholomorphic to $\mathbb{C}^2$. His delicate proof showed that the basin of attraction had a natural fibration as a line bundle with fiber and base both biholomorphic to $\mathbb{C}$.

Weickert ([We1]) investigated this for the case $F'(0) = Id$ and found that in some cases the basin of attraction of the origin contains a biholomorphic copy of $\mathbb{C}^2$. It is however an open question whether there exists a case where a component of the basin of attraction is biholomorphic to $\mathbb{C}^2$. In Weickerts situation it was necessary to make some hypothesis on higher order terms in the Taylor series expansion. It is still an open question in how general a case the result holds. But Hakim generalized the results of Weickert to cover more cases.

Stensønes has an example in $\mathbb{C}^3$ where the basin of attraction of a parabolic boundary point is biholomorphic to $\mathbb{C}^2 \times \mathbb{C}^*$.

In the case of Siegel domains, a Fatou component can be bounded and biholomorphic to a Reinhardt domain (([BBD]). It is an open question which Reinhardt domains can be realized this way.

See ([FS1]) for the case of recurrent Fatou components. A Fatou component Ω is said to be recurrent if there is a point $z \in \Omega$ and a sequence of iterates $F^{n_j}(z) \to w$ where w is an interior point of Ω. These generalize the attracting basins and Siegel domains from one variable. In this case one can have a mix of the two kinds. An example is a Fatou component which is a Siegel cylinder, i.e. biholomorphic to $\Delta \times \mathbb{C}$ and the map is conjugate to $(z, w) \to (e^{i\theta} z, aw)$ for some $|a| < 1$.

See ([FS1]) for further discussions on Fatou components in higher dimension, for example the case of an automorphisms of $\mathbb{C}^2$ with a wandering Fatou component. This was known to exist for entire endomorphisms of the one dimensional complex plane. See also Jupiter-Lilov ([JL]) for a discussion of nonrecurrent periodic Fatou components of automorphisms of $\mathbb{C}^2$. These generalize the parabolic Fatou components in one dimension. In their case the iterates will converge to maps whose image lie in the boundary of the Fatou component. These limits may be points, perhaps uncountable many or higher dimensional complex varieties.

In the endomorphism case not much is known. But Ueda has found a way to construct examples from one variable maps. His idea is to use the classical map $\Phi : \mathbb{P}^1 \times \mathbb{P}^1 \to \mathbb{P}^2$ given by

$$\Phi([z_1 : z_2], [w_1 : w_2]) = ([z_1 w_1 : z_2 w_2 : z_1 w_2 + z_2 w_1]).$$

This is a two to one map and if f is a rational function, then this induces a holomorphic map $F : \mathbb{P}^2 \to \mathbb{P}^2$ given by $F(\Phi(p, q)) = \Phi(f(p), f(q))$. In particular, if U, V are Fatou components of f then $\Phi(U \times V)$ is a Fatou component of F. This gives rise to the similar types of Fatou components as in one variable.

Ueda proved a key theorem about Fatou components for endomorphisms of $\mathbb{P}^k$. We first introduce the notion of Kobayashi hyperbolicity. Let M be a complex manifold, $p \in M$ and $\xi \in T_p(M)$.

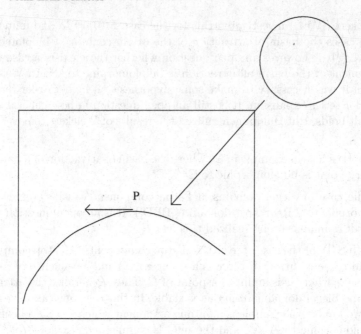

Fig. 9. Kobayashi Metric

Definition 2.14 *The Kobayashi pseudometric, Figure 9, $d_K(p,\xi) := \inf\{\frac{1}{|c|};$ $\exists f : \Delta \to M; f(0) = p, f'(0) = c\xi\}$. If $d_K(p,\xi)$ is locally bounded below for unit vectors ξ, then we say that M is Kobayashi hyperbolic.*

Theorem 2.15 *([Ue2]) Suppose that $F : \mathbb{P}^n \to \mathbb{P}^n$ is a holomorphic endomorphism of degree $d \geq 2$. Then all Fatou components are Kobayashi hyperbolic.*

Proof: We lift F to a homogeneous polynomial map

$$\tilde{F} = (F_1(z_1, \ldots, z_{n+1}), \ldots, F_{n+1}(z_1, \ldots, z_{n+1})).$$

More precisely, if $\pi : \mathbb{C}^{n+1} \setminus (0) \to \mathbb{P}^n$ is the canonical projection mapping a point to the complex line containing it, then $\pi \circ \tilde{F} = F \circ \pi$. Since F is holomorphic, the map $\tilde{F}$ only vanishes at the origin. Hence there exists a C so that $\frac{1}{C}\|z\|^d \leq \|\tilde{F}(z)\| \leq C\|z\|^d$, d the degree.

Let G be the Brolin-Hubbard escape function:

$$G(z) = \lim \frac{\log \|\tilde{F}^m(z)\|}{d^m}$$

This is a continuous plurisubharmonic function on $\mathbb{C}^{n+1}$ with a logarithmic pole at 0 and it is pluriharmonic exactly over the Fatou set. By homogeneity,

$$G(rz) = G(z) + \log r.$$

It follows that $\{G < 0\}$ is the Fatou component of 0 and that this is a bounded set with boundary $\{G = 0\}$. Next observe that every Fatou component Ω of F has a covering $\tilde{\Omega}$ in $\{G = 0\}$. Since $\{G = 0\}$ is bounded it follows that the covering is Kobayashi hyperbolic but then Ω also is Kobayashi hyperbolic. $\qquad\square$

Corollary 2.16 *([Ue2]) Suppose that $F : \mathbb{P}^n \to \mathbb{P}^n$ is a holomorphic endomorphism of degree $d \geq 2$. If Ω is an attracting basin of an attracting fixed point, then Ω contains critical points.*

Proof: Suppose that the basin Ω of the attracting fixed point p does not contain any critical points. Let $B \subset \Omega$ be a small ball containing p. Since there are no critical points in Ω, F has a well defined inverse G on B with $G(p) = p$. Since p is attracting we can assume that $G(B) \supset B$. But we can repeat the process and define inductively G^n. Taking the images under G^n of a small disc through p we get a contradiction to the Kobayashi hyperbolicity of Ω. $\qquad\square$One can also prove:

Theorem 2.17 *([FS3],[Ue2]) Suppose that $F : \mathbb{P}^n \to \mathbb{P}^n$ is a holomorphic endomorphism of degree $d \geq 2$. Then all Fatou components are Stein manifolds.*

Question 3. Are the Fatou components of a nonlinear holomorphic endomorphism of $\mathbb{P}^k$ taut (i.e. is the family of holomorphic maps from the unit disc into a Fatou component a normal family) ?

Question 4. Classify the Fatou components of polynomial skew products,

$$F(z, w) = (P(z), Q(z, w)).$$

For example, can there can be wandering components in this case.

Notice that the above question is related to random iteration of one variable maps.

3 Lecture 3: Saddle Points for Hénon Maps

In this lecture we discuss the basic result by Bedford-Lyubich-Smillie ([BLS]) that saddle points are dense in J^* for Hénon maps. Our focus will be on the ingredients of their proof coming from outside complex analysis. There are many complex analytic estimates using tools like plurisubharmonic functions, which lie outside the scope of this lecture series. Here we refer the reader to the original paper. We follow ([BLS]) closely but stress the general dynamics part.

3.1 Elementary Properties of Hénon Maps

In this section we recall briefly some basic complex analytic properties of Hénon maps.

A (complex) Hénon map on $\mathbb{C}^2$ is a map of the form $F(z,w) = (P(z) + aw, z)$ where P is a polynomial of degree $d \geq 2$. A generalized Hénon map is a composition of finitely many such maps. Let d denote the algebraic degree of the map, which means the degree of the highest order term in z.

We note that it is crucial for the theory that P is a polynomial. It would be interesting to develop dynamics for entire maps such as $H(z,w) = (e^z + aw, z)$.

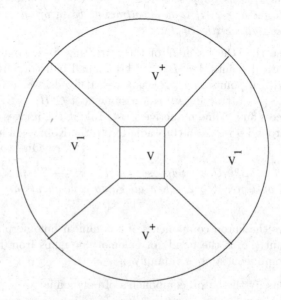

Fig. 10. Hubbard Filtration

The Hubbard-filtration, Figure 10, of $\mathbb{C}^2$ decomposes space into three pieces, $V^{\pm}$ and V.

$$V^- = \{|z| > R, |z| \geq |w|\}$$
$$V^+ = \{|w| > R, |z| < |w|\}$$
$$V = \{|z|, |w| \leq R\}$$

Here R is a large number depending on the generalized Hénon map F.

Lemma 3.1

$$F(V^-) \subset V^-$$
$$F(V) \subset V^- \cup V$$
$$p \in V^+ \Rightarrow F^{n(p)} \in V^- \cup V.$$

Proof: Observe that if $1 << R < |w|$ and $|z| > \frac{|w|}{2}$ then $F(z, w) \in V^-$. On the other hand if $R < |w|$ and $|z| \leq \frac{|w|}{2}$ then the w coordinate drops by a factor of at least 2 as long as the orbit stays in V^+. This implies the third statement of the Lemma. The others are easier. □

The situation is reversed for the inverse map F^{-1}.

Definition 3.2 $K^\pm = \{p; \{(F^\pm)^n(p)\}_{n \geq 0}$ *is a bounded sequence*$\}$

Lemma 3.3 *The sets $K^\pm$ are closed and unbounded. More precisely, $K^\pm \subset V \cup V^\pm$.*

Definition 3.4

$$J^\pm = \partial K^\pm$$
$$K = K^+ \cap K^-$$
$$J = J^+ \cap J^-$$

Following Brolin and Hubbard one defines escape functions:

Definition 3.5 $G^\pm = \lim_{n \to \infty} \frac{1}{d^n} \log^+ \|F^{\pm n}(z, w)\|$.

Lemma 3.6 *$G^\pm$ is a continuous plurisubharmonic function on $\mathbb{C}^2$, pluriharmonic and > 0 outside $K^\pm$ and 0 on $K^\pm$. Moreover $G^\pm \circ F^\pm = dG^\pm$.*

We define $\mu^\pm = dd^c G^\pm$. This is a $(1,1)$ current with measure coefficients since G is plurisubharmonic, ([Le]). Using Lemma 3.5 we obtain:

Lemma 3.7 *$F^*(\mu^+) = d\mu^+$. Support of $\mu^+ \equiv J^+$.*

The function G^- is continuous and μ^+ has measure coefficients, hence $G^- \mu^+$ is a welldefined current, so we can define $\mu := \mu^+ \wedge \mu^- := dd^c(G^- \mu^+)$ (Bedford-Taylor theory ([BT])). Then μ is an invariant probability measure.

3.2 Ergodicity and Measure Hyperbolicity

In this section we start introducing real methods into the theory of Hénon maps. The first basic concept is that of ergodicity of a measure.

Definition 3.8 *Let ν be a probability measure on a space X and let F be a measure preserving map on X. We say that ν (or F) is ergodic if whenever E is a measurable set and $F(E) = E$ up to a set of measure 0, then $\nu(E)$ is 0 or 1.*

Theorem 3.9 *(Corollary 2.2 ([BS3]) The measure μ is ergodic.*

To prove this one actually proves a stronger result, that the measure μ is mixing, Figure 11.

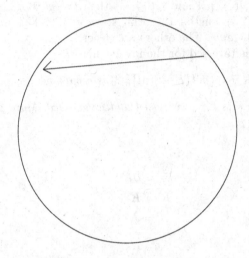

Fig. 11. Irrational Rotation is Ergodic, not Mixing

Definition 3.10 *A Borel measure ν is mixing for a measure preserving map f if for any two Borel sets,*

$$\lim_{n \to \infty} \nu(A \cap f^{-n}(B)) = \nu(A)\nu(B).$$

To see that a mixing measure is ergodic, suppose $F(E) = E$ modulo a set of measure 0. Let $A = B = E$. Then the mixing property says that $\nu(E) = (\nu(E))^2$, hence $\nu(E)$ is 0 or 1.

Theorem 3.11 *(Theorem 2.1 ([BS3]) The measure μ is mixing.*

To prove mixing, it suffices to show that when ϕ and ψ are test functions, then

$$\lim_{n \to \infty} \int (F^n)^* \phi \psi \mu = \int \phi \mu \int \psi \mu.$$

The proof goes by using integration by parts to split $\int (F^n)^*\phi\psi\mu$ into pieces which can be estimated by complex analytic methods, see ([BLS]).

Proof:

$$\int [(F^n)^*\phi]\psi\mu = \int \psi[(F^n)^*\phi][(F^n)^*\mu], \ (\mu = (F^n)^*\mu)$$

$$= \int \psi[(F^n)^*(\phi\mu)]$$

$$= \int \psi[(F^n)^*(\phi\mu^+ \wedge \mu^-)]$$

$$= \int \psi[(F^n)^*(\phi\mu^+)] \wedge [(F^n)^*\mu^-]$$

$$= \int \frac{\psi}{d^n}[(F^n)^*(\phi\mu^+)] \wedge \mu^-, \ \left((F^n)^*\mu^- = \frac{\mu^-}{d^n}\right)$$

$$= \int \frac{\psi}{d^n}[(F^n)^*(\phi\mu^+)] \wedge dd^c G^-]$$

Integration by parts:

$$= \int \frac{dd^c\psi}{d^n}[(F^n)^*(\phi\mu^+)] \wedge G^-]$$

$$+ \int \frac{d\psi}{d^n}[(F^n)^*(d^c\phi \wedge \mu^+)] \wedge G^-]$$

$$- \int \frac{d^c\psi}{d^n}[(F^n)^*(d\phi \wedge \mu^+)] \wedge G^-]$$

$$+ \int \frac{\psi}{d^n}[(F^n)^*(dd^c\phi \wedge \mu^+)] \wedge G^-]$$

Here the main term is the first of the four integrals. One needs to show that $\frac{(F^n)^*(\phi\mu^+)}{d^n} \to c\mu^+$ where $c = \int \phi\mu$. Once this is proved, the first integral converges to

$$\int cdd^c\psi \wedge G^- = \int \phi\mu \int \psi\mu.$$

To prove this one considers all weak limits of $\frac{(F^n)^*(\phi\mu^+)}{d^n}$.

Then one shows by nontrivial complex analytic methods, that this class is invariant under pull-backs and that each limit has a potential which is plurisubharmonic and agrees with cG^+.

To complete the proof one shows that the other three integrals converge to zero. One uses the Schwartz' inequality and complex analytic methods to show that $\partial\psi\mu^+$ has much smaller mass than $\psi\mu^+$.   □

Next we measure the expansion of F. This is conveniently measured by the Lyapunov exponent.

Definition 3.12 *Let $p \in \mathbb{C}^2$ and let v be a tangentvector at p. We define the Lyapunov exponent $\lambda(v,p) := \lim_{n \to \infty} \frac{1}{n} \log |DF^n(v,p)|$ whenever the limit exists.*
Also we define $\Lambda := \lim_{n \to \infty} \frac{1}{n} \int \log \|Df^n(x)\| \mu(x)$.

Since μ is already known to be ergodic, general real dynamics theory (Os-eledets theory) gives that Λ is well-defined. Moreover, the general theory gives that there are two cases: Either there are two distinct Lyapunov exponents (independent of p a.e.$d\mu$), $\lambda_1 > \lambda_2$ (and a corresponding measurable splitting of the tangentbundle, or there is one exponent. Note that $\lambda_1 + \lambda_2 = \log |\delta|$ where $\delta = J(f)$. Since the leading term of the map F is of the form x^d and with the experience from the formula of Przytycki, one might expect that the Λ is bounded below by $\log d$ just by using the expansion along the x direction. Indeed this is the case:

Lemma 3.13 *Suppose that K is a compact subset of $\mathbb{C}$ with Green function $G_K(w) = \log |w| + o(1) = \log |w| + \mathcal{O}(1/w)$. Let $P(w) = w^N + \cdots$ be a monic polynomial. Set $p = \log |P|$. Then $\int p dd^c G_K \geq 0$.*

Proof: We can factor P so we may assume that $p = log|w|$. We can assume that K has smooth boundary, not containing 0.

We get:

$$\int_{\mathbb{C}} \log |w| dd^c G_K = \int_{\partial K} \log |w| d^c G_K$$

$$= \int_{\partial K} (\log |w| - G_K) d^c G_K$$

$$= - \int_{\partial(\mathbb{C}-K)+\infty} (\log |w| - G_K) d^c G_K + \int_{\infty} (\log |w| - G_K) d^c G_K$$

$$= - \int_{\mathbb{C}-K} d(\log |w| - G_K) \wedge d^c G_K$$

$$\text{Recall: } [du \wedge dv = -d^c u \wedge du]$$

$$= \int_{\mathbb{C}-K} d^c(\log |w| - G_K) \wedge dG_K$$

$$= \int_{\mathbb{C}-K} G_K d^c \log |w|$$

$$= G_K(0)$$

$$\geq 0.$$

$\square$

Theorem 3.14 $\Lambda \geq \log d$.

Proof: Let h_p denote the unit horizontal vector. Let $X = \{w = 0\}$ and $K_0^+ = X \cap K^+$. Then

$$\int \log \|DF^k\| \mu \geq \int \log \|DF^k(h_p)\| \mu$$

$$\geq \int \log \|DF_1^k(h_p)\| \mu$$

$$= \lim_{n \to \infty} \int \log \|DF_1^k(h_p)\| \frac{1}{d^n} [F^n X] \wedge \mu^+$$

$$= \lim_{n \to \infty} \int \log \|DF_1^k(h_p)\| F_*^n [X] \wedge F_n^* \mu^+$$

$$= \lim_{n \to \infty} \int (F^n)^* (\log \|DF_1^k(h_p)\|) [X] \wedge \mu^+$$

$$= \lim_{n \to \infty} \int (\log \|DF_1^k(h_p) \circ F^n\|) [X] \wedge \mu^+$$

$$= \lim_{n \to \infty} \int (\log \|DF_1^k(h_p) \circ F^n\|) \mu_{K_0^+}$$

$$\geq k \log d$$

by the previous Lemma, since $DF_1^k(h_p) \circ F^n = d^k(z^{(d^k-1)d^n} + \cdots)$. $\qquad\square$

Theorem 3.15 *(Corollary 3.3 ([BS3])* $\lambda_1 > 0 > \lambda_2$.

Proof: We can suppose that $|J(F)| = \delta \leq 1$ otherwise we use the inverse. Then $\lambda_1 \geq \log d$ and $\lambda_1 + \lambda_2 = \log \delta$, so $\lambda_2 \leq -\log d$. $\qquad\square$

3.3 Density of Saddle Points

Once we have shown that the map is ergodic on μ and that the Lyapunov exponents are nonzero, the powerful theory of Oseledec and Pesin applies. This makes it possible to find compact subsets of the support of μ on which the map is uniformly hyperbolic.

The main result is ([BLS]):

Theorem 3.16 *If F is a generalized Hénon map, then all periodic saddle points are contained in the support of μ.*

We can then combine this result with a general theorem from real dynamics:

Theorem 3.17 *(Katok ([K]), Theorem 4.2) Let f be a $\mathcal{C}^{1+\alpha}$ ($\alpha > 0$) diffeomorphism of a compact manifold M, and μ a Borel probability $f-$ invariant measure with non-zero Lyapunov exponents. If the measure μ is ergodic and not concentrated on a single periodic trajectory then the support of μ is contained in the closure of the set of periodic points of f which are saddle hyperbolic and have a transverse homoclinic point.*

The two above results then imply:

Theorem 3.18 *The support of μ equals the closure of the set of saddle points.*

Let F be a generalized Hénon map. A measurable function $r(x)$ on J^* is called ϵ slowly varying if

$$(1+\epsilon)^{-1}r(x) < r(f(x)) < (1+\epsilon)r(x).$$

Theorem 3.19 *([Pe]) Oseledec Theorem: There is an invariant set $\mathcal{R}$ of full μ measure and invariant measureable distributions E_x^s, E_x^u on $\mathcal{R}$ such that*

$$\lim_{n\to\infty} \frac{1}{n} \log Df^n(x)(v) = \lambda^u, v \in E_x^u \setminus (0).$$

$$\lim_{n\to\infty} \frac{1}{n} \log Df^{-n}(x)(v) = 1/\lambda^s, v \in E_x^s \setminus (0).$$

There is a slowly varying function $s(x) > 0$ on $\mathcal{R}$ which is smaller than the angle between the two bundles.

Next the theory of real dynamics give us corresponding stable and unstable manifolds. We state the result for the stable case. The unstable case is similiar. Let $B^s(x,\delta)$ denote the ball of radius δ in the stable subspace E_x^s.

Theorem 3.20 *([Pe]) For any $\epsilon > 0$ there are ϵ slowly varying positive functions $C(x)$, $r(x)$ on $\mathcal{R}$ and a family $W_{loc}^s(x)$ of smooth manifolds satisfying the following:*
(i) W^s is a graph of a function $B^s(x, r(x)) \to E^u$ tangent to E^s.
(ii) For any $y \in W_{loc}^s$.

$$C(x)e^{-(\lambda^s+\epsilon)n} \leq dist(F^n(x), F^n(y)) \leq C(x)e^{-(\lambda^s-\epsilon)n}$$

$F(W_{loc}^s(x)) \subset W_{loc}^s(F(x))$.

We next fix compact sets of positive μ measure with uniform bounds. We also rotate coordinates suitably: First we give a rough description: Choose a small compact set $K \subset \mathcal{R}$ on which we have uniform estimates. Then for $x \in K$ we can assume that $W_{loc}^u(x)$ is a graph over the $z-$ axis and $W_{loc}^s(x)$ is a graph over the $w-$ axis. Call the local stable and unstable manifolds for $x \in K$, F_x^s, F_x^u. There are pairwise one-point transversal intersections of stable and unstable local leaves. The intersections must be in J since both forward and backward orbits are bounded. The collection of such intersection points is called a Pesin box. After a suitable restriction, the Pesin box is homeomorphic to a product $P^s \times P^u$ where P^s is homeomorphic to the intersection of some $W_{loc}^s(x)$ with the collection of unstable local leaves from K and similiar for P^u.

Next we make a more precise definition:

If ϕ is a plurisubharmonic function and γ is a complex curve, then $dd^c(\phi_{|_\gamma})$ is a measure on γ. We can apply this to the function G^+ and the curves $W_{loc}^u(x)$ for $x \in K$. This gives rise to measures on the local unstable manifolds. We define a set Q_1:

$$Q_1 = \{x \in \mathcal{R}; r(x) \geq r, \text{ and } (i), (ii), (iii) \text{ below hold}\}.$$

The set $\mathcal{R}$ is a set of full measure in the support of μ given by the Oseledec theorem.

(i) For $m_0 > 0, r_0 = r/8$, and $x \in \mathcal{R}$ such that $r(x) \geq r$, we consider the property

$$\mu^+_{|W_r^u(x)} \geq m_0.$$

(ii): dist $(f^n(x), f^n(y)) \leq Ce^{-n\lambda}$ for $n \geq 1$ and $y \in W_r^s(x)$.
(iii) dist $(f^{-n}(x), f^{-n}(y)) \leq Ce^{-n\lambda}$ for $n \geq 1$ and $y \in W_r^u(x)$.

Let $S = \{x \in J; f^n(x) \in Q \text{ for infinitely many } n\}$. Set $Q = Q_1 \cap S$. Let F be a compact subset of Q. We assume that $F \subset \{x \in \mathcal{R}; r(x) \geq r.\}$. Then $W_{loc}^{s/u}(F) := W_r^{s/u}(F) = \cup_{x \in F} W_r^{s/u}(x)$ and $\mathcal{F}^{s/u} = \{W_r^{s/u}(x); x \in F\}$. Assume that the diameter of F is small enough. Then we can assume after an affine coordinate change that every $W_r^s(x)$ is a graph over the vertical axis and every $W_r^u(x)$ is a graph over the horizontal axis. For every $x \in F, W_r^s(x)$ is transversal to $\mathcal{F}^u$, and $W_r^u(x)$ is transversal to $\mathcal{F}^s$. The set $P := W_r^s(F) \cap W_r^u(F)$ is called a Pesin box.

On the Pesin box there is a natural (holonomy) map between any two local unstable manifolds following the local stable manifolds. Using the fact that this holonomy map perserves G^+ one gets, using a complex analysis argument, that the holonomy maps preserve the above measures on the Pesin box.

Lemma 3.21 *If T is a closed current, $0 \leq T \leq \mu^+$, then locally T has a continuous potential.*

Basically this is since μ^+ has a continuous potential and $\mu^+ - T \geq 0$ has a plurisubharmonic potential.

Definition 3.22 *If L_1, L_2 are uniformly laminar currents on Δ^2, then one can define*

$$L_1 \wedge L_2 = \int \lambda_1(a_1) \int \lambda_2(a_2) [\Gamma_{a_1} \cap \Gamma_{a_2}]$$

with $[\Gamma_{a_1} \cap \Gamma_{a_2}]$ defined as the $0-$current which puts unit mass on each point of $\Gamma_{a_1} \cap \Gamma_{a_2}$ with the exception that $[\Gamma_{a_1} \cap \Gamma_{a_2}] = 0$ if $\Gamma_{a_1} = \Gamma_{a_2}$.

Lemma 3.23 *Let L, L' be uniformly laminar currents [such as we have on Pesin boxes] on Δ^2 such that there is a continuous plurisubharmonic function u with $dd^c u = L$. [Such as we have on Pesin boxes.] Then $L \wedge L' = L \dot{\wedge} L'$.*

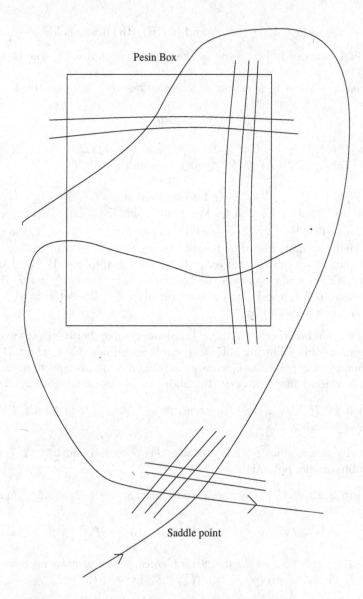

Fig. 12. Support of μ near a saddle point

Lemma 3.24 *Let P be a Pesin box, and Γ^u be the corresponding lamination. If p is a saddle point, then $W^s(p)$ must intersect λ^s almost every disc of Γ^u and the tangential intersections is an isolated subset of $W^s(p)$.*

Proof: Let $D \subset W^s(p)$ be a small disc containing p. Then $\mu^-_{|D}$ is a nonzero measure. This implies that the pullbacks converge to $c\mu^+$ for some $c > 0$. Let E denote a nontrivial collection of unstable manifolds in the Pesin box which avoids W^s. Then $\nu^-_E = \int_{a \in E} \lambda^s(a)[\Gamma^u(a)] = dd^c H$, H continuous. Hence,

$$H \frac{(f^n)^*[D]}{d^n} \to cH\mu^+$$

so,

$$\nu^-_E \wedge \frac{(f^n)^*[D]}{d^n} \to c\nu^-_E \wedge \mu^+.$$

The left side is zero, since the currents have disjoint support and the second is bounded below by a fraction of the mass of μ. □

Proof of Theorem 3.16: The stable manifold enters the Pesin box and there is a compact subset K^u of positive λ^s measure of unstable leaves that it crosses with angle bounded below, Figure 12. The high pullbacks of these unstable leaves converge to a collection of leaves near p by the Λ Lemma. They have positive measure using G^- along W^s to count them. Similarly, the unstable manifold of p must also intersect the stable leaves of P in a set of positive measure of leaves. Pullbacks give again a positive measure using G^+ to count. Here μ^+ is uniformly laminated as $\nu^+ = \int \lambda^s_1(w)[V_w]$ and μ^- is uniformly laminated as $\nu^- = \int \lambda^u_1(z)[V_z]$. Then $\int_{near\,p} \mu \geq \int \nu^+ \wedge \nu^- > 0$. □

Question 5. What can be said about non-saddle points for Hénon maps?

4 Lecture 4: Saddle Hyperbolicity for Hénon Maps

4.1 J and J^*

We say that a generalized Hénon map F is (uniformly) hyperbolic if there exists a continuous splitting of the tangent space of $\mathbb{C}^2$ in two linebundles $E^s_x, E^u_x, x \in J$ where E^s and E^u are stable and unstable repectively. It is an open question whether hyperbolicity is equivalent to hyperbolicity of F on J^*. (It is however known that if F is hyperbolic then $J = J^*$ ([BS1]).)

One can rephrase this question:

Question 6. If F is hyperbolic on J^*, does it follow that F is hyperbolic on J?

A more general question is whether $J = J^*$ for all Hénon maps.

In ([F3]) the author showed that if F is sustainable, it follows that the Hénon map is hyperbolic on J^*. The proof uses very strongly real methods. We show here that for area contracting Hénon maps this condition implies that F is hyperbolic on J. This uses strongly that the map is already known to be hyperbolic on J^* so this is in this case a positive answer to the above question.

In this lecture we will assume that F is a generalized Hénon map which is hyperbolic on J^*. We will also assume that the map is sustainable. We recall the definition:

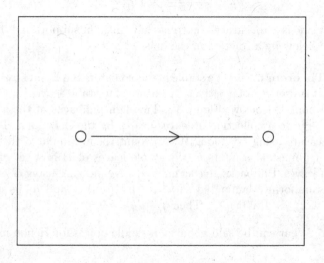

Fig. 13. Desired Orbit

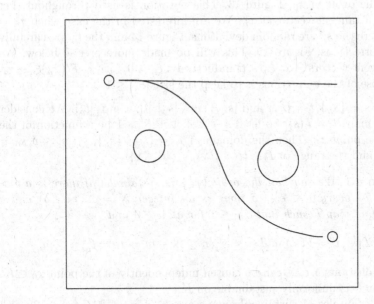

Fig. 14. Random δ Disturbance

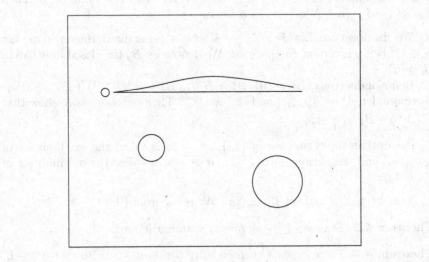

Fig. 15. ϵ corrected orbit

Let $x_0 \in \mathbb{C}^2 \subset \mathbb{P}^2$. Fix a smooth Hermitian metric on $\mathbb{P}^2$. We denote by $\{x_n\}_{n \geq 0}$ the orbit of x_0, Figure 13. This notation is used throughout. Let τ_n denote a tangent vector at x_n. We are interested in the case when $\tau_n = s_n + t_n$ where the s_n are random deviations, Figure 14 and the t_n are carefully selected corrections, Figure 15. This will be made more precise below. We define tangent vectors $\xi_n = \xi_n(s,t)$ inductively: $\xi_0 = 0, \xi_{n+1} = F'(x_n)\xi_n + \tau_n$. The purpose of the corrections is to keep the size $|\xi_n| \leq 1$.

We set $s = \{s_n\}, t = \{t_n\}$ and $|s| = \sup_n |s_n|, |t| = \sup_n |t_n|$. We consider correction maps $t = T(s)$. So let $0 < \epsilon < \delta$. Let $N > 1$ be a function of the ϵ, δ and the point x_0. Then the domain of $T = \{s = \{s_n\}_{n \geq 1}; |s| \leq \delta, s_n = 0, n \leq N\}$ and the range of T is $\{t = \{t_n\}_{n \geq 1}; |t| \leq \epsilon\}$.

Definition 4.1 *We say that the orbit $\{x_n\}$ is sustainable if there is a $\delta > 0$ so that for every $0 < \epsilon < \delta$ there is an integer $N = N(x_0, \epsilon, \delta)$ and a corresponding map T such that $|\xi_n| \leq 1$ for all $n \geq 0$ and*

$$t = T(s), t' = T(s') \text{ and } s_n = s'_n, n \leq N + m \Rightarrow t_n = t'_n, n \leq m.$$

If the constants δ, ϵ, N can be chosen independently of the point $x_0 \in K$, we say that F is uniformly sustainable on K.

Simple examples of sustainable maps are $F_1 : \mathbb{R} \to \mathbb{R}, F_1(x) = 2x$ and $F_2 : \mathbb{R}^2 \to \mathbb{R}^2, F_2(x_1, x_2) = (2x_1, \frac{1}{2}x_2)$. A simple example of a nonsustainable map is $F_3 : \mathbb{R} \to \mathbb{R}, F_3(x) = x + x^2$. For $F_1(x)$, let $t_n = -\frac{s_{n+N}}{2^N}$. In the case of F_2, write $s_n = (s_n^1, s_n^2)$. Then $t_n = (-\frac{s_n^1}{2^N}, 0)$ works. To see that F_3 is not sustainable, choose $s_n = \delta$, $n > N$ for the orbit of $x_0 = 0$. Then $|\xi_{n+N}| \geq n\delta - (n + N)\epsilon \to \infty$ as $n \to \infty$.

We fix notation: Let $F : \mathbb{C}^2 \to \mathbb{C}^2$ be a generalized Hénon map. Let $K \subset \mathbb{C}^2$ be an invariant compact set. We denote by B_p the closed unit ball in $L_p := T_{\mathbb{C}^2}$.

Define inductively, $B_z^0 = \{0\}, B_z^1 = B_{F(z)}, B_z^n = F'(B_z^{n-1}) + B_{F^n(z)}$. Furthermore, let $A_z^n = B_{F^{-n}(z)}^n$ and $A_z^\infty = \cup A_z^n$. Then one can easily show that $F'(A_z^\infty) + B_{F(z)} = A_{F(z)}^\infty$.

Recall that the cluster set of $\{z_n\}_n, n \to \infty$ is called the $\omega-$ limit set of z_0, $\omega(z_0)$ and the cluster set of $\{z_n\}_n, n \to -\infty$ is called the $\alpha-$ limit set of z_0, $\alpha(z_0)$.

Next let $z_0 \in J$ and set $L = \omega(z_0)$. We recall from ([F3]):

Theorem 4.2 *The map F is uniformly sustainable on L.*

Theorem 4.3 *There exists a sustainability constant $\delta > 0$, so that if $z \in L$, $n \geq 1$, then*

$$\|(F^n)'(z)\| \geq \frac{\delta}{4}.$$

Theorem 4.4 *If $z \in L$,*

$$\sup_{n_0, n_1 \geq 1} \|(F^{n_1})'(F^{n_0}(z))\| = \infty.$$

Lemma 4.5 *Let $C > 1$. Then there exists an $M(C)$ so that if $z \in L$ and $n \geq M(C)$, then $Diam(B_z^n) \geq C$.*

Theorem 4.6 *If $z \in L$, then $Vol(A_z^\infty) = \infty$.*

4.2 Proof of Theorem 4.10

The following result is known ([F3]).

Theorem 4.7 *Let F denote a generalized Hénon map on $\mathbb{C}^2$. Suppose that all orbits in J^* are sustainable. Then F is saddle hyperbolic on J^*.*

Our main results are the following:

Theorem 4.8 *Suppose that all orbits in J are sustainable. If $|J(F)| < 1$ then F is hyperbolic on J.*

Theorem 4.9 *Let F be a generalized Hénon map on $\mathbb{C}^2 \subset \mathbb{P}^2$. If F is hyperbolic on J, then F is uniformly sustainable on $\mathbb{C}^2$ with respect to the Fubini Study metric.*

Proof of Theorem 4.10: In this case ([BS1]) $J = J^*$. If $|J(F)| \leq 1$, $\mathbb{C}^2$ is partitioned into the stable set J^+ of J, the attracting basin of $[1 : 0 : 0]$ and at most finitely many attracting periodic basins of bounded attracting periodic orbits. If $|J(F)| > 1$ the similar statement holds for F^{-1} ([BS1]).

Let $R > 1$ be large enough. Set

$$V^- = \{(z, w); |z| > R, |w| \leq |z|\}$$
$$V^+ = \{(z, w); |w| > R, |z| < |w|\}$$
$$V = \{(z, w); |z|, |w| \leq R\}$$

By Lemma 3.1 $F(V^-) \subset V^-$, $F(V) \subset V \cup V^-$ and $F^n(p) \in V \cup V^-$ for every $p \in V^+, n \geq n(p)$.

Lemma 4.10 *There exists a constant $c > 0$ and a neighborhood U of $[1 : 0 : 0]$ in $\mathbb{P}^2$ so that if $p \in U \cap \mathbb{C}^2$ and $\xi \in T_p$ then*

$$\|F'(p)(\xi)\| \leq \frac{c}{2^n}\|\xi\|, n \geq 1.$$

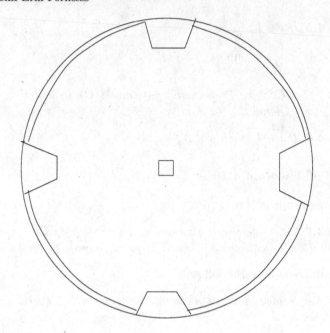

Fig. 16. $\tilde{V}^{\pm}$ and $\tilde{W}$

Proof of the Lemma: Let $H(z,w) = (P(z) + aw, z)$ be a Hénon map, $P(z) = \sum_{j=0}^{d} a_j z^j$, $d \geq 2$, $a_d \neq 0$. We express H in homogeneous coordinates.

$$[z : w : t] \to [\sum_{j=0}^{d} a_j z^j t^{d-j} + awt^{d-1} : zt^{d-1} : t^d].$$

Near $[1 : 0 : 0]$ we can use (w, t) as local coordinates. The map takes the form:

$$\tilde{H}(w,t) = \left(\frac{t^{d-1}}{a_d + a_{d-1}t + \ldots a_0 t^d + awt^{d-1}}, \frac{t^d}{a_d + a_{d-1}t + \ldots a_0 t^d + awt^{d-1}} \right).$$

If $d > 2$ then $\tilde{H}'(0,0) = 0$ so we are done by continuity. If $d = 2$, $\tilde{H}'(0,0)(\xi_1, \xi_2) = \left(\frac{d-1}{a_d}\xi_2, 0 \right)$. Hence if H_1, H_2 are Hénon maps, then $\left(\tilde{H}_1 \circ \tilde{H}_2 \right)'(0,0) = 0$, so we are done again.   □

If $(Z, W) = H(z,w) = (P(z) + aw, z)$ then $z = W$, $aw = Z - P(z)$ so

$$H^{-1}(Z,W) = \left(W, -\frac{P(z)}{a} + \frac{1}{a}Z \right)$$

is of the same form as H. We can hence apply the above Lemma to H^{-1} as well. For $r >> R, 0 < \epsilon << 1$, define, Figure 16,

$\tilde{V}^- := \{|z| > r, |w| \leq \epsilon|z|\}, \tilde{V}^+ := \{|w| > r, |z| < \epsilon|w|\}$. We have the following immediate consequence:

Lemma 4.11 *Let F be a generalized Hénon map. Then there exist r, ϵ, c so that:*

(i) If $p_0 \in \tilde{V}^-, \xi \in T_{p_0}$, then

$$\|(F^n)'(p_0)(\xi)\| \leq \frac{c}{2^n}\|\xi\|, \; n \geq 1.$$

(ii) If $p_n \in \tilde{V}^+, \xi \in T_{p_0}$, then

$$\|(F^n)'(p_0)(\xi)\| \geq \frac{2^n}{c}\|\xi\|, \; n \geq 1.$$

If we pick $r_1 > r$ large enough, then it follows that if $\|(z, w)\| \geq r_1, (z, w)$ not in $\tilde{V}^+$, then $F(z, w) \in \tilde{V}^-$ and if (z, w) is not in $\tilde{V}^-$ then $F^{-1}(z, w) \in \tilde{V}^+$. Let $\tilde{W} = \{\|p\| \geq r_1, p$ not in $\tilde{V}^+ \cup \tilde{V}^-\}$.

By compactness there exists an integer $N_1 > 1$ so that if $p \in V^+ \setminus \tilde{V}^+, \|p\| \leq r_1$, then there exists a minimal $n, 1 \leq n \leq N_1$ so that $F^n(p) \in V \cup V^-$. Moreover, if $p \in V^-$, then there exists a minimal $n, 0 \leq n \leq N_1$ so that $F^n(p) \in \tilde{V}^-$.

Next, we use the hypothesis that F is hyperbolic on J. There exist constants $\lambda > 1, c > 0$ and two continuously varying line bundles $L^s(p), L^u(p), p \in J$, the stable and unstable bundle respectively with the following properties:
(i) The line bundles are subbundles of the tangent space and they span the tangent space at every point.
(ii) $F'(p)(L^s(p)) = L^s(F(p))$
(iii) $F'(p)(L^u(p)) = L^u(F(p))$
(iv) $\|(F^n)'(p)(\xi^u)\| \geq c\lambda^n\|\xi^u\|$ for all $p \in J, \xi^u \in L^u$,
(v) $\|(F^{-n})'(p)(\xi^s)\| \geq c\lambda^n\|\xi^s\|$ for all $p \in J, \xi^s \in L^s$.

In order to find estimates valid on some neighborhood of J it is more convenient to use cone fields. So we replace the line field L^s (L^u) be a small cone in the tangent space centered at L^s (L^u). Then we get:

Lemma 4.12 *There exists a neighborhood $U \subset V$ of J and for every $p \in U$ small closed cones $C^s_p, C^u_p, C^s_p \cap C^u_p = (0)$ in the tangent space varying continuously with p and containing the stable and unstable line fields respectively in their interior on J. There exists an integer $m > 1$ and a constant $\lambda > 1$ so that:*
(i) If $p \in U, F^n(p) \in U, n \geq m$, then $(F^n)'(p)(C^u_p) \subset C^u_{F^n(p)}, \|(F^n)'(p)(\xi)\| \geq \lambda^n\|\xi\| \; \forall \, \xi \in C^u_p$.
(i) If $p \in U, F^{-n}(p) \in U, n \geq m$, then $(F^{-n})'(p)(C^s_p) \subset C^s_{F^{-n}(p)}, \|(F^{-n})'(p)(\xi)\| \geq \lambda^n\|\xi\| \; \forall \, \xi \in C^s_p$.

Case (i). $|J(F)| \leq 1$:

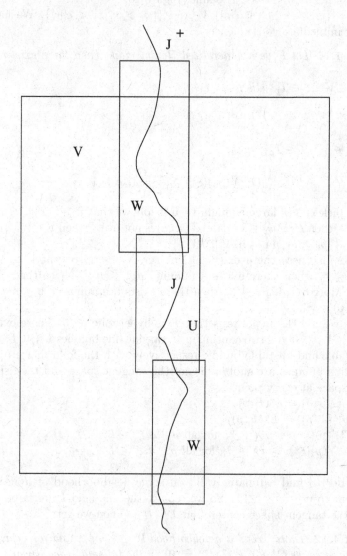

Fig. 17. Neighborhoods near J

There is a neighborhood W, Figure 17, of $[\partial K^+ \cap V] \setminus U$ and an integer $\ell \geq 1$ so that for any point $p \in W$, $F^\ell(p) \in U$ and $F^{-\ell}(p) \subset V^-$. Next we observe that $V \setminus [U \cup W] = V_1 \cup V_2$ where V_1 consists of points in attracting basins of finitely many attracting periodic orbits in V and V_2 belongs to the attracting basin of $[1:0:0]$.

Let $N > 1$ be some integer. We give a prescription for how to find t as a function of s. After that the uniform sustainability of F on $\mathbb{C}^2$ will be clear.

The easiest orbits to deal with are those in V^- where F is expanding, the orbits in J which are saddles and the orbits in V^+ which are contracting. The complications arise when orbits pass from one region to another.

Suppose at first that $p_{n+N} \in \tilde{V}^-$. Then we define $t_n = 0$. The same applies if $p_{n+N} \in \tilde{W}$. If $p_{n+N} \in \tilde{V}^+$, we let t_n be the vector given by

$$(F^N)'(p_n)(t_n) = -s_{n+N}.$$

From now on we can assume that the orbit $\{p_n\}$ never enters $\tilde{W}$.

Suppose next that $p_{n+N} \in V^+$. We assume that $N > N_1$. Then $p_n, \ldots, p_{n+N-N_1} \in \tilde{V}^+$ where the map is uniformly expanding. We define t_n by $(F^N)'(p_n)(t_n) = -s_{n+N}$. Then

$$\|t_n\| \leq c \frac{\|s_{n+N}\|}{2^N}.$$

Moreover for $1 \leq j \leq N$,

$$
\begin{aligned}
\|(F^j)'(p_n)t_n\| &= \|(F^j)'(p_n)[(F^{-N})'(p_{n+N})]s_{n+N}\| \\
&= \|(F^{-(N-j)})'(p_{n+N})(s_{n+N})\| \\
&\leq c \frac{\|s_{n+N}\|}{2^{N-j}}.
\end{aligned}
$$

Hence the corrections add up to something small. We can next use the same definition of t_n if $p_{n+N-2\ell-2m} \in V^-$. The last $2\ell + 2m$ iterates will not seriously affect the sizes of the corrections.

Next we restrict to orbits for which $p_{n+N-2\ell-2m}$ is not in V^+.

We divide into two cases depending on whether or not $p_{n+N+j} \in W \cup U, -2\ell - 2m \leq j \leq m$.

Assume at first that $p_{n+N+j} \in W \cup U, -2\ell - 2m \leq j \leq m$ so we are close to J. Then $p_{n+N+j} \in U, -\ell - 2m \leq j \leq m$. Let $0 \leq j_1 \leq N - \ell - 2m$ be minimal so that $p_{n+j} \in U$ for any $j, j_1 \leq j \leq N$. Let $j_2, m \leq j_2 \leq \infty$ denote the sup over all integers so that $p_{n+N+j} \in U$ for all integers $1 \leq j \leq j_2$.

Let ξ^s be a unit vector in $C^s_{p_{n+N}}$ so that $(F^j)'(p_{n+N})(\xi^s) \in C^s_{p_{n+N+j}}$ for any integer $j, m \leq j \leq j_2$. Also let ξ^u be a unit vector in $C^u_{p_{n+N}}$ so that $\xi^u = (F^{N-j_1})'(p_{n+j_1})\tilde{\xi}^u$ for some $\tilde{\xi}^u \in C^u_{p_{n+j_1}}$. Then we can write $s_{n+N} = a\xi^s + b\xi^u$. Next we define $t_{n+j_1, n+N} := b\tilde{\xi}^u$. Then $\|t_{n+j_1, n+N}\| \leq C \frac{\delta}{\mu^{N-j_1}}$ for some constants $C, \mu > 1$. Hence the size decreases geometrically.

We call $t_{n+j_1, n+N}$ a *contribution* to t_{n+j_1}. The value of t_{n+j_1} will be the sum of all contributions to t_{n+j_1}. The key point why these values of t_n work is that the contributions to the random disturbances from the components $a\xi^s$ decrease exponentially under iteration.

Finally we deal with the case when $p_{n+N-2\ell-2m}$ is not in V^+ and for some $j, -2\ell - 2m \le j \le m$, p_{n+N+j} is not in $W \cup U$. There is a uniform constant $k \ge 1$ (independent of N) so that if q is not in $V^+ \cup W \cup U$, then $F^j(q)$ is in V^- or in any prescribed small neighborhood of the finitely many attracting periodic orbits of F for any $j \ge k$. In particular this holds for $F^j(p_{n+N})$ for any $j \ge m + k$. In this case it suffices to choose $t_n = 0$.

Case (ii). $|J(F)| > 1$:

In this case there might be finitely many repelling periodic orbits. These are all in $J^+ = \partial K^+$. In this case we let W be the same as before after removing small neighborhoods of the repelling periodic points. The rest of the argument is the same.   □

4.3 Proof of Theorem 4.9

Here we prove Theorem 4.9, improving Theorem 4.8 from ([F3]). Some of the arguments are similar to the ones in ([F3]), but we include them in this basic course, for the ease of the reader. Let $z_0 \in J$ and set $L = \omega(z_0)$ or $L = \alpha(z_0)$. We assume that the orbit of z_0 is sustainable.

Lemma 4.13 *If F is saddle hyperbolic on L, then $L \subset J^*$.*

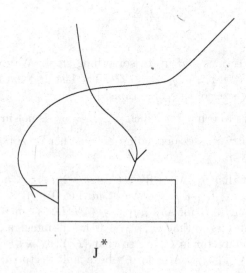

J^*

Fig. 18. Homoclinic intersections belong to J^*

Proof: Since L is an ω or α limit set, it follows that L is chain recurrent, i.e., for every point $z_0 \in L$ and every $\epsilon > 0$ there is an ϵ pseudo orbit $\{z_n\} \subset L$

with $z_n = z_0$ for some $n \geq 1$. But then it follows from ([S]) that saddle points are dense in K. [They might not be in L.] Since all saddle points of F are in J^* ([BLS]), it follows that $L \subset J^*$.   □

Lemma 4.14 *If F is saddle hyperbolic on J^*, then J^* has local product structure.*

Proof: Suppose that $p \in W_{loc}^s(J^*) \cap W_{loc}^u(J^*)$, $p \notin J^*$. Then set $K := J^* \cup \{p_n\}_{n=-\infty}^\infty$. Then F, F^{-1} are saddle hyperbolic on K as we see using local cone fields, and every point in K is chain recurrent. Hence there are periodic saddle points arbitrarily close to p. This is impossible since all periodic saddles are contained in J^* by ([BLS]). This shows that in fact J^* has local product structure.   □

A result of Shub ([S]) then implies that

Lemma 4.15 *If F is saddle hyperbolic on J^*, $W^s(J^*) = \cup_{x \in J^*} W^s(x)$ and similiar for the unstable set.*

Lemma 4.16 *([BLS], Lemma 6.4) Suppose a stable and an unstable manifold for points in J^* are tangent at p. Then after an arbitrarily small change in p, the intersection will be transverse.*

Proof: We remind the reader of the proof in ([BLS]). Assume at first that all nearby intersections are tangential. We minimize the local tangency. Then we have a stable manifold of the form $w = 0$ and an unstable manifold of the form $w = z^k$, $k \geq 2$ minimal in some local coordinate system. Next, consider the unstable manifold of a close by point in J^*. In the local coordinate system, this will by minimality still be tangent to $w = 0$ of order k. So it has the form $w = (z - a)^k \phi$. Then ϕ must be close to one on a given fixed disc. Hence we can take roots and write $w = [(z - a)\psi(z)]^k$, ψ close to one. Now since the unstable manifolds $w = z^k$ and $w = [(z - a)\psi]^k$ are disjoint, there can be no roots of $z^k = [(z - a)\psi]^k$. But if $\omega \neq 1$ is a primitive root of unity, we get $z = \omega(z - a)\psi$ which must have a root.   □

Lemma 4.17 *Suppose that $p \in W^s(J^*) \cap W^u(J^*)$ and that F is hyperbolic on J^*. Then $p \in J^*$.*

Proof: See Figure 18. By the previous Lemma we can assume after a small change in p that p is a transverse intersection between a stable and an unstable manifold for points in J^*. But then $K := J^* \cup \{p_n\}_{n \in Z}$ is a compact invariant set. Moreover, F is saddle hyperbolic on K and K is chain recurrent. Hence by Shub, $K \subset J^*$.   □

The next result is the analogue of Proposition 4.2 of ([F3]). The condition that saddle points are dense in J^* is here replaced by contraction of the mapping.

Fig. 19. B_z^n along the unstable direction

Proposition 4.18 *Let F be a generalized Hénon map with $|J(F)| = \lambda < 1$. Suppose that F is uniformly sustainable on $L = \omega(z_0), z_0 \in J$, with uniform constants $0 < \delta < 1$, $N(\epsilon, \delta)$. Let $r_n(z)$ denote the radius of the largest ball $\mathbb{B}(0, r) \subset B_z^n$. Then $r_n(z) \leq \frac{1}{5}$ for all $n \geq 1, z \in L$, Figure 19.*

Proof: If not, pick $w \in L, m \geq 1$ so that $r_m(w) > \frac{1+\sigma}{\delta}, \sigma > 0$. Let $\Lambda > 1$, to be chosen below. Let $\epsilon, 0 < \epsilon < \delta$, to be chosen below and set $N = N(\epsilon, \delta)$. Let $n \geq 1$ be the smallest integer so that $\frac{4\lambda^n}{\delta} < \frac{\delta}{\Lambda}$. Set $z_0 = w_{-n-N}$.

Before we proceed we need the following immediate Lemma:

Lemma 4.19 *We can diagonalize $(F^n)'(w_{-n})$ by using suitable orthonormal bases $\{A, B\}$ at w_{-n} and $\{\xi_1, \xi_2\}$ at w_0. The largest diagonal entry satisfies $|E_z^n| \geq \frac{\delta}{4}$ (Theorem 4.4), the other satisfies $|F_z^n| \leq \frac{4\lambda^n}{\delta}$. $[(F^n)'(z)(A) = E_z^n \xi_1, (F^n)'(z)(B) = F_z^n \xi_2.]$*

We diagonalize $(F^n)'(w_{-n})$ as in the Lemma.

Let $\tilde{\xi}$ be the image $(F^m)'(w_0)(\xi_1)$. We choose $\tilde{v} \perp \tilde{\xi}$ of norm $\frac{1+\sigma}{\delta}$. Then there exist vectors $\tilde{s}_1, \ldots, \tilde{s}_m$ of norm ≤ 1 so that

$$\tilde{v} = F'(w_{m-1}) \cdots F'(w_1)\tilde{s}_1$$
$$+ \cdots + F'(w_{m-1})\tilde{s}_{m-1} + \tilde{s}_m.$$

We set

$$s_j = 0, j \leq n+N, s_j = \delta\tilde{s}_{j-n-N}, n+N < j \leq n+N+m, s_j = 0, j > n+N+m.$$

By sustainability there is $|t| \leq \epsilon$, so that $|\xi_\ell(s, t)| \leq 1$ for the orbit of z_0.

At first, we observe that $|\xi_N(s, t)| \leq 1$. Next we observe that

$$\xi_{N+n}(s, t) = [(F^n)'(z_N(s, t))(\xi_N)]$$
$$+ [F'(z_{n+N-1}) \cdots F'(z_N)t_N + \cdots + t_{n+N}]$$

$$\xi_{N+n+m}(s,t) = [(F^m)'(w)(F^n)'(z_N)(\xi_N)]$$
$$+ [F'(z_{n+N+m-1})\cdots F'(z_N)t_N + \cdots + t_{n+N+m}]$$
$$+ \delta\tilde{v}$$

Let $[\cdot]_1$ denote the component of a vector parallel to $\tilde{v}$. Then we get

$$1 \geq \|\xi_{N+n+m}(s,t)\|$$
$$\geq \|[\xi_{N+n+m}(s,t)]_1\|$$
$$\geq \delta\|\tilde{v}\| - \|[(F^m)'(w)(F^n)'(z_N)(\xi_N)]_1\|$$
$$- \|[F'(z_{n+N+m-1})\cdots F'(z_N)t_N + \cdots + t_{n+N+m}]_1\|$$
$$\geq 1 + \sigma - \|(F^m)'(w)\|\frac{4\lambda^n}{\delta} - \sum_{j=0}^{n+m}(\|F'\|_L)^j|t|$$

Let $C := \max\{2, \sup_{p\in L}\|F'(p)\|\}$. We get:

$$1 \geq 1 + \sigma - C^m\frac{\delta}{\Lambda} - C^{n+m}\epsilon.$$

Next we choose Λ so that $C^m\frac{\delta}{\Lambda} < \frac{\sigma}{2}$. This determines n. Next we choose ϵ so that $C^{n+m}\epsilon < \frac{\sigma}{2}$. It then follows that $\|\xi_{N+n}(s,t)\| > 1$, a contradiction. $\square$

Proposition 4.20 *Assume that* $|J(F)| < 1$. *Then there is a unique continuous invariant line field* $\{L_z\}, z \in L, L_z \subset A_z^\infty$.

Proof:
$$A_z^\infty = \cup_{n\geq 1}A_z^n \supset \cdots A_z^k \supset \cdots \supset A_z^1$$

where $A_z^n = B_{F^{-n}(z)}^n$.

By construction $B_{F^{-n}(z)}^n \supset \mathbb{B}_2(0,1)$. However, by Proposition 4.19, $B_{F^{-n}(z)}^n$ has a boundary point in $\overline{\mathbb{B}}_2(1,1/\delta)$. Each A_z^n is a convex, complete Reinhardt domain. Hence A_z^∞ is a convex, complete Reinhardt domain and A_z^∞ has a boundary point in $\overline{\mathbb{B}}_2(1,1/\delta)$ while $A_z^\infty \supset \mathbb{B}_2(0,1)$. Moreover, by Theorem 4.7, $\text{Vol}(A_z^\infty) = \infty$. Therefore A_z^∞ is an unbounded set. Let $\{z_n\} \subset A_z^\infty, z_n \to \infty$. Then there exist radii $R_n \to \infty$ so that

$$\{\tau\frac{z_n}{\|z_n\|}, \tau \in \mathbb{C}, |\tau| < R_n\} \subset A_z^\infty.$$

By passing to a limit we find that there exists a complex line L_z through 0 so that $L_z \subset \overline{A}_z^\infty$. Since A_z^∞ contains an open neighborhood of the origin, it follows from convexity that $L_z \subset A_z^\infty$.

The continuity of the line field follows from the fact that A_z^n varies continuously with z for all n and the fact that the volume of $A_z^n = B_{F^{-n}(z)}^n$ grows uniformly by Lemma 4.6. $\square$

We prove a contraction estimate similar to ([FS3, Proposition 4.6]) and include the details for the readers convenience.

Define $\{\tilde{\xi}^s\}_{z \in L}, \|\tilde{\xi}_z^s\| = 1, \tilde{\xi}_z^s \perp L_z$. Then for ξ_z^u a unit vector in L_z:

Proposition 4.21 *Assume* $|J(F)| < 1$. *Write* $(F^k)'(z)(\tilde{\xi}_z^s) = \alpha_z^k \tilde{\xi}_{F^k(z)}^s + \beta_z^k \xi_{F^k(z)}^u$. *Then* $|\alpha_z^k| \leq C\mu^k, C > 1, \mu < 1, z \in L$ *arbitrary,* $k \geq 1$.

We will first prove some Lemmas.

We know from Proposition 4.21 that $A_z^\infty = \{\tau \tilde{\xi}_z^s; |\tau| < r_z\} + L_z, 1 \leq r_z \leq \frac{1}{\delta}$. We have that $F'(z)(\tilde{\xi}_z^s) = \alpha_z \tilde{\xi}_{F(z)}^s + \beta_z \xi_{F(z)}^u$. Hence $F'(z)(r_z \tilde{\xi}_z^s) = r_z \alpha_z \tilde{\xi}_{F(z)}^s + r_z \beta_z \xi_{F(z)}^u$.

This implies

Lemma 4.22 $r_{F(z)} = r_z |\alpha_z| + 1$.

We can apply this Lemma repeatedly to the orbit $\{z_k\}_{k \geq 0}$:

$$r_{z_1} = r_{z_0} |\alpha_{z_0}| + 1$$
$$r_{z_2} = r_{z_1} |\alpha_{z_1}| + 1$$
$$= [r_{z_0} |\alpha_{z_0}| + 1]|\alpha_{z_1}| + 1$$
$$= |\alpha_{z_1}||\alpha_{z_0}|r_{z_0} + |\alpha_{z_1}| + 1$$

Lemma 4.23

$$r_{z_k} = |\alpha_{z_{k-1}}| \cdots |\alpha_{z_0}|r_{z_0} + |\alpha_{z_{k-1}}| \cdots |\alpha_{z_1}| + \cdots + |\alpha_{z_{k-1}}| + 1.$$

Since $1 \leq r_{z_0}$ and $r_{z_n} \leq 1/\delta$ we get after replacing k by $k + 1$:

Corollary 4.24 *If* $k \geq 0$,

$$|\alpha_{z_k}| \cdots |\alpha_{z_0}| + |\alpha_{z_k}| \cdots |\alpha_{z_1}| + \cdots + |\alpha_{z_k}| + 1 \leq 1/\delta.$$

Let $j = \left[\frac{2}{\delta}\right] + 1$. Then $j \geq \frac{2}{\delta}$.

Lemma 4.25 *If* $k \geq j$, *then there is an* $m(k), k - j < m(k) \leq k$ *so that* $|\alpha_{z_k}| \cdots |\alpha_{z_{m(k)}}| \leq \frac{1}{2}$.

Proof:

$$|\alpha_{z_k}| + |\alpha_{z_k}||\alpha_{z_{k-1}}| + \cdots + |\alpha_{z_k}| \cdots |\alpha_{z_{k-j+1}}| \leq \frac{1}{\delta}.$$

There are j terms, so the smallest one is at most $\frac{1}{\delta}\frac{1}{j} \leq \frac{1}{\delta}\frac{\delta}{2} = \frac{1}{2}$.   □

Set $C = \sup_{z \in L} |\alpha_z| \vee 1$.

Lemma 4.26 *Let* $k = \ell j + r, 0 \leq r < j$. *Then* $|\alpha_{z_k}| \cdots |\alpha_{z_0}| \leq 2C^j \left(\frac{1}{2}\right)^{\frac{k}{j}}$.

Proof: Set $k = k_1, k_2 = m(k_1) - 1, \ldots, k_n = m(k_{n-1}) - 1, k_n < j$. Then $n \geq \ell$ and we get

$$
\begin{aligned}
|\alpha_{z_k}| \cdots |\alpha_{z_0}| &= \left[|\alpha_{z_{k_1}}| \cdots |\alpha_{z_{m(k_1)}}| \right] \\
&* \left[|\alpha_{z_{k_2}}| \cdots |\alpha_{z_{m(k_2)}}| \right] \\
&\cdots \left[|\alpha_{z_{k_{n-1}}}| \cdots |\alpha_{z_{m(k_{n-1})}}| \right] \\
&* |\alpha_{z_{k_n}}| \cdots |\alpha_{z_0}| \\
&\leq \left(\frac{1}{2} \right)^n C^j \\
&\leq \left(\frac{1}{2} \right)^\ell C^j
\end{aligned}
$$

Since $\ell j \geq k - j, \ell \geq \frac{k-j}{j} = \frac{k}{j} - 1$. $\qquad\square$

Proof of the Proposition: The estimate follows from the identity $|\alpha_z^k| = |\alpha_{F^k(z)}| \cdots |\alpha_z|$ and by the above Lemma. $\qquad\square$

We set $(F^k)'(\xi_z^u) = \gamma_z^k \xi_{F^k(z)}^u$. Theorem 4.4 implies:

Lemma 4.27 *Let $z \in L$. Assume that $|J(F)| < 1$ Then, $|\alpha_z^n| \leq C\mu^n, C > 1, \mu < 1, \max\{|\beta_z^n|, |\gamma_z^n|\} \geq \tau > 0$ for each $n \geq 1$. The constants C, μ, τ are independent of z.*

The next Lemma is the same as ([FS3], Lemma 4.11). We still include the proof here for the benefit of the reader.

Lemma 4.28 *There exists an increasing sequence $\{C_n\}, 1 < C_1 < C_2 < \cdots < \lim C_n < \infty$ so that $|\beta_z^n| \leq C_n |\gamma_z^n|$ for all $z \in L$.*

Proof: We get if $n \geq 1$ and $C' := \sup_{q \in L} |\beta_q^1|$:

$$
\begin{aligned}
\beta_p^{n+1} &= \alpha_p^n \beta_{F^n(p)}^1 + \beta_p^n \gamma_{F^n(p)}^1 \\
\gamma_p^{n+1} &= \gamma_{F^n(p)}^1 \gamma_p^n
\end{aligned}
$$

Hence:

$$
\begin{aligned}
|\beta_p^{n+1}| &\leq C\mu^n |\beta_{F^n(p)}^1| + C_n |\gamma_p^n| |\gamma_{F^n(p)}^1| \\
|\beta_p^{n+1}| &\leq CC'\mu^n + C_n |\gamma_p^{n+1}|.
\end{aligned}
$$

Suppose at first that $|\gamma_p^{n+1}| \geq \tau$. Then we have

$$CC'\mu^n = \left(\frac{CC'}{\tau}\right)\mu^n\tau$$

$$\leq \left(\frac{CC'}{\tau}\right)\mu^n|\gamma_p^{n+1}|$$

Hence:

$$|\beta_p^{n+1}| \leq \left[C_n + \frac{CC'}{\tau}\mu^n\right]|\gamma_p^{n+1}|$$

$$\leq C_n\left[1 + \frac{CC'}{\tau}\mu^n\right]|\gamma_p^{n+1}|$$

Assume next that $|\gamma_p^{n+1}| < \tau$. Then by Lemma 4.27, $|\beta_p^{n+1}| \geq \tau$. We get:

$$|\beta_p^{n+1}| \leq C_n|\gamma_p^{n+1}| + \left(\frac{CC'}{\tau}\right)\mu^n\tau$$

$$\leq C_n|\gamma_p^{n+1}| + \left(\frac{CC'}{\tau}\right)\mu^n|\beta_p^{n+1}|$$

Hence, if $n \geq n_0, n_0$ large enough,

$$|\beta_p^{n+1}| \leq \frac{C_n}{1 - \frac{CC'}{\tau}\mu^n}|\gamma_p^{n+1}|.$$

Hence we first define $C_1, \ldots, C_{n_0}$ so that $1 < C_1 < \cdots < C_{n_0}$ and $|\beta_p^j| \leq C_j|\gamma_p^j|$ for all $j = 1, \ldots, n_0, p \in L$: This is possible since the functions $|\gamma_p^j|$: $L \to \mathbb{R}^+$ are strictly positive continuous functions. Then for $n \geq n_0$, we define

$$C_{n+1} = \max\left\{\frac{C_n}{1 - \frac{CC'}{\tau}\mu^n}, C_n\left[1 + \frac{CC'}{\tau}\mu^n\right]\right\} = \frac{C_n}{1 - \frac{CC'}{\tau}\mu^n}.$$

Since $\mu < 1$ the Lemma follows.   □

Let $C_\infty = \lim C_n$.

$$|\beta_z^n| \leq C_\infty|\gamma_z^n| \; \forall \, z \in L, n \geq 1.$$

Since $\max\{|\beta_p^n|, |\gamma_p^n|\} \geq \tau$, we have $|\gamma_p^n| \geq \frac{\tau}{C_\infty}$. From this we immediately get:

$$\|(F^n)'(z)(\xi_z^u)\| \geq \frac{\tau}{C_\infty}\|\xi_z^u\|$$

for any $z \in L$ and any $n \geq 1$.

Proposition 4.29 *Suppose that F is a generalized Hénon map,*
$|J(F)| < 1$. *Suppose $z_0 \in J$ is sustainable and let $L := \omega(z_0)$. Then F is saddle hyperbolic on L.*

Proof: We show first that F is uniformly expanding on the line field L_z. If not, there is a point $z \in L$ so that

$$\|(F^n)'(z)(\xi_z^u)\| \le \|\xi_z^u\|$$

for all $n \ge 1$. Let $T > C$ be a constant to be fixed later, C as in Lemma 4.28. Then there exists by Theorem 4.5 integers $n_0, n_1 \ge 1$ so that the norm of the matrix $(F^{n_1})'(F^{n_0}(z))$ is at least $3T$. Since $|\alpha^{n_1}_{F^{n_0}(z)}| \le C\mu^{n_1} \le C < T$ we obtain that

$$\max\{|\beta^{n_1}_{F^{n_0}(z)}|, |\gamma^{n_1}_{F^{n_0}(z)}|\} \ge T.$$

We also know that $|\gamma^{n_1}_{F^{n_0}(z)}| \ge \frac{T}{C_\infty}$ by Lemma 4.29. Since we know by Lemma 4.28 that $|\gamma^{n_0}_z| \ge \frac{\tau}{C_\infty}$, we conclude that

$$|\gamma^{n_1+n_0}_z| = |\gamma^{n_1}_{F^{n_0}(z)}||\gamma^{n_0}_z| \ge \frac{T\tau}{C_\infty^2}.$$

This contradicts that $|\gamma^{n_1+n_0}_z| = \|(F^{n_1+n_0})'(z)\xi_z^u\| \le 1$ if we choose $T > \frac{C_\infty^2}{\tau}$.

Hence we have shown that F is uniformly expanding on $\{L_z\}, z \in L$, i.e. there are constants $C_3 > 0, \lambda > 1$ so that

$$|\gamma_z^n| \ge C_3\lambda^n, z \in L, n \ge 1.$$

Next we find a stable invariant continuous line field on L by a standard argument. Fix a $k \ge 1$ so that $|\gamma_z^k| > 2, |\alpha_z^k| < \frac{1}{2}$ for all $z \in L$. Also, let $A = \sup_L |\beta_z^k|$.

We start for any z with a disc of candidates for stable vectors,

$$\Delta_0^z = \{\tilde{\xi}_z^s + \tau\xi_z^u; |\tau| < 2A\}.$$

After k iterates this disc expands to cover $\{\alpha_z^k\tilde{\xi}_{F^k(z)}^s + \sigma\xi_{F^k(z)}^u; |\sigma| < 3A\}$. We let Δ_1^z consist of the relatively compact subdisc of Δ_0^z whose image covers $\{|\sigma| < 2A\}$. We repeat the process. The intersection of the discs Δ_j^z define the line field L_z^s. It is easy to see that L_z^s is invariant and varies continuously and is stable. $\qquad\square$

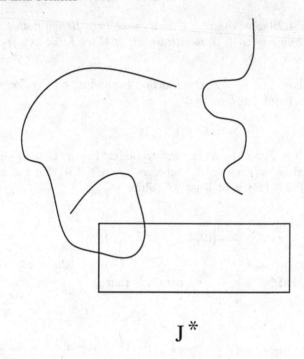

$$J^*$$

Fig. 20. α and ω limit sets

Completion of the Proof of Theorem 4.9: Let $p \in J \setminus J^*$. By the above Lemma F is saddle hyperbolic on $\omega(p)$. By Shub ([S]) there are saddle points clustering everywhere on $\omega(p)$. This implies by Lemma 4.14 that $\omega(p) \subset J^*$. By Lemma 4.18 it suffices to show that $\alpha(p) \subset J^*$ because this would imply that $p \in J^*$, a contradiction, Figure 20.

By the shadowing Lemma it follows that there is a neighborhood $U(J^*)$ so that if $p \in J \setminus J^*$ and $\alpha(p)$ is not contained in J^* then $\alpha(p)$ is not contained in U.

Let $p \in J \setminus J^*$, $\alpha(p)$ not a subset of J^*. Let $q \in \alpha(p) \setminus U$, then $\alpha(q) \subset \alpha(p)$. Notice that if $\alpha(q) \subset U$, then again it follows that $q \in J^*$, a contradiction, so $\alpha(q)$ can never be contained in U. Next pick $q \in \alpha(p) \setminus U$ so that $\alpha(q)$ is minimal. Then for every $z \in \alpha(q) \setminus U$, $\alpha(z) = \alpha(q)$. It follows that $F(\alpha(q)) \subset \alpha(q)$ since if $z \in \alpha(q)$, then there is a sequence n_j so that $F^{-n_j}(z) \to z$.

Let $z \in \alpha(q)$. Since $\omega(z) \subset J^*$, there is a minimal $n(z)$ so that for every $n \geq n(z)$, we have $F^n(z) \subset \overline{U}$. Let $K_m = \{z \in \alpha(q); n(z) \leq m\}$. Then K_m is closed. Moreover K_m is nowhere dense in $\alpha(q)$ outside $\overline{U}$. By a category argument there is a point z in $\alpha(q) \setminus \cup K_m$. But this contradicts that $\omega(z) \subset J^*$.

References

[BBD] Barrett, D., Bedford, E., Dadok, J; T^n *Actions on Holomorphically Separable Complex Manifolds*, Math.Z. 202 (1989), 65–82.

[BLS] Bedford, E., Lyubich, M., Smillie, J; *Polynomial diffeomorphisms of* $\mathbf{C}^2$. *IV. The measure of maximal entropy and laminar currents*, Inv. Mat. 112 (1993), 77–125.

[BS1] Bedford, E., Smillie, J; *Polynomial diffeomorphisms of* $\mathbf{C}^2$: *Currents, equilibrium measure and hyperbolicity*, Inv. Math. 103 (1991), no. 1, 69–99.

[BS2] Bedford, E., Smillie, J; *Polynomial diffeomorphisms of* $\mathbf{C}^2$. *II. Stable manifolds and recurrence*, J. Amer. Math. Soc. 4 (1991), no. 4, 657–679.

[BS3] Bedford, E., Smillie, J; *Polynomial diffeomorphisms of* $\mathbf{C}^2$. *III. Ergodicity, exponents and entropy of the equilibrium measure*, Math. Ann. 294 (1992), no. 3, 395–420.

[BS5] Bedford, E., Smillie, J; *Polynomial diffeomorphisms of* $\mathbf{C}^2$. *V. Critical points and Lyapunov exponents*, J. Geom. Anal. 8 (1998), no. 3, 349–383.

[BS6] Bedford, E., Smillie, J; *Polynomial diffeomorphisms of* $\mathbf{C}^2$. *VI. Connectivity of J*, Ann. of Math. (2) 148 (1998), no. 2, 695–735.

[BS7] Bedford, E., Smillie, J; *Polynomial diffeomorphisms of* $\mathbf{C}^2$. *VII. Hyperbolicity and external rays*, Ann. Sci. Ecole Norm. Sup. (4) 32 (1999), no. 4, 455–497.

[BS8] Bedford, E., Smillie, J; *Polynomial Diffeomorphisms of* $\mathbf{C}^2$. *VIII. Quasiexpansion*, Preprint.

[BT] Bedford, E., Taylor, B. A; *The Dirichlet problem for a complex Monge-Ampére equation*, Invent. Math. 37 (1976), 1–44.

[BF] Bonifant, A., Fornæss, J. E; *Attractors*, In Complex Geometry, Springer Verlag 2002. Editors, Bauer, Catanese, Kawamata, Peternell and Siu, 73–84.

[BD] Briend, J. Y., Duval, J; *Exposants de Lyapunov et points periodiques d'endomorphismes de* $\mathbb{P}^k$, Acta Math. 182 (1999), no. 2, 143–157.

[B] Brjuno, A. D; *Convergence of transformations of differential equations to normal form*, Dokl. Akad. Nauk USSR, 165 (1965), 987–989.

[Bo] Böttcher, L; *The principle laws of convergence of iterates and their application to analysis*, Bull. Kasan Math. Soc. 14 (1904), 155–234.

[CG] Carleson, L., Gamelin, T.W; Complex Dynamics, Springer, (1993)

[Cr] Cremer, H; *Über die Häufigkeit der Nichtzentren*, Math. Ann. 15 (1938), 573–580.

[DJ] Diller, J., Jonsson, M; *Topological Entropy on saddle sets in* $\mathbb{P}^2$, Duke Math. J. 103 (2000), 261–278.

[FJ] Favre, C., Jonsson, M; *Manuscript*

[F1] Fornæss, J. E; Dynamics in Several Complex Variables, CBMS 87, (1996).

[F2] Fornæss, J. E; *Infinite dimensional complex dynamics*, Discrete and Continuous Dynamical Systems, Millenium Issue, 6 (2000), 51–60.

[F3] Fornæss, J. E; *Random Iteration and Shadowing*, preprint.

[F4] Fornæss, J. E; *Short* $\mathbb{C}^k$, preprint.

[FGa] Fornæss, J. E., Gavosto, E; *Existence of Generic Homoclinic Tangencies for Henon mappings*, Journal of Geometric Analysis, 2 (1992), 1–16.

[FGr] Fornæss, J. E., Grellier, S; *Exploding Orbits of Hamiltonian and Contact Structures*, Lecture notes in pure and applied math, 173 (1996), 155–172.

[FS1] Fornæss, J. E., Sibony, N; *Random Iterations of Rational Functions*, Ergodic Theory and Dynamical Systems, 11 (1991), 687–708.

[FS2] Fornæss, J. E., Sibony, N; *Complex Hénon mappings in* $\mathbb{C}^2$ *and Fatou-Bieberbach domains,* Duke Math. J. 65 (1992), 345–380.

[FS3] Fornæss, J. E., Sibony, N; *Complex Dynamics in higher dimension II,* Ann. Math. Studies 137 (1995), 135–187.

[FS4] Fornæss, J. E,, Sibony, N; *Holomorphic Symplectomorphisms in* $\mathbb{C}^2$, World Sc. Series, Appl. Anal., Vol 4, Dynamical Systems and Applications, (1995), 239–262.

[FS5] Fornæss, J. E., Sibony, N; *Oka's inequality for currents and applications,* Math. Ann. 301 (1995), 399–419.

[FS6] Fornæss, J. E,, Sibony, N; *Holomorphic Symplectomorphisms in* $\mathbb{C}^{2p}$, Duke Math J. 82 (1996), 521–533.

[FS7] Fornæss, J. E., Sibony, N; *Hyperbolic maps on* $\mathbb{P}^2$, Math. Ann. 311 (1998), 305–333.

[FS8] Fornæss, J. E., Sibony, N; *Fatou and Julia sets for entire mappings in* $\mathbb{C}^k$, Math. Ann. 311 (1998), 27–40.

[FS9] Fornæss, J. E., Sibony, N. *Localization for Unitary Operators,* Preprint.

[FW1] Fornæss, J. E., Weickert, B; *Attractors on* $\mathbb{P}^2$, Several complex variables (Berkeley, CA, 1995–1996), 297– 307, Cambridge Univ. Press, Cambridge, (1999).

[FW2] Fornæss, J. E., Weickert, B; *Random iteration in* $\mathbb{P}^k$, Ergodic Theory and Dynamical Systems 20 (2000), no. 4, 1091–1109.

[FW3] Fornæss, J. E,, Weickert, B; *A quantized Henon map,* Discrete Contin. Dynam. Systems 6 (2000), no. 3, 723–740.

[FLM] Freire, A., Lopes, A., Mañe, R; *An invariant measure for rational maps,* Bol. Soc. Bras. Mat 6 (1983), 45–62.

[FM] Friedland, S., Milnor, J; *Dynamical Properties of plane polynomial automorphisms,* Ergodic Theory and Dynamical Systems. 9 (1989), 67–99.

[Ga] Gavosto, E; *Attracting basins in* $\mathbb{P}^2$, J. Geometric Analysis, 8 (1998), no. 3, 433–440.

[Gu] Guedj, V; *Approximation of currents on complex manifolds,* Math. Ann. 313 (1999), 437–474.

[HH] Hénon, M., Heiles, C; *The Applicability of the Third Integral of Motion: Some Numerical Experiments,* The Astronomical Journal, 69, Number 1 (1964), 73–79.

[He] Herman, M. R; *Recent results and some open questions on Siegel's linearization theorem of germs of complex-analytic diffeomorphisms of* $\mathbb{C}^n$ *near a fixed point,* Proceedings on Mathematical Physics, Marseille (1986), Mebkhout, M and Sénéor, R., editors, World Sc., Singapore (1987), 138–184.

[HP] Hubbard, J., Papadopol; *Superattractive fixed points in* $\mathbb{C}^n$, Indiana Univ. Math. J. 43 (1994), 321–365.

[JV] Jonsson, M., Varolin, D; *Stable manifolds of holomorphic diffeomorphisms,* Invent. Math., 149 (2002), 409–430.

[JW] Jonsson, M., Weickert, B; *A nonalgebraic attractor in* $\mathbb{P}^2$, Proc. Amer. Math. Soc. 128 (2000), no. 10, 2999–3002.

[Ko] Koenigs, G; *Recherches sur les integrales de certaines équations fonctionelles,* Ann. Sci. de l'École Normale Supérieure, Series 3, 1 (1884), 1–41.

[LM] Laskar, J., Marmi, S; *Chaotic behaviour in the solar system,* Seminaire Bourbaki, Vol. 1998/99. Asterisque, No. 266 Exp. No. 854, 3 (2000), 113–136.

[Le] Lelong, P; Fonctions plurisousharmoniques et formes différentielles positive, New York, Gordon and Breach, Paris, (1968).

[Ly]   Lyubich, M; *Entropy properties of rational endomorphisms of the Riemann sphere*, Erg. Th. Dyn. Syst. 3 (1983), 351–385.

[MSS]   Mañe, R., Sad, P., Sullivan, D; *On the dynamics of rational maps*, Ann. Éc. Norm. Sup. 16 (1983), 193–217.

[M]   McMullen, C; Complex Dynamics and Renormalization, Ann. of Math. Studies, 135 (1994) Princeton Univeristy Press.

[Mi]   Milnor, J; Dynamics in one complex variable: Introductory lectures, Friedr. Vieweg and Sohn, Braunschweig, (1999), viii+257 pp.

[MNTU] Morosawa, S., Nishimura, Y., Taniguchi, M., Ueda, T; Holomorphic dynamics. Cambridge University Press, Cambridge,(2000), xii+338 pp.

[Pe]   Pesin, Y; *Characteristic Lyapunov exponents and smooth ergodig theory*, Russian Math. Surveys 32 (1977), 55–114.

[Po]   Poincaré, H; *Sur les proprietés des Fonctions Définis par les Equations aux Différences Partielles*, (1987), in Oeuvres, Tome 1, Gauthier-Villars, Paris (1928), 49–129.

[Pr]   Przytycki, F; *Riemann map and holomorphic dynamics*, Inv. math 85 (1986), 439–455.

[R]   Rees, M; *Positive measure sets of ergodic rational maps*, Ann. Sci. Éc. Norm Sup. 19 (1986), 383–407.

[RR]   Rosay, J. P., Rudin, W; *Holomorphic maps from $\mathbb{C}^n$ to $\mathbb{C}^n$*, Trans. Amer. Math. Soc. 310 (1988), no. 1, 47–86.

[Ru]   Ruelle, D; Elements of differentiable dynamics and bifurcation theory, Academic Press (1989).

[Sc]   Schröder, E; *Ueber iterierte Funktionen*, Math. Ann. 3 (1871), 296–322.

[S]   Shub, M; Global stability of Mappings, Springer Verlag, (1987).

[S1]   Sibony, N; *Dynamique des applications rationelles de $\mathbb{P}^k$*, Panoramas et Synthèses, In Dynamique et geometrie complexes (Lyon, 1997), ix–x, xi–xii, 97–185, Soc. Math. France, Paris, (1999).

[S2]   Sibony, N; *Valeurs au bord de fonctions holomorphes et ensembles polynomialement convexes*, Springer-Verlag, Lecture Notes in Math. 578 (1977), 300–312.

[Si1]   Siegel, C. L; *Iteration of analytic functions,* Ann. Math. 43 (1942), 807–812.

[Si2]   Siegel, C. L; *Über die Normalform analytischer Differentialgleichungen in der Nahe einer Gleichgewichtslösung,* Nachr. Akad. Wiss. Göttingen, Math. Phys. K1, (1952), 21–30.

[Ste]   Sternberg, S; Celestial Mechanics, Part II, Benjamin, New York (1969).

[T]   Tsuji; Potential theory in modern function theory, Tokyo Mazuren, (1959).

[Ue1]   Ueda, T; *Local structure of analytic transformations of two complex variables, I,* J. Math. Kyoto Univ, 26-2 (1986), 233–261.

[Ue2]   Ueda, T; *Fatou sets in complex dynamics on projective spaces,* J. Math. Soc. Japan 46, no.3 (1994), 545–556.

[Ue3]   Ueda, T; *Critical orbits of holomorphic maps on projective spaces,* J. Geom. Anal. 8 (1998), no. 2, 319–334.

[We1]   Weickert, B; *Attracting basins for automorphisms of $\mathbb{C}^2$,* Invent. Math. 132 (1998), no. 3, 581–605.

[We2]   Weickert, B; *Infinite-dimensional complex dynamics: a quantum random walk,* Discrete Contin. Dynam. Systems 7 (2001), no. 3, 517–524.

[Y]   Yoccoz, J. C; *Linéarisation des germes de difféomorphisms holomorphes de* $(\mathbb{C}, 0)$, C. R. Acad. Sci. Paris 306 (1988), 55–58.

Local Equivalence Problems
for Real Submanifolds in Complex Spaces

Xiaojun Huang *

Department of Mathematics, Rutgers University at New Brunswick, USA
huangx@math.rutgers.edu

1 Global and Local Equivalence Problems

There is a classical theorem in complex analysis, called the Riemann mapping theorem, which states that any simply connected domain in $\mathbf{C}$ is either holomorphically equivalent to $\mathbf{C}$ or to the unit disk. For more general domains in $\mathbf{C}$, He-Schramm showed [HS] that if ∂D has countably many connected components, then D is holomorphically equivalent to a circle domain whose boundaries are either points or circles. These results give a nice picture on the holomorphic structures for domains in $\mathbf{C}$. When one goes to higher dimensions, a natural question is then to investigate the complex structure for domains in $\mathbf{C}^n$ for $n \geq 2$. More precisely, given two domains in $\mathbf{C}^n$, one would like to know if there is a biholomorphic map between them. This the so-called global equivalence problem in several complex variables. Along these lines of investigations, substantial progress has been made in the past 30 years ([Fe], [CM], [BSW], etc.). However, we are still a certain big distance away from getting a relatively complete picture as in the one complex variable.

An approach to the study of the equivalence problem is to attach holomorphic invariants to each given domain. Since domains in $\mathbf{C}^n$ are open complex manifolds, many (interior) invariants which are crucial for the study of compact complex manifolds are difficult even to define. As already observed by Poincaré about 100 years ago, the interior complex structure of a domain D in $\mathbf{C}^n$ for $n > 1$ is closely related to the partial complex structure in its boundary, which is the so-called CR structure. Hence, the classification of

---

* Supported in part by NSF-0200689 and a grant from the Rutgers University Research Council

the complex structures for domains in $\mathbf{C}^n$ may be reduced to the equivalence problem for the boundary CR structures. Indeed, this idea has been proved to be fundamental through the work of Cartan, Tanaka, Chern-Moser, etc. And it indeed led to the solutions to many questions.

To illustrate what we said above, we give the following classical example of Poincaré:

Proposition 1.1: Let $\mathbf{B}^n = \{z \in \mathbf{C}^n : |z| < 1\}$ and $\Delta^n = \Delta \times ... \times \Delta := \{(z_1, \cdots, z_n) : |z_j| < 1\}$. $\mathbf{B}^n$ and Δ^n are diffeomorphic to each other. But $\mathbf{B}^n$ is not biholomorphic to Δ^n.

Proof of Proposition 1.1: Suppose that there is a biholomorphic map f from Δ^n to $\mathbf{B}^n$. Let $p = (p_1, ..., p_n) \in \Delta^n$ be such that $f(p) = 0$. Let $\sigma_j \in Aut(\Delta)$ be such that $\sigma_j(0) = p_j$. Write $\sigma(z_1, ..., z_n) = (\sigma_1(z_1), ..., \sigma(z_n))$ and $F = f \circ \sigma$. Then F is also a biholomorphic map from Δ^n to $\mathbf{B}^n$ with $F(0) = 0$.

Write $F(z) = zA + \sum a_\alpha z^\alpha$. Write τ_θ for the map sending z to $e^{i\theta}z$, and define $F_\theta = \tau_\theta^{-1} \circ F \circ \tau_\theta : \Delta^n \rightarrow \mathbf{B}^n$. Then F_θ has the following Taylor expansion at 0: $Az + \sum e^{i(|\alpha|-1)\theta} a_\alpha z^\alpha$. Since $\mathbf{B}^n$ is convex, the map $\frac{1}{2\pi} \int_0^{2\pi} F_\theta(z)d\theta = Az$ still maps Δ^n to $\mathbf{B}^n$. Applying the same argument to F^{-1}, we similarly conclude that the $A^{-1}z$ maps $\mathbf{B}^n$ to Δ^n. Hence, $\mathbf{B}^n$ and Δ^n are holomorphically equivalent through the linear map Az. This yields a contradiction; for $\mathbf{B}^n$ has a smooth boundary, while the boundary of Δ is only Lipschitz continuous. $\square$

The key step in the proof of the above proposition is to find a better behaved map so that it induces a nice boundary map. Hence the existence of the holomorphic equivalence map imposes the 'match-up' of certain boundary geometry. In the case considered above, the group structure of the domains allows us to get a very rigid map, which can be actually made to be linear. In general, since most domains have trivial automorphism groups (see [GK], for instance), it is unrealistic to conjecture that the holomorphic equivalence of two domains must induce the linear equivalence of their boundary. A fundamental result by the work of C. Fefferman [Fe] and Bochner (see, e.g. [Ho] or [Kr]) asserts that for two bounded smooth strongly pseudoconvex domains, they are holomorphic equivalent if and only if their boundaries are CR equivalent. To state precisely the result of Bochner and Fefferman, we recall the following definition [Kr].

Let $D \subset\subset U$ be a bounded domain in $\mathbf{C}^n$ with defining function $r \in C^\alpha(U)$, where $\alpha \geq 2$. Namely, we assume that $r < 0$ in D, $r > 0$ in $U \setminus \bar{D}$ and $dr|_{\partial D} \neq 0$. (We call D a domain with C^α-smooth boundary.) Define the Levi form of r by

$$\mathcal{L}_{r,p}(\xi, \xi) = \sum \frac{\partial^2 r}{\partial z_j \partial \bar{z}_k}|_p \xi_j \overline{\xi_k}.$$

We call D is pseudoconvex (or strongly pseudoconvex) at $p \in \partial D$ if $\mathcal{L}_{r,p}(\xi, \xi) \geq 0$ (or, $\mathcal{L}_{r,p}(\xi, \bar{\xi}) \geq C|\xi|^2$ with $C > 0$, respectively) for any $\xi = (\xi_1, \cdots, \xi_n)$ with

$\sum_{j=1}^{n} \xi_j r_{z_j}(p) = 0$. D is called a pseudoconvex domain (or, a strongly pseudo-convex domain) if D is pseudoconvex (or, strongly pseudoconvex, respectively) at any boundary point p.

More generally, for a real submanifold $M \subset \mathbf{C}^n$ of real codimension k. We define CT_pM to be the collection of vectors: $\mathcal{L} = \sum(a_j \frac{\partial}{\partial z_j} + b_j \frac{\partial}{\partial \overline{z_j}})|_p$, such that $\mathcal{L}(g) = 0$ for any function g, which is smooth in a neighborhood of M and is constant along M. CT_pM is called the complex tangent vector space of M at p. We define the holomorphic and conjugate holomorphic tangent vector space of M at p to be

$$T_p^{(1,0)}M = CT_pM \cap T_p^{(1,0)}\mathbf{C}^n, \quad T_p^{(0,1)}M = CT_pM \cap T_p^{(0,1)}\mathbf{C}^n, \text{ respectively.}$$

Write $CR_p(M) = \dim_{\mathbf{C}} T_p^{(1,0)}M$. $CR_p(M)$ is called the CR dimension of M at p. It is easy to see that $CR_p(M)$ is an upper semi-continuous function in p, which is the simplest holomorphic invariant that one can attach to the germ of a real submanifold in $\mathbf{C}^n$. When $CR_p(M)$ is identically 0, we call M a totally real submanifold. When $CR_p(M)$ is a positive constant, we call M a CR submanifold of $\mathbf{C}^n$. Notice that in case M is a real hypersurface, $CR_p(M) \equiv n - 1$ and thus M must be a CR submanifold for $n \geq 2$.

Let M and M' be two CR submanifolds of $\mathbf{C}^n$ for $n \geq 2$. We call M and M' to be CR equivalent if there is a smooth diffeomorphism F from M into M' such that $F_*(T_Z^{(1,0)}M) = T_{F(Z)}^{(1,0)}M'$ for any $Z \in M$. Such a map F is called a smooth CR diffeomorphism from M to M'. We call two germs of real submanifolds (M_j, p_j) with $p_j \in M_j$ to be CR equivalent if there is a CR diffeomorphism from a small neighborhood of p_1 in M_1 to a small neighborhood of p_2 in M_2, which maps p_1 to p_2. The following theorem, called the Bochner-Fefferman theorem, is crucial to reduce the equivalence problem for domains to the study of the boundary CR equivalence problem:

Theorem 1.2 (Bochner-Fefferman [Fe] [Ho]): Let D_1 and D_2 be bounded strongly pseudoconvex domains in $\mathbf{C}^n$ with C^∞ boundaries. Then D_1 and D_2 are biholomorphically equivalent if and only if there is a smooth CR equivalence map from ∂D_1 to ∂D_2.

For any real submanifolds M and M' in $\mathbf{C}^n$, we call M and M' to be holomorphically equivalent if there is a biholomorphic map Φ from a neighborhood of M to a neighborhood of M' in $\mathbf{C}^n$ such that $\Phi(M) = M'$. Apparently, when M and M' are holomorphically equivalent, then they are automatically CR equivalent. By the work of many people (see [CM], [Le], [Pi], [BJT], etc.), it is now clear that when M and M' are real analytic CR submanifolds with some extra geometric restrictions, the CR equivalence of M with M' implies their holomorphic equivalence. For instance, the following is a special case of the Baouendi-Jacobowitz-Treves theorem: (For more references on this matter, we refer the reader to the book of Baouendi-Ebenfelt-Rothschild [BER1] or the survey paper [Hu1]):

Theorem 1.3 (Baouendi-Jacobowitz-Treves [BJT]): Let M_1 and M_2 be two real analytic hypersurfaces in $\mathbf{C}^n$. Suppose that M_1 and M_2 do not contain any non-trivial holomorphic curves. Then any smooth CR equivalence map from M_1 to M_2 is actually a holomorphic equivalence map from M_1 to M_2.

We notice that by a result of Diederich-Fornaess [DF], any compact real-analytic submanifold in $\mathbf{C}^n$ does not contain any non-trivial germs of complex analytic curves. We also mention that Theorem 1.3 follows from the more general theory of Chern-Moser, when both M_j are Levi non-degenerate. (See the next section for more notation on this matter.)

Different from the situation in one complex variable, in the 70's, Pinchuk and Vitushkin first showed that germs of local holomorphic equivalences between strongly pseudoconvex hypersurfaces can be extended to the global holomorphic equivalence maps under certain geometric assumptions for the hypersurfaces. (See [Vit] for references). This gives the evidence that for many important classes of domains, the local CR structures of their boundaries essentially determine their interior global complex structures. There have been many developments along these lines of research. Here, we only state the following theorem recently obtained in [HJ1] and refer the reader to [HJ1] for more references on this matter:

Thereon 1.4 (Huang-Ji [HJ1]): Let D be a bounded strongly pseudoconvex domain in $\mathbf{C}^n$ defined by a real polynomial. If there is a point $p \in \partial D$ such that a small piece of ∂D near p is CR equivalent to a small piece of the unit sphere $\partial \mathbf{B}^n$, then D must be biholomorphic to the unit ball $\mathbf{B}^n$.

With the above discussions, it is also natural to study the local holomorphic equivalence problem for real submanifolds in complex spaces. Namely, one can consider the following two problems:

Question 1.5: Let (M_j, p_j) be real submanifolds in $\mathbf{C}^n$. When is there a biholomorphic map F from a neighborhood U_1 of $p_1 \in M_1$ in $\mathbf{C}^n$ into a neighborhood of $p_2 \in M_2$ in $\mathbf{C}^n$ such that $f(M_1 \cap U_1) \subset M_2$?

Question 1.6: Let (M_j, p_j) be CR submanifolds in $\mathbf{C}^n$ and $\mathbf{C}^N$, respectively, with $N \geq n$. Classify all CR embeddings from (M_1, p_1) to (M_2, p_2) up to the CR automorphism groups: $\mathrm{Aut}(M_1)$ and $\mathrm{Aut}(M_2)$.

Question 1.6 is more along the lines of CR rigidity problems, which, unfortunately, we can only briefly touch in §2 of this lecture notes, due to the time limit. In the following sections, we will mainly address some of the recent work on Question 1.5.

Acknowledgment: The present notes grew out of the lectures I gave in the summer graduate school held at Martina Franca, Italy, June-July, 2002. The materials discussed are from the research work carried out in [CM] [Ch] [Mos] [MW] [HJ2] [EHZ1], etc.. Also, we mention to the reader the books by Krantz [Kr], Baouendi-Ebenfelt-Rothschild [BER1] and R. Gardner [Ga], where one can find most of the necessary prerequisites. We thank G. Zampieri and D. Zaitsev for their invitation and hospitality during my pleasant stay in Italy in the summer of 2002. Thanks are also due to S. Ji and D. Zaitsev for their generous help provided during the preparation of the notes.

2 Formal Theory
for Levi Non-degenerate Real Hypersurfaces

Let (M, p) be the germ of a real hypersurface in $\mathbf{C}^n$ $(n > 1)$ near p. We will construct the holomorphic invariants of M at p such that we can distinguish the hypersurfaces by reading off their invariants. In this section, we will use the formal power series method. There is a more geometric approach based on the ideas of E. Cartan, that we will address in §5. In the power series method, we will try to find good representation for the hypersurfaces, called their normal form. The invariants are then embedded in the coefficients of their normal form. This section is based on the papers [CM] [EHZ1].

2.1 General Theory for Formal Hypersurfaces

We let (M, p) be a germ of real (formal) hypersurface in the complex n-space with $n \geq 2$. First, after a local change of coordinates, we assume that $p = 0$, $TM = \{v = 0\}, T^{(1,0)}M = \{w = 0\}$, where we use $(z, w) \in \mathbf{C}^{n-1} \times \mathbf{C}$ for the coordinates of $\mathbf{C}^n$ and write $w = u + iv$. Then M near 0 is defined by an equation of the form: $v = \rho$ with $\rho(0) = d\rho(0) = 0$. Notice that ρ is real-valued. Write

$$\rho = \sum a_{k\bar{l}} z_k \overline{z_\ell} + \sum b_{kl} z_k z_l + \sum \overline{b_{kl} z_k z_l}$$
$$+ \sum \overline{e_k z_k} u + \sum e_k z_k u + du^2 + O(|(z, w)|^3).$$

Then we have on M:

$$Re\{-iw - 2\sum_{kl} b_{kl} z_k z_l - 2\sum_k c_k z_k u\} = \sum a_{k\bar{l}} z_k \overline{z_\ell} + du^2 + O(|(z, w)|^3).$$

Define

$$\begin{cases} w' = w - 2i\sum_{kl} b_{kl} z_k z_l - 2i\sum_k c_k z_k w - diw^2, \\ z' = z. \end{cases}$$

In the (z', w') coordinates, M can be expressed as the graph of the following function:

$$v' = \sum a_{k\bar{l}} z'_k \overline{z'_l} + O(3) = z' A \overline{z'}^t + O(3)$$

where $A = \overline{A}^t$ is a matrix. Write

$$A = \Gamma \begin{pmatrix} \lambda_1 & 0 & \cdots & 0 \\ 0 & \lambda_2 & \cdots & \cdots \\ \vdots & \vdots & & \vdots \\ 0 & 0 & \cdots & \lambda_{n-1} \end{pmatrix} \overline{\Gamma}^t = \Gamma \Lambda \overline{\Gamma}^t,$$

Then

$$v' = z' \Gamma \Lambda \overline{(z'\Gamma)}^t + O(3).$$

Let $z'' = z'\Gamma$, $w'' = w'$. We have $v'' = \sum \lambda_j |z_j''|^2 + O(3)$. We say that $p = 0$ is a Levi non-degenerate point of M if $\lambda_j \neq 0$ for each j.

Assume, for the rest of this section, that M is Levi non-degenerate at 0. Then without loss of generality, we can assume that

$$v'' = \sum \epsilon_j \left| \sqrt{|\lambda_j|} z_j'' \right|^2 + O(3),$$

where $\epsilon_j = -1$ if $j \leq \ell$; and $\epsilon_j = 1$ if $j > \ell$. With $z_j''' = \sqrt{|\lambda_j|} z_j''$, $w''' = w''$. Then in the (z''', w''') coordinates, M is the graph of the following function:

$$v''' = \sum \epsilon_j |z_j'''|^2 + O(3).$$

Still write z for z''' and w for w'''. Then M is defined by:

$$(2.0) \qquad v = -\sum_{j=1}^{\ell} |z_j|^2 + \sum_{j=\ell+1}^{n-1} |z_j|^2 + O(|(z,w)|^3).$$

In the above expression and for the rest of this section, when $\ell = 0$, we regard the first term after the equality sign to be zero. Replacing (z, w) by $(z_{\ell+1}, \cdots, z_{n-1}, z_1, \cdots, z_\ell, -w)$ if necessary, we can assume that $\ell \leq \frac{n-1}{2}$. The pair $(\ell, n-1-\ell)$ is called the signature of M at 0. The model of Levi non-degenerate hypersurfaces with signature $(l, n-1-l)$ is the hyperquadric defined as follows:

$$(2.0)' \qquad \mathbf{H}_\ell^n = \{ v = -\sum_{j=1}^{\ell} |z_j|^2 + \sum_{j=\ell+1}^{n-1} |z_j|^2 \}.$$

Notice that the pair $(\ell, n-1-\ell)$ is completely determined by ℓ. Hence, in what follows, for brevity, we call ℓ the signature of the above hypersurface M.

When $\ell = 0$, we call M strongly pseudoconvex. Also, when $\ell = 0$, $\mathbf{H}_0^n = \mathbf{H}^n$ reduces to the classical Heisenberg hypersurface. Let

$$(2.0)'' \qquad \mathbf{S}_\ell^n := \{ v > -\sum_{j=1}^{\ell} |z_j|^2 + \sum_{j=\ell+1}^{n-1} |z_j|^2 \},$$

which is called the Siegel upper half-space and has $\mathbf{H}_\ell^n$ as its real analytic boundary. Let

$(2.0)'''\qquad \mathbf{B}_\ell^n := \{1 + |z_1|^2 + ... + |z_\ell|^2 \geq |z_{\ell+1}|^2 + ... + |z_{n-1}|^2 + |w|^2\}.$

Define $\Psi_n = \left(\frac{2z}{i+w}, \frac{i-w}{i+w}\right)$. Then

$$\Phi_n = \Psi_n^{-1} = \left(\frac{2z}{1+w}, \frac{i-iw}{1+w}\right).$$

Both Ψ_n and Φ_n are called the Cayley transformations. It is easy to verify the following properties:

Lemma 2.1: Ψ_n is a bimeromorphic map from $\mathbf{S}_\ell^n$ to $\mathbf{B}_\ell^n$; and Ψ_n bimeromorphically maps $\mathbf{H}_\ell^n = \partial\mathbf{S}_\ell^n$ to $\partial\mathbf{B}_\ell^n$. In particular, Ψ_n is a holomorphic equivalence map from $(\mathbf{H}_\ell^n, 0)$ to $(\partial\mathbf{B}_\ell^n, 0)$.

For convenience of the discussion, we set up some notation to be used for the rest of this section.

For two m-tuples $x = (x_1, \cdots, x_m), y = (y_1, \cdots, y_m)$, we write $< x, y >_\ell = \sum_{j=1}^m \delta_{j,\ell} x_j y_j$, and $|x|_\ell^2 = \sum_{j=1}^n \delta_{j,\ell} |x_j|^2$. Here $\delta_{j,\ell}$ is defined to be -1 for $j \leq \ell$ and to be 1 otherwise. We define the matrix $E_{\ell,n-1}$ to be the diagonal matrix with its first ℓ diagonal elements -1 and the rest 1.

Parameterize $\mathbf{H}_\ell^n$ by $(z, \bar{z}, u)$ through the map $(z, \bar{z}, u) \to (z, u + i|z|_\ell^2)$. In what follows, we will assign the weight of z and u to be 1 and 2, respectively. For a nonnegative integer m, a function $h(z, \bar{z}, u)$ defined over a small ball M of 0 in $\mathbf{H}_\ell^n$ is said to be of quantity $o_{wt}(m)$, if $\frac{h(tz, t\bar{z}, t^2 u)}{|t|^m} \to 0$ uniformly for (z, u) on any compact subset of U as $t(\in \mathbf{R}) \to 0$. (In this case, we write $h = o_{wt}(m)$). By convention, we write $h = o_{wt}(0)$ if $h \to 0$ as $(z, \bar{z}, u) \to 0$). For a smooth function $h(z, \bar{z}, u)$ defined over U, we use $h^{(k)}(z, \bar{z}, u)$ for the sum of terms of weighted degree k in the weighted expansion of h up to order k. If h is not specified, we use it to denote a weighted homogeneous polynomial of weighted degree k. For a weighted homogeneous holomorphic polynomial of degree k, we use the notation: $(\cdot)^{(k)}(z, w)$, or $(\cdot)^{(k)}(z)$ if it depends only on z.

Next returning to (2.0), we would like to simplify terms in $O(|(z, w)|^3)$ by further changes of coordinates. These changes of coordinates should have the following properties:

(i). Preserves the origin and the real tangent space $\{v = 0\}$ of the hypersurfaces at the origin.

(ii). Preserves the complex tangent space $\{w = 0\}$ at the origin.

(iii). Preserve the hyperquadric $\mathbf{H}_\ell^n$ up to weighted order 3.

Let $(z', w') = F = (f, g)$ be such a map. Then the general form that F can take, with the properties in (i)-(iii), is as follows:

(2.1) $f = zA + \mathbf{a}w + O(|(z,w)|^2), \quad g = \lambda w + O(|(z,w)|^2)$

with $\lambda \in \mathbf{R} \setminus \{0\}$.

Since F preserves $v = |z|_\ell^2$ up to the third order, it follows that $\lambda > 0$ if $\ell < \frac{n-1}{2}$ and $AE_{\ell,n-1}\overline{A}^t = \lambda E_{\ell,n-1}$ in general. The following proposition indicates that we can further limit down our transformation group to make calculations more accessible:

Proposition 2.2: For any transformation F of the form in (2.1), there is a unique $T \in Aut_0(\mathbf{H}_\ell^n)$ such that $F = T \circ F_0$ with $F_0 = (f_0, g_0)$ having the following properties:

(2.2) $f_0 = z + O(|(z,w))|^2), \quad g_0 = w + O(|(z,w))|^2), \quad Re\left(\frac{\partial^2 g}{\partial w^2}(0)\right) = 0.$

In fact, by a straightforward verification, we can set $T = T_1 \circ T_2$. Here,
(a) If $l = \frac{n-1}{2}$ and $\lambda < 0$, then $T_2(z,w) = (z_{\ell+1}, \cdots, z_{n-1}, z_1, ..., z_\ell, -w)$. Otherwise T_2 is always set to be the identity.
(b) When $\lambda > 0$, we have

$$T_1 = \left(\frac{(z + \mathbf{a} \cdot w)A}{q(z,w)}, \frac{\lambda w}{q(z,w)}\right)$$

where $q(z,w) = 1 - 2i\langle z, \overline{\mathbf{a}}\rangle_\ell + (r - i|\mathbf{a}|_\ell^2)w$. For $\lambda < 0$, one can similarly define T_1.

We next normalize $(M, 0)$ by transformations satisfying (2.2). Of course, the invariants we get in this way are still subject to the action of $Aut_0(\mathbf{H}_\ell^n)$, which is a finite dimensional Lie group.

By induction, suppose that we have found a coordinates system: (z, w), in which M has been normalized up to the weighted order s. We then want to see how to choose the new coordinates (z', w') to get the invariant form at the level of weighted order $(s + 1)$.

We first mention that for a formal power series $N(z, \overline{z}, w, \overline{w}) = N(z, \overline{z}, u, v)$, we have the decomposition $N = \sum_{s=0}^{\infty} N^{(s)}(z, \overline{z}, u, v)$, where $N^{(s)}(tz, \overline{tz}, t^2 u, t^2 v) = t^s N^{(s)}(z, \overline{z}, u, v)$ for $t \in \mathcal{R}$.

Suppose that in (z, w)-coordinates, M is given implicitly by

(2.3) $v = |z|_\ell^2 + N_1(z, \overline{z}, u, v)$

and in $(z', w') = (f(z,w), g(z,w))$ coordinates system , M is given by

$$v' = |z'|_\ell^2 + N_2(z', \overline{z}', u', v').$$

By what we mentioned above, we want to keep what we have already achieved. Namely, we want to have $N_2^{(\sigma)}(z, \overline{z}, u, |z|_\ell^2) = N_1^{(\sigma)}(z, \overline{z}, u, |z|_\ell^2)$ for $\sigma \leq s$. Since we have assumed that (f, g) satisfies the normalization in (2.2), we have

(2.4) $$(f - z)^{(1)} = 0, \quad (g - w)^{(2)} = 0.$$

Write the weighted expansion of (f, g) as follows:

(2.5) $$f = z + \sum_{\sigma \geq 2} f^{(\sigma)}(z, w), \quad g = w + \sum_{\sigma \geq 3} g^{(\sigma)}(z, w),$$

where $f^{(\sigma)}(tz, t^2 w) = t^\sigma f^{(\sigma)}(z, w)$; $g^{(\sigma)}(tz, t^2 w) = t^\sigma g^{(\sigma)}(z, w)$. Then we have

(2.6)
$$\operatorname{Im}\left(w + \sum_{\sigma \geq 3} g^{(\sigma)}(z, w)\right) =$$
$$= |z|_\ell^2 + 2\operatorname{Re}\sum_{\sigma_1 \geq 2} < \overline{z}, f^{(\sigma_1)}(z, w) >_\ell + \sum_{\sigma_1, \sigma_2 \geq 2} < f^{(\sigma_1)}, \overline{f^{(\sigma_2)}} >_\ell +$$
$$+ N_2\left(z + \sum f^{(\sigma)}, z + \sum f^{(\sigma)}, u + \operatorname{Re}(\sum_{\sigma \geq 3} g^{(\sigma)}), \operatorname{Im}(w + \sum_{\sigma \geq 3} g^{(\sigma)}(z, w))\right)$$

where (z, w) satisfies (2.3). Suppose that $f^{(\tau-1)}$ and $g^{(\tau)}$ have been determined for $\tau < \sigma \leq s + 1$, we want to find $f^{(\sigma-1)}$ and $g^{(\sigma)}$ for any $\sigma \leq s + 1$. In particular, we would like to find $f^{(s)}$, $g^{(s+1)}$, and $N_2^{(s+1)}$. Substituting $w = u + iv$ and $v = |z|_\ell^2 + N_1(z, \overline{z}, u, v)$ in (2.6), we have

(2.7)
$$\operatorname{Im}(g^{(\sigma)}(z, u + i|z|_\ell^2)) = 2\operatorname{Re} < \overline{z}, f^{(\sigma)}(z, u + i|z|_\ell^2) >_\ell$$
$$+ N_2^{(\sigma)}(z, \overline{z}, u, |z|_\ell^2) - N_1^{(\sigma)}(z, \overline{z}, u, |z|_\ell^2) + G^{(\sigma)}(z, \overline{z}, u),$$

where $G^{(\sigma)}$ is completely determined by $f^{(\tau-1)}$ and $g^{(\tau)}$ for $\tau \leq \sigma - 1$ and is zero if $(f - z)^{(\tau-1)}$, $(g - w)^{(\tau)} = 0$ for $\tau \leq \sigma - 1$. To proceed further, we make the following definition:

Definition 2.3: Let $f = (f_1, \cdots, f_{n-1})$ and g be (formal) holomorphic functions in a neighborhood of 0 in $\mathbf{C}^n$. Suppose that

(2.8) $$(f, g) = O(|(z, w)|^2) \quad \text{with } \operatorname{Re}(g''_{ww}(0)) = 0.$$

Then the map $\mathcal{L}_\ell$, which sends the above (f, g) to the real-valued (formal) analytic function over $\mathbf{H}_\ell^n$ defined below:

(2.8)' $$\mathcal{L}_\ell(f, g) := \operatorname{Im}\left(g(z, w) - 2i < \overline{z}, f(z, w) >_\ell |_{w = u + i|z|_\ell^2}\right)$$

is called the Chern-Moser operator.

Returning to (2.7), we get for $\sigma \geq 3$

(2.9) $$\mathcal{L}_\ell(f^{(\sigma-1)}, g^{(\sigma)}) = N_2^{(\sigma)}(z, \overline{z}, u, |z|_\ell^2) - N_1^{(\sigma)}(z, \overline{z}, u, |z|_\ell^2) + G^{(\sigma)}(z, \overline{z}, u).$$

Since our ℓ is always fixed, we will write $\mathcal{L}$ instead of $\mathcal{L}_\ell$ to simplify the notation. Later we will see that $\mathcal{L}$ has the following uniqueness property: If

$\mathcal{L}(f^{(\sigma-1)}, g^{(\sigma)}) = 0$ with $\text{Re}(\frac{\partial^2 g^{(\sigma)}}{\partial w^2})|_0 = 0$ and $\sigma \geq 3$, then it follows that $(f^{(\sigma-1)}, g^{(\sigma)}) \equiv 0$.

Since we assumed that $N_2^{(\sigma)} = N_1^{(\sigma)}$ for $\sigma \leq s$, we conclude that $(f^{(\sigma-1)}, g^{(\sigma)}) \equiv 0$ for $\sigma \leq s$ and thus $G^{(s+1)} \equiv 0$. At the level of weighted degree $s+1$, we have

$$(2.10) \qquad \mathcal{L}(f^{(s)}, g^{(s+1)}) = N_2^{s+1}(z, \overline{z}, u, |z|_\ell^2) - N_1^{(s+1)}(z, \overline{z}, u, |z|_\ell^2).$$

Notice that $N_1^{(s+1)}(z, \overline{z}, u, |z|_\ell^2)$ is known from the induction assumption. Our purpose is then to choose

$$f^{(s)}(z, w), \quad g^{(s+1)}(z, w)$$

appropriately so that we can make $N_2^{(s+1)}$ as simple as possible. (2.10) suggests us to pick $N_2^{(s+1)}(z, \overline{z}, u, v)$ so that $N_2^{(s+1)}(z, \overline{z}, u, |z|_\ell^2)$ is in the 'complement' of the range of the Chern-Moser operator $\mathcal{L}$.

Definition 2.4: Let $\mathcal{A}^{(s)}$ be a collection of real-valued polynomials of weighted degree s in (z, w) for $s \geq 4$. Let $\mathcal{A} = \oplus_{s \geq 4} \mathcal{A}^{(s)}$. Assume that $0 \in \mathcal{A}$.
(a) We call $\mathcal{A}$ a uniqueness set for the Chern-Moser operator $\mathcal{L}$ if $\mathcal{L}(f, g) = G|_{w=u+i|z|_\ell^2}$ with $G \in \mathcal{A}$ is only solvable when $G = 0$ and $(f, g) = 0$. (As in Definition 2.3, for (f, g) in the domain of the Chern-Moser operator, we always assume that $(f, g) = O(|(z, w)|^2)$ with $\text{Re}(g''_{ww}(0)) = 0$.)
In (b) and (c), we assume further that $\mathbf{H}_\ell^n$ is a uniqueness set for $\mathcal{A}$ in the sense that for any $G_1, G_2 \in \mathcal{A}$, $G_1 \equiv G_2$ if and only if $G_1|_{w=u+i|z|_\ell^2} = G_2|_{w=u+i|z|_\ell^2}$.
(b) $\mathcal{A}$ is called an admissible space for $\mathcal{L}$ if for any $G_1, G_2 \in \mathcal{A}$, the equation $\mathcal{L}(f, g) = (G_2 - G_1)|_{w=u+i|z|_\ell^2}$ has solution (f, g) only when $(f, g) \equiv 0$ and $G_1 \equiv G_2$.
(c). $\mathcal{A}$ is called a normal space if $\mathcal{A}$ is admissible and for any real-valued polynomial $B^{(s)}(z, \overline{z}, u)$ of degree $s \geq 4$, there is a unique $G^{(s)} \in \mathcal{A}^{(s)}$ such that $\mathcal{L}(f^{(s-1)}, g^{(s)}) = G^{(s)}|_{w=u+i|z|_\ell^2} - B^{(s)}(z, \overline{z}, u)$ is solvable.

We remark that it can be easily proved that any weighted homogeneous polynomial of weighted degree 3, when restricted to $\mathbf{H}_\ell^n$, is in the range of the Chern-Moser operator. Hence, in Definition 2.4, we take $s \geq 4$.

Summarizing the above, we have the following:

Theorem 2.5: (a) Suppose $\mathcal{A}$ is a normal space for the Chern-Moser operator. Then any formal real hypersurface $(M, 0)$ can be transformed by a formal power series to a formal hypersurface defined by $v = |z|_\ell^2 + N$ with $N \in \mathcal{A}$.
(b) Suppose that $\mathcal{A}$ is an admissible space. Let $(M_j, 0)$ be formal hypersurfaces which are in the $\mathcal{A}$-normal form, namely, M_j are defined by an equation of the form $v = |z|_\ell^2 + N_j$ with $N_j \in \mathcal{A}$. Let F be a formal holomorphic map from $(M_1, 0)$ to $(M_2, 0)$ satisfying the normalization condition (2.2). Then $F \equiv \text{Id}$ and $N_1 \equiv N_2$.

2.2 $\mathcal{H}_k$-Space and Hypersurfaces in the $\mathcal{H}_k$-Normal Form

To be able to make good use of Theorem 2.5, we need to construct the normal space for the Chern-Moser operator. Apparently, the normal space associated to the Chern-Moser operator is not unique. And it is the case that for different problems, one has to use different normal or admissible spaces. In the following, we present two different admissible spaces for the Chern-Moser operator, following the work in [CM] and [EHZ1]. Unfortunately, the one obtained in [EHZ1] is not a normal space and the normal form obtained in terms of that is in the implicit form. However, it is invariant under the action of the group $\text{Aut}_0(\mathbf{H}_\ell^n)$. This makes it very convenient to use in working on certain problems.

We first discuss the space $\mathcal{S}_k^0$ (The $\mathcal{S}_k$ defined in [EHZ1] is slightly more general than the one defined below):

Definition 2.6: For $s \geq 4$, $\mathcal{S}_k^{0(s)}$ is the collection of all real-valued weighted homogeneous polynomials of degree s in $(z, \overline{z}, w, \overline{w})$ with the following property: For each $A(z, \overline{z}, w, \overline{w}) \in \mathcal{S}_k^{0(s)}$, there is a set of weighted homogeneous holomorphic polynomials

$$E = \{\phi_j(z, w), \ \psi_j(z, w)\}_{j \leq k^*} \text{ with } k^* < \infty,$$
$$\deg_{wt}(\phi_j) = p_j \leq s/2, \ \deg_{wt}(\psi_j) = q_j \geq s/2 \text{ and } p_j + q_j = s$$

$(j \leq k^*)$ such that
(A): $\phi_j(z, w)$ and $\psi_j(z, w)$ have no linear and constant terms in (z, w) for any j;
(B): for each $\tau \leq s/2$, there are at most k $\phi_j's$ in E with $\deg_{wt}(\phi_j) = p_j = \tau$
(C): $A(z, \overline{z}, w, \overline{w})$ is real valued for any $(z, w) \in \mathbf{C}^n$, and has the following decomposition:

$$(2.11) \quad A(z, \overline{z}, w, \overline{w})) = \sum_{q_j = p_j} \phi_j(z, w)\overline{\psi_j(z, w)} + 2 \sum_{q_j > p_j} \text{Re}(\phi_j(z, w)\overline{\psi_j(z, w)}).$$

We define $\mathcal{S}_k^0 := \oplus_{s=4}^\infty \mathcal{S}_k^{0(s)}$

What makes $\mathcal{S}_k^0$ convenient to use is the so-called $\mathcal{H}_k$-class contained in $\mathcal{S}_k^0$, which is defined to be the collection of all real-valued formal power series $A(z, \overline{z}, w, \overline{w})$ (for $(z, w) \in \mathbf{C}^{n-1} \times \mathbf{C}$) such that

$$(2.11)' \qquad A(z, \overline{z}, w, \overline{w}) = \sum_{j=1}^k \phi_j(z, w)\overline{\psi_j(z, w)}$$

where ϕ_j, ψ_j are formal holomorphic power series in (z, w) which do not contain any constant and linear terms.

As we will see, the $\mathcal{H}_k$-normal form is invariant under the action of $\mathrm{Aut}_0(\mathbf{H}_\ell^n)$. This makes it very convenient to apply in applications. More precisely, let $T \in \mathrm{Aut}_0(\mathbf{H}_\ell^n)$. Then we can write

(2.12)
$$T(z,w) = (\tfrac{\lambda(z-aw)U}{q(z,w)}, \tfrac{\sigma\lambda^2 w}{q(z,w)}),$$
$$\text{with } q(z,w) = 1 + 2i < z, \overline{a} >_\ell + (r - i < a, \overline{a} >_\ell)w,$$

where λ is a non-zero real number, $a \in \mathbf{C}^{n-1}$ and U is a certain $(n-1) \times (n-1)$ matrix such that

(2.13)
$$U E_{\ell,n-1} \overline{U}^t = \sigma E_{\ell,n-1}, \quad \sigma = \pm 1.$$

Let M be a formal real hypersurface which is in the $\mathcal{H}_k$-normal form. Namely, M is defined by an equation of the form:

$$v = |z|_\ell^2 + N(z, \overline{z}, w, \overline{w}), \text{ with } N \in \mathcal{H}_k.$$

The following lemma, which can be proved easily, makes the $\mathcal{H}_k$-normal form convenient to apply:

Lemma 2.7: Under the above notation and assumption, $T(M)$ is also in the $\mathcal{H}_k$-normal form. In fact, $T(M)$ is defined by an equation of the form:
(2.14)
$$v = |z|_\ell^2 + N_2(z, \overline{z}, w, \overline{w}), \text{ with } N_2(z, \overline{z}, w, \overline{w}) = \frac{\sigma\lambda^2}{|q \circ T^{-1}|^2} N_1 \circ T^{-1}(z, w) \in \mathcal{H}_k.$$

The following result from [EHZ1] is basic for the application of Lemma 2.7 and Theorem 2.5 to work on various local equivalence problems:

Theorem 2.8 (Ebenfelt-Huang-Zaitsev)([EHZ1]) (a): $\mathcal{S}_k^0$ is a uniqueness set for the Chern-Moser operator for $k \le n - 2$. (b). $\mathcal{S}_k^0$ is an admissible space for the Chern-Moser operator for $k \le \frac{n-2}{2}$.

The $\mathcal{S}_k^0$ (or the $\mathcal{S}_k$ in [EHZ1]) is far from being a normal space. It is an open problem how to complete $\mathcal{H}_k$ or $\mathcal{S}_k^0$ for $k \le \frac{n-2}{2}$ into a normal space. This problem is closely related to the study of the embeddability problem for real analytic Levi non-degenerate hypersurfaces into hyperquadrics.

We refer the reader to the paper [EHZ1] for a proof of Theorem 2.8. Here, we give a proof of the part that $\mathbf{H}_\ell^n$ is a uniqueness set for $\mathcal{S}_k^0$ when $k \le \frac{n-2}{2}$. We notice that

$$c_1 \mathcal{S}_{k_1}^0 + c_2 \mathcal{S}_{k_2}^0 \subset \mathcal{S}_{k_1+k_2}^0$$

for any complex numbers c_1 and c_2.

Proposition 2.8': Let $A(z, \overline{z}, w, \overline{w}) \in \mathcal{S}_k^0$ with $k \le n - 2$. Assume that

$$A^0(z, \overline{z}, u) := A(z, \overline{z}, u + i < z, \overline{z} >_\ell, u - i < z, \overline{z} >_\ell) \equiv 0$$

as a formal power series in $(z, \overline{z}, u)$. Then $A(z, \overline{z}, w, \overline{w}) \equiv 0$ as a formal power series in $(z, \overline{z}, w, \overline{w})$. In particular, $\mathbf{H}_\ell^n$ is a uniqueness set for $\mathcal{S}_k^0$ with $k \leq \frac{n-2}{2}$.

We first observe that if $A(z, \overline{z}, w, \overline{w})$ is weighted homogeneous of degree σ, then $A^0(z, \overline{z}, u)$ is weighted homogeneous of degree σ. Hence, if we decompose a formal power series $A(z, \overline{z}, w, \overline{w})$ into its weighted homogeneous components

$$A(z, \overline{z}, w, \overline{w}) = \sum_\sigma A^{(\sigma)}(z, \overline{z}, w, \overline{w})$$

then the decomposition of $A^0(z, \overline{z}, u)$ is given by

$$A^0(z, \overline{z}, u) = \sum_\sigma (A^0)^{(\sigma)}(z, \overline{z}, u),$$

where, in the terminology introduced above, $(A^0)^{(\sigma)} = (A^{(\sigma)})^0$. Moreover, if $A \in \mathcal{S}_k^0$, then $A^{(\sigma)} \in \mathcal{S}_k^0$.

Proof of Proposition 2.8': It is enough to prove the lemma when A is a weighted homogeneous polynomial of degree $s \geq 4$ and

$$A(z, \overline{z}, w, \overline{w}) = \sum_{j=1}^{k^*} \phi_j(z, w)\overline{\psi_j(z, w)},$$

where ϕ_j, ψ_j are weighted homogeneous holomorphic polynomials, without constant or linear terms, of weighted degree p_j and $q_j = s - p_j$, respectively, and where for each τ there are at most k terms with $p_j = \tau$. Let us expand ϕ_j and ψ_j as follows:

$$\phi_j = \sum_{\nu^j + 2\nu_w^j = p_j} a_j^{(\nu^j)}(z)w_w^{\nu_w^j}, \quad \psi_j = \sum_{\mu^j + 2\mu_w^j = s - p_j} b_j^{\mu^j}(z)w_w^{\mu_w^j}.$$

Then, if we expand $A(z, \overline{z}, w, \overline{w})$ in powers of $w, \overline{w}$, we can write

$$A(z, \overline{z}, w, \overline{w}) = \sum_{m,l} c_{m,l}(z, \overline{z})w^m \overline{w}^l,$$

where

$$c_{m,l}(z, \overline{z}) = \sum_{j=1}^{k^*} a_j^{(p_j - 2m)}(z)\overline{b_j^{(q_j - 2l)}(z)}$$

and $p_j + q_j = s$. By isolating the terms in $c_{m,l}(z, \overline{z})$ of bidegree (α, β) in $(z, \overline{z})$ (denoted $c_{m,l,\alpha,\beta}(z, \overline{z})$), we conclude that $A(z, \overline{z}, w, \overline{w}) \equiv 0$ if and only if, for every 4-tuple of nonnegative integers (m, l, α, β),

$$c_{m,l,\alpha,\beta}(z,\bar\xi) := \sum_{j \in J(m,l,\alpha,\beta)} a_j^{(\alpha)}(z)\overline{b_j^{(\beta)}(z)} \equiv 0,$$

where the index set $J(m,l,\alpha,\beta)$ consists of those $j \in \{1,\ldots,k^*\}$ for which

$$p_j = \alpha + 2m, \quad q_j = \beta + 2l.$$

Observe that, since $A \in S_k^0$, there are at most $k \le n-2$ indices in the set $J(m,l,\alpha,\beta)$ for each (m,l,α,β).

Now, we use the fact that $A^0(z,\bar z, u) \equiv 0$ is equivalent to $A(z,\bar z, w, \bar w)$ vanishing on the quadric $\mathbf{H}_\ell^n$, and the usual complexification argument, to conclude that

$$A(z,\bar\xi,w,\bar\eta) = 0$$

whenever $w = \eta + 2i < z,\bar\xi >_\ell$; or, equivalently,

$$\sum_{m,l} c_{k,l}(z,\bar\xi)(\eta + 2i < z,\bar\xi >_\ell)^m \bar\eta^l \equiv 0.$$

Assume, in order to reach a contradiction, that $A(z,\bar z, w, \bar w) \not\equiv 0$. Then, there is a smallest nonnegative integer l_0 such that $c_{m,l_0,\alpha,\beta}(z,\bar\xi) \not\equiv 0$ for some m,α,β. Hence, we can factor out η^{l_0} (of course, if $l_0 = 0$, then we do not need to factor anything) from the identity above and get

$$\sum_{m,l \ge l_0} c_{m,l}(z,\bar\xi)(\eta + 2i < z,\bar\xi >_\ell)^m \bar\eta^{l-l_0} \equiv 0.$$

By setting $\eta = 0$, we conclude that

$$\sum_m c_{m,l_0}(z,\bar\xi)(2i < z,\bar\xi >_\ell)^m \equiv 0.$$

Isolating the terms of bidegree (α,β) above, we deduce

$$\sum_m c_{m,l_0,\alpha-m,\beta-m}(z,\bar\xi)(2i < z,\bar\xi >_\ell)^m \equiv 0.$$

It now follows from [Lemma 3.2, Hu2], [Lemma 3.2, EHZ1] that, for every m,γ,μ,

$$c_{m,l_0,\gamma,\mu}(z,\bar\xi) \equiv 0,$$

which contradicts the choice of l_0. This completes the proof of Lemma 3.3. $\square$

Making use of Lemma 2.7 and Theorem 2.8, one has the following:

Corollary 2.9 ([EHZ1]): Let $(M_1,0)$ and $(M_2,0)$ be two germs of formal real hypersurfaces in the $\mathcal{H}_{k_1}$ and $\mathcal{H}_{k_2}$-normal form defined, respectively, by

$$v = |z|_\ell^2 + N_j.$$

Assume that $k_1 + k_2 \leq n - 2$. Then $(M_1, 0)$ is equivalent to $(M_2, 0)$ by a formal holomorphic map if and only if there is an automorphism $T \in \text{Aut}_0(\mathbf{H}_\ell^n)$ such that

$$(2.15) \quad N_2 = \frac{\sigma\lambda^2}{|q \circ T^{-1}|^2} N_1 \circ T^{-1}, \text{ or } N_1 = \sigma\lambda^{-2}|q(z, w)|^2 N_2 \circ T(z, w).$$

Here, we write $T(z, w) = (\frac{\lambda(z - aw)U}{q(z,w)}, \frac{\sigma\lambda^2 w}{q(z,w)})$, with $q(z, w) = 1 + 2i < \bar{z}, \bar{a}^t >_\ell$ $+(r - i < a, \bar{a} >_\ell)w$, $UE_{\ell,n-1}\overline{U}^t = \sigma E_{\ell,n-1}$, $\sigma = \pm 1$, $\lambda > 0, r \in \mathbf{R}$, and with a a certain $(n - 1)$-tuple. In particular, when M_1 is equivalent to M_2, it must hold that $N_1 \in \mathcal{H}_{k_2}$. And the set of all equivalence maps from $(M_1, 0)$ to $(M_2, 0)$ is precisely the collection of the automorphisms T of $\text{Aut}_0(\mathbf{H}_\ell^n)$ which make (2.15) hold. (Hence, any formal equivalence map from $(M_1, 0)$ to $(M_2, 0)$ is given by a convergent power series.)

Let M be in the $\mathcal{H}_k$-normal form, namely, let M be defined by $v = < z, \bar{z} >_\ell$ $+N$ with $N = \sum_{j=1}^k \phi_j\overline{\psi_j} \in \mathcal{H}_k$, where ϕ_j, ψ_j have no constant and linear terms in (z, w). Let $R(M, N)$ be the minimum k to get such an expression for N. Then as an application of Corollary 2.9, we have the following weak invariant property for $R(M, N)$:

Corollary 2.10: (I). Let M_1 and M_2 be in the $\mathcal{H}_k$-normal form for a certain k. If $R(M_1, N_1) + R(M_2, N_2) \leq n - 2$ and $(M_1, 0)$ is equivalent to $(M_2, 0)$, then $R(M_1, N_1) = R(M_2, N_2)$. (II). If in the expression $N = \sum_{j=1}^k \phi_j\overline{\psi_j} \in \mathcal{H}_k$, both $\{\phi_j\}$ and $\{\psi_j\}$ are linearly independent over $\mathbf{C}$, then $R(M, N) = k$, where $M := \{v = < z, \bar{z} >_\ell +N\}$. Moreover, if $N = \sum_{j=1}^k A_j\overline{B_j}$ for A_j, B_j satisfying the same property as ϕ_j, ψ_j do, then there is an invertible constant $k \times k$ matrix C such that $(\phi_1, \cdots, \phi_k) = (A_1, \cdots, A_k)C$, and $(\psi_1, \cdots, \psi_k) = (B_1, \cdots, B_k)\overline{(C^t)^{-1}}$.

Proof of Corollary 2.10: The first part apparently follows from Theorem 2.9 and Equation (2.15). Let $\{\phi_j, \psi_j\}$ be as in Part (II) of the corollary and assume that $\sum_{j=1}^k \phi_j\overline{\psi_j} = \sum_{j=1}^{k'} A_j\overline{B_j}$, where A_j, B_j are holomorphic in their arguments $Z = (z, w)$. Since $\{\phi_j\}$ is a linearly independent finite set, it is easy to see that the set $\{\phi_j^l\}$, where ϕ_j^l are the truncation of ϕ_j up to order l, must be also independent for k sufficiently large. Hence, there exits $\{Z_j\}_{j=1}^k$ such that the matrix $D := ((\{\phi_j^l(Z_1)\})^t, \cdots, (\{\phi_j^l(Z_k)\})^t)$ is invertible. Since $\sum_{j=1}^k \phi_j^l\overline{\psi_j} = \sum_{j=1}^{k'} A_j^l\overline{B_j}$, it follows clearly that $\{\psi_1, \cdots, \psi_k\}$ is a linear combination of $\{B_1, \cdots, B_{k'}\}$. Hence, $k' \geq k$. The last statement can also be similarly seen. $\qquad\square$

Remark 2.11: Corollary 2.10 can be further used to simplify the equation (2.14). To see this, let $M_1 := \{v = < z, \bar{z} >_\ell +\sum_{j=1}^{k_1} \phi_j\overline{\psi_j}\}$ and $M_2 :=$

$\{v = <z, \overline{z}>_\ell + \sum_{j=1}^{k_2} \widetilde{\phi}_j \overline{\widetilde{\psi}_j}\}$ be in the $\mathcal{H}_k$-normal form ($k = \max\{k_1, k_2\}$) such that $R(M_1, N_1) = k_1$ and $R(M_2, N_2) = k_2$, where $N_1 = \sum_{j=1}^{k_1} \phi_j \overline{\psi_j}$ and $N_2 = \sum_{j=1}^{k_1} \widetilde{\phi}_j \overline{\widetilde{\psi}_j}$. Assume that $k_1 + k_2 \leq n - 2$ and $(M_1, 0)$ is equivalent to $(M_2, 0)$. Then, by Corollary 2.10, it holds that: (I) $k_1 = k_2 = k$; (II) There are a $k \times k$ constant invertible matrix C, an automorphism $T \in \text{Aut}_0(\mathbf{H}_\ell^n)$ with its associated data q, σ, λ as given in Theorem 2.9 such that

$$(2.16) \quad \begin{aligned} \sigma\lambda^{-2} q(z, w) \cdot (\widetilde{\phi}_1, \cdots, \widetilde{\phi}_{k_0}) \circ T &= (\phi_1, \cdots, \phi_{k_0}) \cdot C, \\ \sigma\lambda^{-2} q(z, w) \cdot (\widetilde{\psi}_1, \cdots, \widetilde{\psi}_{k_0}) \circ T &= (\psi_1, \cdots, \psi_{k_0}) \cdot \overline{(C^t)^{-1}}. \end{aligned}$$

Immediately, we have from (2.16) the following conclusions:

(A). If all $\widetilde{\phi}_j$, $\widetilde{\psi}_j$, ϕ_j ψ_j are polynomials, and at least one of them is not zero, then $q \equiv 1$ and $T = (\lambda(z - aw)U, \sigma\lambda^2 w)$.

(B). If $\{\widetilde{\phi}_j, \widetilde{\psi}_j\}$ are rational functions, then so are $\{\phi_j, \psi_j\}$.

(C). If at least one of $\{\widetilde{\phi}_j, \widetilde{\psi}_j\}$ is a transcendental function, then at least one of ϕ_j and ψ_j is transcendental, too.

Example 2.12: Let $M_1 = \mathbf{H}_0^3$ and let $M_2 := \{(z, w = u + iv) \in \mathbf{C}^3 : v = |z|^2 + |\frac{2z_1^2(1-iw)}{(1+iw)^2}|^2 + |\frac{2z_1 z_2}{(1+iw)}|^2\}$. Then $R(M_1) = 0$, $R(M_2) = 2$. Also, M_1 is equivalent to M_2. Notice that $R(M_1) + R(M_2) = 2 > n - 2 = 1$. Hence, the assumption that $R(M_1) + R(M_2) \leq n - 2$ in Corollary 2.10 can not be weakened.

2.3 Application to the Rigidity and Non-embeddability Problems

we now first present a discussion on how to apply the materials in §2.2 for the study of the rigidity problem for mappings between the hyperquadrics.

Theorem 2.13 ([EHZ1]): Let $F = (f_1, \cdots, f_{n-1}, \phi_1, \cdots, \phi_{N-n}, g)$ be a formal holomorphic mapping sending $\mathbf{H}_{\ell_1}^n$ into $\mathbf{H}_{\ell_2}^N$ with $F(0) = 0$, $\frac{\partial g}{\partial w}|_0 \neq 0$, where g is the normal component of F and $N \geq n > 2$. Suppose that $\ell_2 \geq \ell_1$ and $\ell_1 + \ell_2 \leq n - 1$. Suppose that $N \leq 2n - 2$. Then there is a linear fractional holomorphic embedding T from $\mathbf{H}_{\ell_2}^N$ to $\mathbf{H}_{\ell_1, \ell_2}^N := \{(Z, W) \in \mathbf{C}^N : \text{Im}(W) = -\sum_{j \leq \ell_1} |Z_j|^2 + \sum_{\ell_1 < j \leq n-1} |Z_j|^2 - \sum_{n-1 < j \leq n-1+\ell_2-\ell_1} |Z_j|^2 + \sum_{n-1+\ell_2-\ell_1 < j \leq N-1} |Z_j|^2\}$ and $T_0 \in \text{Aut}_0(\mathbf{H}_{\ell_1}^n)$ such that $T \circ F \circ T_0(z, w) = (z, \phi^*, w)$ with $\phi^* = O(|(z, w)|^2)$. Moreover, when $\ell_2 = \ell_1$, T is a self-map and $\phi^* = 0$. (For $\ell_1 = \ell_2 = \ell$, $\mathbf{H}_{\ell_1, \ell_2}^N$ is understood as $\mathbf{H}_\ell^N$.) Also, when $\ell_1 < \frac{n-1}{2}$, T_0 can be taken to be the identity map.

Proof of Theorem 2.13: Let $F = (f, \phi, g) = (\widetilde{f}, g)$ be a formal holomorphic mapping from $(\mathbf{H}_{\ell_1}^n, 0)$ into $(\mathbf{H}_{\ell_2}^N, 0)$ with $\frac{\partial g}{\partial w}|_0 \neq 0$. Then $\text{Im}(g) = <\widetilde{f}, \overline{\widetilde{f}}>_{\ell_2}$ along $\mathbf{H}_{\ell_1}^n$ as a formal power series. Collecting the coefficients of weighted degree 1 and 2, we see that $g = \sigma\lambda^2 w + o_{wt}(2)$, $\widetilde{f} = \lambda z U + \sigma\lambda^2 aw + O(|zw| + |z|^3 +$

$|w|^2$). Here $\sigma = \pm 1$, $\lambda > 0$, a is a certain complex vector, $U = (E_1^t, \cdots, E_{n-1}^t)^t$ with $E_j = \frac{1}{\lambda}\frac{\partial \tilde{f}}{\partial z_j}(0)$, and $< E_j, \overline{E_k} >_{\ell_2} = \sigma \delta_j^k \delta_j$. Since $< E_j, \overline{E_j} >_{\ell_2} \neq 0$ for $j \leq n - 1$, we can extend $\{E_j\}_{j=1}^{n-1}$ to an orthogonal basis $\{E_j\}_{j=1}^{N-1}$ (with respect to the Hermitian product $< \cdot, \overline{\cdot} >_{\ell_2}$). Let $\tilde{U} = (E_1^t, \cdots, E_{N-1}^t)^t$, then $\tilde{U}E_{\ell_2, N-1}\overline{\tilde{U}} = \text{diag}(< E_1, \overline{E_1} >_{\ell_2}, \cdots, < E_{N-1}, \overline{E_{N-1}} >_{\ell_2})$, where $E_{\ell_2, N-1}$ is defined, as before, by $< X, \overline{X} >_{\ell_2} = XE_{\ell_2, N-1}\overline{X^t}$. In particular, we see that $< E_j, \overline{E_j} >_{\ell_2} \neq 0$ for any j. Without loss of generality, we can assume that $< E_j, \overline{E_j} >_{\ell_2} = c_j = \pm 1$ for $j \geq n$, too. (Notice that $c_j = -\sigma$ for $j \leq \ell_1$ and $c_j = \sigma$ for $\ell_1 < j \leq n - 1$.) After changing the position of $E_j's$ for $j > n - 1$, we can assume that $\tilde{U}E_{\ell_2, N-1}\tilde{U}^t = \sigma B^*$, where σB^* is determined by the following Hermitian product:

$$< Z, Z >_{\ell_1, \ell_2} := -\sum_{j=1}^{\ell_1} Z_j\overline{Z_j} + \sum_{\ell_1 < j \leq n-1} Z_j\overline{Z_j} - $$
$$- \sum_{n-1 < j \leq \ell_2 - \ell_1 + n - 1} Z_j\overline{Z_j} + \sum_{N-1 > j \geq \ell_2 - \ell_1 + n - 1} Z_j\overline{Z_j},$$

Apparently, when $\ell_1 + \ell_2 < \frac{n-1}{2}$, σ must be 1. Otherwise, $\ell_1 = \frac{n-1}{2} = \ell$. In this case, composing F with $T_0(z, w) = (z_{\ell+1}, \cdots, z_{n-1}, z_1, \cdots, z_\ell, -w) \in \text{Auto}(\mathbf{H}_\ell^n)$, if necessary, we can also make $\sigma = 1$.

In the following, we assume that $\sigma = 1$. Letting

$$(2.17) \qquad T(z, w) = (\frac{\lambda^{-1}(Z - aW)\tilde{U}^{-1}}{q(Z, W)}, \frac{\lambda^{-2}W}{q(Z, W)}),$$

where $q(Z, W) = 1 + 2iZE_{\ell_2, N-1}\overline{a}^t + (r - i < a, \overline{a} >_{\ell_2})W$, with $r = \frac{1}{2}\lambda^{-2}\text{Re}(\frac{\partial^2 g}{\partial w^2}|0)$. Write $F^* = T \circ F := (f^*, \phi^*, g^*)$. Then (f^*, g^*) satisfies the normalization condition (2.2), and $\phi^* = O(|(z, w)|^2)$. Notice that T biholomorphically maps $\mathbf{H}_{\ell_2}^n$ to $\mathbf{H}_{\ell_1, \ell_2}^N$. Namely, $\text{Im}(g^*) = \sigma f^* B^* \overline{f^*}^t$ along $\mathbf{H}_{\ell_1}^n$. Now, we can inductively apply Theorem 2.8 to prove that $f^* = z_{,}$ $g^* = w$. Indeed, we first notice that by collecting terms of weighted degree ≤ 4 in the equation $\text{Im}(g^*) = f^*B^*\overline{f^*}^t$, we see by Theorem 2.8 and the normalization condition that $f^{*(j-1)} = 0$, $g^{*(j)} = 0$, for $3 \leq j \leq 4$. Suppose that $f^{*(\tau-1)}$, $g^{*(\tau)} = 0$ for $\tau \leq K_0$. Collecting terms of weighted degree $K_0 + 1$,

$$(2.17)' \qquad \mathcal{L}(f^{*(K_0)}, g^{*(K_0+1)}) = 2\sum_{\kappa=1}^{k}\sum_{j=2}^{[K_0/2]} \epsilon_\kappa \text{Re}(\phi_\kappa^{*(j)}\overline{\phi_\kappa^{*(K_0-j)}}),$$

where ϵ_j is the $(n - 1 + j)$-th element in the diagonal matrix σB^*. Since $k \leq n - 2$, the right hand side of (2.17') is in $\mathcal{S}_{n-2}^0$. Hence, it follows from Theorem 2.8, that $f^{*(K_0)} = 0$, $g^{*(K_0+1)} = 0$. By induction, we see that $f^* = 0$, $g^* = 0$.

Returning to ϕ^*, we get $\sum_{j=1}^k \epsilon_j|\phi^*_j|^2 \equiv 0$. Assume that $\ell_2 = \ell_1$. Since we assumed that $\ell_1 \leq (n - 1)/2$, all ϵ_j then must have fixed sign. Hence,

$\phi^* \equiv 0$. as remarked above, σ must be 1 when $\ell_1 = \ell_2 < (n-1)/2$; and σ can be made to be 1 by replacing F with $F(z_{\ell+1}, \cdots, z_{n-1}, z_1, \cdots, z_\ell, -w)$, if necessary, when $\ell_1 = \ell_2 = (n-1)/2$.

When $\ell_2 > \ell_1$, write

(2.18). $\Phi_I = (\phi_1^*, \cdots, \phi_\kappa^*)$ with $\kappa = \ell_2 - \ell_1$ and $\Phi_{II} = (\phi_{\kappa+1}^*, \cdots, \phi_{N-1}^*)$.

We also see $\|\Phi_I\|^2 = \|\Phi_{II}\|^2$ over $\mathbf{H}_{\ell_1}^n$.   □

We give some applications of Corollary 2.9 to the problem of embedding a non-degenerate formal hypersurface $M \subset \mathbf{C}^n$ of signature ℓ into $\mathbf{H}_{\ell'}^N$ with $N \le 2n - 2$ ($\frac{N-1}{2} \ge \ell' \ge \ell$, $\ell + \ell' \le n - 1$).

Let $M = \{v = <z, \overline{z}>_\ell + N\}$ be a formal non-degenerate hypersurface of signature ℓ with $N = o_{wt}(3)$. Assume that F is a formal holomorphic embedding from $(M, 0)$ into $(\mathbf{H}_{\ell'}^N, 0)$. As we see above, after replacing F by $F \circ T_0$, if necessary, and then composing it with a certain holomorphic linear fractional map from $(\mathbf{H}_{\ell'}^N, 0)$ to $(\mathbf{H}_{\ell,\ell'}^N, 0)$, we can write $F = (f, \Phi_I, \Phi_{II}, g)$, where (f, g) satisfies the normalization condition (2.2) and $\Phi_I, \Phi_{II} = O(|(z, w)|^2)$ as defined in (2.18). Applying the implicit function theorem, we conclude that M is equivalent through $F_0 = (f, g)$ to the following hypersurface:

$$\widetilde{M} = \{v = <z, z>_\ell - \|\Phi_I \circ F_0^{-1}\|^2 + \|\Phi_{II} \circ F_0^{-1}\|^2 = <z, z>_\ell + H_{N-n}\}.$$

Notice that $H_{N-n} \in \mathcal{H}_{N-n}$ and F_0 satisfies (2.2). Conversely, by Corollary 2.9, we have the following

Proposition 2.14: Let $M := \{v = <z, z>_\ell + N\}$ where $N \in \mathcal{H}_k$. Suppose that $N \le 2n - 2 - k$. Then $(M, 0)$ can be formally embedded into $\mathbf{H}_{\ell,\ell'}^N$ if and only if there are vector valued holomorphic functions $\Phi_I(z, w)$, $\Phi_{II}(z, w) = O(|(z, w)|^2)$ with $\ell' - \ell$ and $N - n - \ell' + \ell$ components, respectively, such that

(2.19) $\sigma^* N(z, \overline{z}, w, \overline{w}) = -\|\Phi_I(z, w)\|^2 + \|\Phi_{II}(z, w)\|^2,$

where σ^* is either identically 1 or identically -1. In particular, when $\ell = \ell'$, then M can be embedded into $\mathbf{H}_\ell^N$ with $N \le 2n - 2 - k$ if and only there are $(N - n)$ formal holomorphic functions $\{\phi_j\}_{j=1}^{N-n}$ such that

(2.20) $$N(z, \overline{z}, w, \overline{w}) = \sigma^* \sum_{j=1}^{N-n} |\phi_j(z, w)|^2.$$

where σ^* must be 1 when $\ell < \frac{n-1}{2}$.

More generally, assume that $\ell' = \ell$ and let M be given by $M := \{v = <z, \overline{z}>_\ell + N^{(s)} + o_{wt}(s)\}$ with $N^{(s)}(\not\equiv 0) \in \mathcal{S}_k^{0(s)}$. Let $F = (f, \phi, g)$ be a formal embedding of M into $\mathbf{H}_\ell^N$ with (f, g) satisfying (2.2) and $\phi = O(|(z, w)|^2)$.

When $N \leq 2n - 2$, an inductive use of Theorem 2.8 shows that $(f, g) = (z + f^{(s-1)} + o_{wt}(s), w + g^{(s)} + o_{wt}(s+1))$ and $\phi = (\phi_1^{(\sigma)}, \cdots, \phi_{N-n}^{(\sigma)}) = 0$ for $2\sigma < s$. In particular, it follows from Theorem 2.8 that $s = 2s'$ must be even if $N \leq 2n-2$ and $k \leq n-2$. Assume this. For terms of weighted degree s, we have $\mathcal{L}(f^{(s-1)}, g^{(s)}) = \|\phi^{(s')}\|^2 - N^{(s)}$. Since $\|\phi^{(s')}(z, u + i < z, \overline{z} >_{\ell_1})\|^2 \in \mathcal{S}_{N-n}^0$, it follows that if $k + N - n \leq n - 2$ then

$$(2.21) \qquad N^{(s)}(z, \overline{z}, u, v)|_{v=|z|_\ell^2} \equiv \|\phi^{(s')}(z, u + i < z, \overline{z} >_\ell)\|^2.$$

Therefore, we have

Corollary 2.15: Let $M = \{v =< z, \overline{z} >_\ell +N^{(s)} + o_{wt}(s)\}$ be a formal non-degenerate hypersurface of signature ℓ with $N^{(s)}(\neq 0) \in \mathcal{S}_k^{0(s)}$, $k \leq n-2$, $s \geq 4$. Assume that $k \leq n - 2$ and $N \leq 2n - 2 - \delta_s^e k$ with $\delta_s^e = 0$ for s odd and equal to 1 otherwise. Suppose that there is no holomorphic solution $\phi^{(s')}$ to (2.21). Then $(M, 0)$ cannot be formally embedded into $\mathbf{H}_\ell^N$, when $\ell < \frac{n-1}{2}$. For $\ell = \frac{n-1}{2}$, if there is no solution to

$$N^{(s)}(z, \overline{z}, u, v)|_{v=|z|_\ell^2} \equiv \pm \|\phi^{(s')}(z, u + i < z, \overline{z} >_\ell)\|^2.$$

Then $(M, 0)$ cannot be formally embedded into $\mathbf{H}_\ell^N$.

Example 2.16: Let $M(\subset \mathbf{C}^n) := \{v = |z|^2 + \mathrm{Re}(w^{s-1}\overline{h(z)}) + o_{wt}(2s)\}$ be the germ of a formal non-degenerate hypersurface of signature 0, where $s > 2$ and $h(z)$ is a non-zero homogeneous polynomial of degree 2. Then there is no vector valued weighted holomorphic polynomial $\phi^{(s)}$ of weighted degree s such that $\mathrm{Re}((u + i|z|^2)^{s-1}\overline{h(z)}) = \|\phi^{(s+1)}(z, u + i|z|^2)\|^2 \geq 0$ over $w = u + i|z|^2$. Notice that $k = 2$. Hence, when $N \leq 2n - 4$, $(M, 0)$ can never be formally holomorphically embedded into $\mathbf{H}^N$. Also notice that $M_0(\subset \mathbf{C}^n) := \{v = |z|^2 + \mathrm{Re}(w^{s-1}\overline{h(z)})\}$ can be holomorphically embedded into $\mathbf{H}_1^{n+2}$ through the map $F = (\frac{1}{2}(w^{s-1} - h(z)), z, \frac{1}{2}(w^{s-1} + h(z)), w)$.

To conclude this subsection, we present one more application to the study of a rigidity problem, which asks if two CR embeddings of a strongly pseudoconvex hypersurface M in $\mathbf{C}^n$ into the Heisenberg hypersurface $\mathbf{H}^N$ are the rigid motion of each other. Namely, if F, Ψ are two C^l-smooth CR embeddings from M into $\mathbf{H}^N$, is there a $T \in \mathrm{Aut}(\mathbf{H}^N)$ such that $T \circ F = \Psi$? Here l is a certain positive number. This problem has been answered in the work of Webster [We2] when $N = n + 1 \geq 4$. The reader can find a geometric approach along the lines of Webster [We2] on this problem in [EHZ2] when $N - n \leq \frac{n-2}{2}$. The arguments here are essentially those in [EHZ1].

Let $M = \{v =< z, \overline{z} >_\ell +N\}$ be a formal non-degenerate hypersurface of signature ℓ with $N = o_{wt}(3)$. Assume that F, Ψ are formal holomorphic embeddings from $(M, 0)$ into $(\mathbf{H}_{\ell'}^N, 0)$ and $(\mathbf{H}_{\ell''}^{N'}, 0)$, respectively. (Assume that $N' \geq N$. Also, for simplicity, assume that $\ell + \ell' < n-1$). After composing F, Ψ

with certain holomorphic linear fractional maps from $(\mathbf{H}^N_{\ell'}, 0)$ to $(\mathbf{H}^N_{\ell,\ell'}, 0)$ and from $(\mathbf{H}^{N'}_{\ell'}, 0)$ to $(\mathbf{H}^{N'}_{\ell,\ell'}, 0)$, respectively, we can write $F = (f, \Phi_I, \Phi_{II}, g)$ and $\Psi = (f^*, \Phi_I^*, \Phi_{II}^*, g^*)$ where (f, g) (f^*, g^*) satisfy the normalization condition (2.2) and $\Phi_I, \Phi_{II}, \Phi_I^*, \Phi_{II}^* = O(\|(z, w)|^2)$ as defined in (2.18). Therefore, M is equivalent through $F_0 = (f, g)$ or $\Psi_0 = (f^*, g^*)$ to the following hypersurfaces $\widetilde{M}$, M^*, defined, respectively by:

$$v = <z, z>_\ell - \|\Phi_I \circ F_0^{-1}\|^2 + \|\Phi_{II} \circ F_0^{-1}\|^2,$$
$$v = <z, z>_\ell - \|\Phi_I^* \circ \Psi_0^{-1}\|^2 + \|\Phi_{II}^* \circ \Psi_0^{-1}\|^2.$$

Notice that $F_0 \circ \Psi_0^{-1}$ is a normalized formal biholomorphic map from $(M^*, 0)$ to $(\widetilde{M}, 0)$ satisfying (2.2), and $\widetilde{M}$, M^* are in the $\mathcal{H}_{N-n}$, $\mathcal{H}_{N'-n}$-normal form, respectively. By Theorem 2.8, we see that when $N + N' \leq 4n - 2$, $F_0 \equiv \Psi_0$ and $-\|\Phi_I\|^2 + \|\Phi_{II}\|^2 \equiv -\|\Phi_I^*\|^2 + \|\Phi_{II}^*\|^2$ along M as formal power series. In particular, when $\ell' = \ell'' = \ell$, there is a constant matrix U with $U \cdot \overline{U}^t = \mathrm{Id}$ such that $\Phi_{II}^* = \Phi_{II} \cdot U$ by a result of D'Angelo [Da] and by noting that $\Phi_I^* = \Phi_I = 0$. Hence, after applying another $T \in \mathrm{Aut}_0(\mathcal{H}^{N'}_l)$ to Ψ, we see that the new F and Ψ satisfy the relation: $\Psi = (F, 0)$.

2.4 Chern-Moser Normal Space $\mathcal{N}_{CH}$

The space $\mathcal{H}_k$ we presented in the above subsections is indeed very convenient to apply due to its invariant property under the action of $\mathrm{Aut}_0(\mathbf{H}^n_\ell)$. However, it is not a normal space and thus can only be used to model a very limited class of germs of real hypersurfaces. For the study of general Levi non-degenerate hypersurfaces, we need to make use of the normal space $\mathcal{N}_{CM}$ discovered by Chern-Moser in [CH]. The Chern-Moser normal space is not invariant under the action of $\mathrm{Aut}_0(\mathbf{H}^n_\ell)$. Thus a hypersurface which is in the $\mathcal{N}_{CH}$-normal form is still subject to the action of this group. However, it can be used to model any germ of hypersurface.

Since the discussion on the Chern-Moser normal form is available in many nice expositions ([Vit], [BER2], etc.), we here just give a brief account on this theory. Define
(2.23)
$$\Delta_\ell := -\sum_{j \leq \ell} \frac{\partial^2}{\partial z_j \overline{z_j}} + \sum_{j \geq \ell+1}^{n-1} \frac{\partial^2}{\partial z_j \partial \overline{z_j}},$$
$$\mathcal{N}_{CH} := \{h = \sum_{k,l \geq 2} F_{k\bar{l}}(z, \overline{z}, u), \text{ with } F_{k\bar{l}} = \sum_{|\alpha|=k, |\beta|=l} a_{\alpha\overline{\beta}}(u) z^\alpha \overline{z}^\beta$$
$$\overline{h} = h, \quad \Delta_\ell F_{2\overline{2}} = \Delta_\ell^2 F_{2\overline{3}} = \Delta_\ell^3 F_{3\overline{3}} = 0\}$$

The following is a fundamental result of Chern-Moser in this subject:

Theorem 2.17(Chern-Moser [CM]): Assume the above definition and notation. Then (a). $\mathcal{N}_{CM}$ is a normal space. (b). Any germ of Levi non-degenerate real analytic hypersurface $(M, 0)$ with signature ℓ can be transformed by the germ of a biholomorphic map to a convergent Chern-Moser normal form. (c).

Let $(M_j, 0)$ be germs of formal real hypersurfaces at 0 defined by $v = |z|_\ell^2 + N_j$ with $N_j \in \mathcal{N}_{CM}$. Then $(M_1, 0)$ and $(M_2, 0)$ are equivalent by a formal holomorphic map F, satisfying the normalization (2.2), if and only if $F \equiv \mathrm{Id}$ and $N_1 \equiv N_2$.

The proof of Theorem 2.17 can be found in [§3, 4, CM], which we skip here. However, we mention that one of the significant features in the above theorem is that a convergent germ of hypersurface has a convergent Chern-Moser normal form.

In terms of the above theorem, the general procedure to see if two germs $(M_j, 0)$, which are already in the Chern-Moser normal form, are equivalent to each other, is as follows: First apply $T \in \mathrm{Aut}_0(\mathbf{H}_\ell^n)$ to M_2 to obtain $T(M_2)$. Then by solving infinitely many times the Chern-Moser equation (2.10) to find a new normal form for $T(M_2) : v = |z|_\ell^2 + N_{2,T}$. Finally, $(M_1, 0)$ is equivalent to $(M_2, 0)$ if and only if $N_1 \equiv N_{2,T}$ for a certain T. The major difficulty here is that it is extremely difficult in general to find $N_{2,T}$ from the defining equation of $T(M_2)$. Indeed, it is the purpose to get rid of this difficulty that motivated us to find an invariant normal form (with respect to $\mathrm{Aut}_0(\mathbf{H}_\ell^n)$) in [EHZ1]. Unfortunately, the admissible space we obtained in [EHZ1] only works for a very small class of real hypersurfaces, which are actually those which can be formally embedded into the hyperquadrics with restricted codimension. The interested reader is referred to the paper [EHZ1] for more on this matter.

We notice that $\mathcal{S}_k^0$ is not a subclass of the Chern-Moser normal space. For instance, for $\sigma > 1$, $h = \mathrm{Re}(z_1^{2\sigma}\overline{w^\sigma})$ contains a term of the form $u^\sigma z_1^{2\sigma}$. While h is in $\mathcal{H}_2$, it is not in the Chern-Moser normal space.

3 Bishop Surfaces with Vanishing Bishop Invariants

In this section, we study the holomorphic equivalence problem for submanifolds in $\mathbf{C}^n$ with higher codimension. There have been many generalizations of the Chern-Moser theory to the so-called generic strongly pseudoconvex CR submanifolds. (See the survey paper [BER2] for some references in this regard and the recent paper [BRZ] for some other related studies). In this notes, we would like to focus on the normal form problem for Bishop surfaces [Bis] in $\mathbf{C}^2$. The study of Bishop surfaces has attracted considerable attention since the work of E. Bishop in 1965. (See [BG], [KW], [Mos], [MW] [HK]). These surfaces are interesting, due to the following reasons: First, from the point of view of complex analysis, they can be viewed as the simplest higher codimensional analogy of strongly pseudoconvex hypersurfaces; secondly, they have a rich complex structure at the complex tangent and have trivial complex structure elsewhere, namely they can also be viewed as the simplest models where one sees the CR singularity; thirdly, from the work of Moser-Webster [MW], which we will discuss in the next section, one sees a tremendous interaction of complex analysis with the classical dynamics problems encountered in Mechanics [SM]– An understanding of such a problem may provide useful information and motivation to many converge problems in Mechanics. The basic references to this section include the papers [MOS] [MW] and [HK].

To be more specific, we let M be a real surface in $\mathbf{C}^2$. Then for any $p \in M$, $CR_M(p)$ can be only $0, 1$. When $CR_M(p) = 0$, we say M is totally real at p. By the semi-continuity of the CR dimension function, we conclude that M must be totally real in a neighborhood of p in M. When M is further real analytic, then an easy application of the complexification shows that (M, p) is holomorphically equivalent to $(\mathbf{R}^2, 0)$, where $\mathbf{R}^2 := \{(x, y) \in \mathbf{C}^2, \ x, y \in \mathbf{R}\}$. On the other hand, if $CR_M(q) \equiv 1$ for $q \approx p$, then apparently $(M, p) \approx (\mathbf{C} \times \{0\}, 0)$. Hence, from the equivalence point of view, only points with CR dimension 1 but not constantly 1 nearby are interesting. Among such points, only those which have CR dimension 1 but 0 nearby are stable under small perturbation. Such points are called isolated CR singular points.

Now, let $p \in M$ be a point with a non-trivial complex tangent. Namely, we assume that $CR_M(p) = 1$. After a holomorphic change of coordinates, we can assume that $p = 0$ and $T^{(1,0)} = CT_pM = \{w = 0\}$, where we use (z, w) for the coordinates of $\mathbf{C}^2$. Then M near 0 can be defined by an equation of the form: $w = h(z, \overline{z}) + o(|z|^2)$. Here $h(z, \overline{z}) = az^2 + bz\overline{z} + c\overline{z}^2$. Replacing w by $w - (a - c)z^2$, if necessary, we can assume that $a = c$. Assume that $b \neq 0$. Replacing w by w/b and replacing z by $ze^{i\theta}$ for a suitable θ, we can assume that $h = z\overline{z} + \lambda(z^2 + \overline{z}^2)$ with $\lambda \geq 0$. By a straightforward verification, one can see that λ is a biholomorphic invariant, called the Bishop invariant.(See Lemma 3.2 below). When $\lambda < \frac{1}{2}$, we call $p = 0$ an elliptic complex tangent of M. When $\lambda > \frac{1}{2}$, we call $p = 0$ a hyperbolic complex tangent point of M. When $\lambda = 1/2$ or when $b = 0$ but $c \neq 0$, we say $p = 0$ is a parabolic complex tangent. An elliptic, parabolic or hyperbolic complex tangent point is called a non-degenerate complex tangent point. In the other case, we say 0 is a degenerate complex tangent point. A real surface M is called a Bishop surface if all of its complex tangents are non-degenerate. In this notes, we are mainly concerned with the equivalence problem of M at an elliptic complex tangent point. Hence, we have $\lambda \in [0, 1/2)$. In this section, we discuss the formal theory of Moser [Mos] when the surface is formally equivalent to the model surface $M_0 := \{w = |z|^2\}$. In the next section, we discuss the Moser-Webster theory for Bishop surfaces with non-vanishing Bishop invariants.

The understanding to the general Bishop surfaces with vanishing Bishop invariant is still not complete. It is an open question to get a complete set of invariants for analytic Bishop surfaces with vanishing Bishop invariant.

We first state a general result along these lines proved in [HK]:

Theorem 3.1 (Huang-Krantz [HK]): Let M be a real analytic Bishop surface with vanishing Bishop invariant at 0. Then $(M, 0)$ can be flattened in the sense that there is a biholomorphic change of coordinates such that in the new coordinates, it holds that $M \subset \mathbf{C} \times \mathbf{R}$. More precisely, in the new coordinates, M near 0 can be defined by an equation of the form:

(3.1) $$w = |z|^2 + E(z, \overline{z}), \ E(z, \overline{z}) = \overline{E(z, \overline{z})}.$$

We start with the following statement on invariance of the Bishop invariant.

Lemma 3.2: Suppose that M_j for $j = 1, 2$ are Bishop surfaces with only CR singular point at p_j, respectively. Then the Bishop invariant of M_1 at p_1 is the same as the Bishop invariant of M_2 at p_2, if M_1 is biholomorphically equivalent to M_2.

Proof of Lemma 3.2: Without loss of generality, we can assume that $p_j = 0$. Let $F = (f, g)$ be a biholomorphic map from M_1 to M_2. Then $F(0) = 0$, for F preserves the CR dimension. After a change of coordinates, we can assume that

$$M_j : w = z\overline{z} + \lambda_j(z^2 + \overline{z}^2) + O(|z|^3), \ 0 \le \lambda_j \le \infty.$$

When $\lambda_j = \infty$, we regard M_j as a surface defined by an equation of the form: $w = z^2 + \overline{z}^2 + o(|z|^2)$. For simplicity of calculation, we assume, in the following, that $\lambda_j < \infty$.

Notice that F must preserve the complex tangent space of M_j at 0. We can write $F = (f, g)$ with $f = az + bw + O(|(z, w)|^2)$ and $g = cw + d^{(2)}(z) + O(|w|^2 + |zw| + |z|^3)$. Using the equation of M_2, we get

$$c(z\overline{z} + \lambda_1(z^2 + \overline{z}^2)) + d^{(2)}(z) = |az + bw|^2 + \lambda_2 2Re(az + bw)^2 + O(|z|^3),$$

where $(z, w) \in M_1$. Collecting the coefficients of $z\overline{z}, z^2, \overline{z}^2$, we get

(3.2) $$c = |a|^2, \ d^{(2)} = d_2 z^2, \ c\lambda_1 + d_2 = \lambda_2 a^2, \ c\lambda_1 = \overline{a}^2 \lambda_2.$$

Hence it follows that

(3.3) $$c > 0, \lambda_1 = \lambda_2, a \in \mathbf{R}, \ d^{(2)} = 0$$

This completes the proof of Lemma 3.2.   □

3.1 Formal Theory for Bishop Surfaces with Vanishing Bishop Invariant

We now focus on the case $\lambda = 0$ and present the formal theory of Moser [Mos]. Let M be a real analytic Bishop surface with vanishing Bishop invariant at 0. By Theorem 3.1, after a change of coordinates, we can assume that M is defined by an equation of the form:

$$w = |z|^2 + E(z, \overline{z}) \ \text{with} \ \overline{E(z, \overline{z})} = \overline{E}(\overline{z}, z) = E(z, \overline{z}).$$

We notice that M near 0 bounds a family of holomorphic disks defined by

$$\{(z, w): \ v = 0, u = r^2, r^2 \geq |z|^2 + E(z, \overline{z})\}.$$

Namely, let σ_r be a Riemann mapping from the unit disk in $\mathbf{C}$ to the domain

(3.4) $$D_r := \{z \in \mathbf{C} : r^2 > |z|^2 + R(z, \overline{z})\}.$$

Then the map ϕ_r from the unit disk $\Delta := \{z \in \mathbf{C} : |z| < 1\}$, which sends z to $(\sigma_r(z), r^2)$, is holomorphic in Δ, real analytic up to the unit circle and maps the unit circle to M. Such a ϕ_r is called a holomorphic disk attached to M.

Conversely, for any holomorphic map $\phi = (\phi_1, \phi_2)$ from the unit disk to $\mathbf{C}^2$, which is continuous up to $\partial\Delta$, if it is attached to M (namely, $\phi(\partial\Delta) \subset M$) and if $\|\phi\| << 1$, then $\phi(\Delta) = D_r$ for a certain r. This can be seen easily by noticing that for such a map, ϕ_2 must be constant; for its imaginary part has boundary value 0.

Next, let $(M_j, 0)$ $(j = 1, 2)$ be two real analytic surfaces defined, respectively, by an equation of the form:

$$w = |z|^2 + E_j(z, \overline{z}) \ \text{ with } \ \overline{E_j(z, \overline{z})} = \overline{E_j}(\overline{z}, z) = E_j(z, \overline{z}).$$

And let $F = (f, g)$ be a biholomorphic map from $(M_1, 0)$ to $(M_2, 0)$. Then F must send a holomorphic disk attached to M_1 to a holomorphic disk attached to M_2. From this, it follows easily that $g(z, w) = g(w)$ with $g(r^2) > 0$ for $0 < r << 1$. Also, $f(z, r^2)$ for each fixed r must be a conformal map from the disk $|z|^2 + E_1(z, \overline{z}) \leq r^2$ to the disk $|z|^2 + E_2(z, \overline{z}) \leq g(r^2)$.

In particular, when both $M_1 = M_2 = M_\lambda = \{w = |z|^2 + \lambda(z^2 + \overline{z}^2)\}$ with $\lambda = 0$, $f(z, r^2)$ must be a conformal map from $|z|^2 \leq r^2$ to $|z| \leq \sqrt{g(r^2)}$. Hence

$$f(z, r^2) = \sqrt{g(r^2)} e^{i\theta(r)} \frac{z - ra(r)}{r - a(r)z}$$

for certain $\theta(r), a(r)$.

Since f is analytic in (z, w), we can conclude that $f(0, u) = -\sqrt{g(u)} e^{i\theta(r)} a(r)$ is real analytic in u. Write $g(w) = w(g^*(w))^2$ with $g^*(r^2) > 0, g^*(0) > 0$. Then

$$f(0, u) = -\sqrt{u} g^*(u) e^{i\theta(\sqrt{u})} a(\sqrt{u}),$$

we see that $\sqrt{u} a(\sqrt{u}) e^{i\theta(\sqrt{u})}$ is analytic.

In particular, we see that $u|a(\sqrt{u})|^2$ and thus $|a(\sqrt{u})|^2$ is analytic.

Next, $\frac{\partial f}{\partial z}(0, u) = g^*(u) e^{i\theta(\sqrt{u})}(1 - |a(\sqrt{u})|^2)$. We conclude that $e^{i\theta(\sqrt{u})}$ is also analytic, and thus $\sqrt{u} a(\sqrt{u})$ is analytic too. In this manner, we can write

$$f(z, u) = g^*(u) \Lambda(u) \frac{z - c(u)u}{1 - \overline{c}(u)z}$$

where $c(u) = \frac{a(\sqrt{u})}{\sqrt{u}}$ is analytic in u with $|c(u)| \leq \frac{1}{\sqrt{u}}$,or $|c(u)u| < \sqrt{u}$; $g^*(u)$ and $\Lambda(u)$ are analytic in u with $g^*(0) > 0$. Summarizing what we did and with a further straightforward verification, we have

Proposition 3.3([MW] [Mos]): $Aut_0(M_\lambda)$ with $\lambda = 0$ consists of the following transformations:

$$(3.5) \qquad \begin{cases} w' = wa(w)\overline{a}(w), \\ z' = a(w)\frac{z - wb(w)}{1 - \overline{b}(w)z} \end{cases}$$

with $a(0) \neq 0$, $a(w), b(w)$ holomorphic functions in w.

Still let M be defined by $w = |z|^2 + E(z, \overline{z})$ with $E(z, \overline{z} = \overline{E(z, \overline{z})} = O(|z|^3)$ real analytic in z. We subject to M a transformation of the form: $F = (f, g)$ where $f = az + bw + O(|z, w|^2)$, $g = g(w)$ with $g(r^2) > 0$ for $r > 0$.

Lemma 3.4: There is a unique $T \in Aut_0(M_\lambda)$ such that $T \circ F = (\widetilde{f}, \widetilde{g})$ satisfies the following normalization condition:

$$(3.6) \qquad \widetilde{f} = \sum_{j=0}^{\infty} z^j f_j(w) \;\; with \; f_0 = 0, \;\; f_1(u) > 0 \;\; f_1(0) = 1 \;, \widetilde{g} = w.$$

Proof of Lemma 3.4: First, we can easily make $F = (f, g) = (z + O(w + |z|^2), w + O(w^2))$. Choose $T_0 \in Aut_0(M_0)$: $T_0(z, w) = (a(w)z, a^2(w)w)$, with $a(0) \neq 0$, $a(u) > 0$ for $u \geq 0$. We like to have $T_0 \circ F = (\cdot, w)$. For this, we need the function relation: $a^2(g(w))g(w) = w$. Hence, $a(g(w)) = \frac{1}{g^*(w)}$, where, as before, $g(w) = wg^*(w)$ with $g^*(0) \neq 0$. Apparently, such an $a(w)$ can be uniquely solved.

Still write F for $T_0 \circ F$. Let

$$T_1 = \left(\Lambda(w)\frac{z - c(w)w}{1 - \overline{c}(w)z}, w \right) \in Aut_0(M_0).$$

Write $F = (\sum_{j=0}^{\infty} f_j(w)z^j, w)$ and letting $c(w) = \frac{f_0(w)}{w}$. Then

$$T_1 \circ F = \left(\Lambda(w)\frac{\sum_{j=1}^{\infty} f_j(w)z^j}{1 - \overline{c}(w)f(z, w)}, w \right) = (\sum_{j=1}^{\infty} \widetilde{f}_j(w)z^j, w).$$

Then

$$\widetilde{f}_1(w) = \frac{\Lambda(w)f_1(w)}{(1 - \overline{c}(w)wc(w))}.$$

Notice that $f_1(0) = 1$. Also, we can apparently choose Λ such that $\widetilde{f}_1(u) > 0$ for $u \geq 0$. We proved the existence of $T \in Aut_0(M_\lambda)$ such that the $T \circ F$ satisfies the normalization (3.6) in the lemma.

We next prove the uniqueness of T. Suppose that there are $T_1 = (\phi_1, \psi_1)$ and $T_2 = (\phi_2, \psi_2)$ such that both $T_1 \circ F$ and $T_2 \circ F$ satisfy the normalization condition in (3.6). Then one can see easily that it must hold $\psi_1 = \psi_2$ when restricted to M_0. We leave it to the reader to verify that $\phi_1 = \phi_2$ along M_0.

There is another normalization used in [Mos] for F:

Lemma 3.4′([Mos]): There is a unique $T \in Aut_0(M_\lambda)$ with $\lambda = 0$ such that
(3.7)
$$T \circ F = \left(\sum_{j=0}^{\infty} z^j f_j(w), g(w) \right) \text{ with } f_0(w) = 0, \ f_1(w) \equiv 1, \ g(w) = w + o(|w|).$$

Proof of Lemma 3.4′: We choose T_1 to be of the form $\left(\frac{z - c(w)w}{1 - \bar{c}(w)z}, w \right)$. Let $c(w) = \frac{f_0(w)}{w}$. Then $T_1 \circ F = \left(\sum_{j=1}^{\infty} f_j(w)z^j, g(w) \right)$.

Next we take $T_2 = (g^*(w)\Lambda(w)z, w(g^*(w))^2)$ with $g^*(u) > 0$ if $u \geq 0$. Then, we can choose $\Lambda(u)$ with $|\Lambda(u)| \equiv 1$ such that $g^*(u)\Lambda(u)f_1(u) \equiv 1$. Then $T_2 \circ T_1 \circ F$ has the normalization as in Lemma 3.4′. The uniqueness part can also be done easily.   □

We now derive the Moser pseudo-formal norm for $(M, 0)$, where M is defined as in (3.1). We will subject to M the transformation of the following form:

(3.8) $\begin{cases} z' = F = z + f(z, w), \ w' = w \text{ with} \\ f(z, w) = \sum_{l=1}^{\infty} f_l(w)z^l, \ f_1(w) > 0, \ f_1(0) = 1. \end{cases}$

Proposition 3.5: With the above notation, there is a unique formal holomorphic transformation $(z', w') = F(z, w)$ as in (3.8) such that in the (z', w') coordinates, $F(M)$ is given by the following pseudo-normal form:

(3.9) $$w' = z'\overline{z'} + \phi(z') + \overline{\phi(z')}$$

where $\phi(z') = \sum_{j=s\geq 3}^{\infty} a_j(z')$.

In the above lemma, if all $a_j = 0$, then M is formally equivalent to the model M_0. Otherwise, we can assume that $a_s \neq 0$. In fact, replacing (z', w') by $(\kappa z, \kappa^2 w)$ for a suitable κ, we can further make $a_s = 1$. It can be verified that s is then also a biholomorphic invariant of $(M, 0)$, which we call the s-invariant.

When $(M, 0)$ is formally equivalent to the model, we say the s-invariant of $(M, 0)$ is ∞.

Proof of Proposition 3.5: Substituting (3.8) into (3.9), we have

$$w = (z + f(z, w))(\overline{z} + \overline{f}(\overline{z}, \overline{w})) + \phi(z + f(z, w)) + \overline{\phi}(\overline{z} + \overline{\phi(z, w)}),$$

for $w = |z|^2 + E(z, \overline{z})$. Collecting terms of degree s in $(z, \overline{z})$, we get

$$E^{(s)} = z\overline{f^{(s-1)}}(z, z\overline{z}) + f^{(s-1)}(z, z\overline{z})\overline{z} + \phi^{(s)}(z) + \overline{\phi^{(s)}(z)} + G^{(s)}(z, \overline{z})$$

where $G^{(s)}$ is completely determined by $f^{(\sigma-1)}(z, z\overline{z}), g^{(\sigma)}(z, z\overline{z})$ and $\phi^{(\sigma)}$ for $\sigma < s$. Moreover, $G^{(s)}$ is 0 when $f^{(\sigma-1)}(z, z\overline{z}) = g^{(\sigma)}(z, z\overline{z}) = \phi^{(\sigma)}(z) = 0$ for $\sigma < s$. We will also assign the weight of u to be 2.

We will inductively determine F and ϕ. Suppose $F^{(\sigma)}$ and $\phi^{(\sigma)}$ have been solved for $\sigma < s$. Write $\Gamma(z, \overline{z}) = E^{(s)} - G^{(s)}$. We then see that $\phi^{(s)} = \Gamma(z, 0)$. Write $\Gamma(z, \overline{z}) - \Gamma(z, 0) - \Gamma(0, \overline{z}) = \Gamma_0(z\overline{z}) + \sum_{l=1}^{\infty}(z^l \Gamma_l(z\overline{z}) + \overline{z}^l \Gamma_l(z\overline{z}))$ with $\overline{\Gamma}_l = \Gamma_l$, $\Gamma_0 = \overline{\Gamma}_0$. Since $f_1(u) > 0, f_0 = 0$, we obtain

(3.10)
$$\begin{cases} f_1^{(2s')}(u) = \dfrac{\Gamma_0^{(2s'+2)}(u) - \Gamma_0^{(2s'+2)}(0)}{2u}, \\ f_l^{(2s')}(u)u = \Gamma_{l-1}^{(2s'+2)}(u) \text{ or } f_l^{(2s')}(u) = \dfrac{\Gamma_{l-1}^{(2s'+2)}(u) - \Gamma_{l-1}^{(2s'+2)}(0)}{u}, \quad l > 1. \end{cases}$$

Let $f^{(s)}, g^{(s)}$ be the unique solutions given as above. Let $F^{(s)} = (z + \sum f^{(s)}, w)$. Then F satisfies the normalization as in (3.6). Now, the composition of such a map formally transforms $(M, 0)$ into a special form as in (3.9).

Similarly, one also has the following:

Proposition 3.5′ ([Mos]): Let $(M, 0)$ be given as in (3.1). Then there is a unique formal holomorphic transformation $(z', w') = F(z, w)$, that satisfies the normalization in (3.7), such that in the (z', w') coordinates, $F(M)$ is given by a pseudo-normal as in (3.9).

A surface defined by an equation of the form in (3.9) is said to be presented in the Moser pseudo-normal form. It should be mentioned that the coefficients embedded in the Moser pseudo-normal form are far from being holomorphic invariants. Indeed, the Moser pseudo-normal form is still subject to the action of a huge group: $\text{Aut}_0(M_0)$, which, different from the real hypersurface case, is of infinite dimension. It has been an open question how to simplify the Moser peudo-normal form further to get a more invariant representation for Bishop surfaces with vanishing Bishop invariant. It is also an open question if a real analytic $(M, 0)$ can be transformed into a convergent Moser pseudo-normal form through a convergent power series. In the following subsection, we will show that if M is formally equivalent to the model M_0, then it is biholomorphic to M_0. We will follow essentially the argument in [Mos] for this purpose.

3.3. Bishop surfaces which are formally equivalent to $(M_\lambda, 0)$ with $\lambda = 0$: In this section, we give the proof of the following theorem of Moser:

Theorem 3.6(Moser)[Mos]: Suppose $(M, 0)$ is formally equivalent to $(M_\lambda, 0)$ with $\lambda = 0$. Then $(M, 0)$ is biholomorphic equivalent to $(M_\lambda, 0)$.

Let $M : w = z\overline{z} + E(z, \overline{z})$ with $E = O(|z|^3)$ real valued be formally equivalent to M_λ with $\lambda = 0$. $E(z, \xi)$ can be assumed to be holomorphic in the polydisc $|z|, |\xi| \le 1$, $\sup_{|z|, |\xi| < 1} |E(z, \overline{z})| < \eta_0$. Replacing (z, w) by $(\epsilon, \epsilon^2 w)$ for $\epsilon << 1$, we can always make η_0 sufficiently small.

We will seek the transformation of the form $z' = z + f(z, w)$, $w' = w$ as in (3.6), such that

$$w = (z + f(z, w))(\overline{z} + \overline{f}(\overline{z}, \overline{w})), \text{ or}$$

$$\overline{z}f(z, w) + z\overline{f}(\overline{z}, \overline{w}) = E - |f(z, w)|^2, \ w = z\overline{z} + E(z, \overline{z}).$$

Consider its lineariztion:

$$\overline{z}f(z, z\overline{z}) + z\overline{f}(\overline{z}, z\overline{z}) = E$$

which may not be solvable in general. However, as what we did above, we can solve the following

$$\overline{z}f(z, z\overline{z}) + z\overline{f}(\overline{z}, z\overline{z}) + \phi(z) + \overline{\phi(z)} = E(z, \overline{z})$$

where $f(z, w) = \sum_{j=1}^{\infty} f_j(w)z^j$ with $f_0 = 0$, $f_1(u) > 0$ and $f_1(0) = 1$. Still write

$$(3.11) \qquad E = E_0(z\overline{z}) + \sum_{l=1}^{\infty} (E_\ell(z\overline{z})z^l + E_l(z\overline{z})\overline{z}^l) + E(z, 0) + E(0, \overline{z}).$$

Then as in (3.10), we have the following

$$(3.12) \qquad \begin{cases} \phi(z) = E(z, 0), \\ f_1(u) = \frac{E_0(u) - E_0(0)}{2u} \\ f_\ell(u) = \frac{E_{l-1}(u) - E_{l-1}(0)}{u}, \ l = 2, 3, \cdots. \end{cases}$$

For the rest of this section, for $1/2 < r < 1$, we write

$$(3.12)' \ \Delta_r = \{(z, w) : |z| < r, \ |w| < r^2\}, \ D_r = \{(z, w) : |z| < r, \ |w| < r\}.$$

We will also use c_j, c_j' to denote certain absolute constant.

Proposition 3.7: Suppose that $E(z, \xi) \in \text{Hol}(D_r)$. Let $\rho \in (1/2, r)$. Write

$$\|E\|_r = \sup_{|z| < r, |\xi| < r} |E(z, \xi)|, \ |f|_r = \sup_{|z| < r, |w| < r^2} |f(z, w)|.$$

Then f, ϕ are holomorphic over D_r with following estimates:

(3.13)
$$\begin{cases} |f|_\rho < c_1(r-\rho)^{-1}\|E\|_r; \\ |f_z|_\rho + |f_w|_\rho \le c_1(r-\rho)^{-2}\|E\|_r, \\ \sup_{|z|<r}|\phi(z)| \le \|E\|_r. \end{cases}$$

Proof: Note that $z^\ell E_\ell(z\xi) = \frac{1}{2\pi i}\int_0^{2\pi} E(e^{i\phi}z, e^{-i\phi}\xi)e^{-il\theta}d\theta$. By the maximum principle,

$$\sup_{|w|\le r^2}|E_\ell(w)| \le \sup_{|w|=r^2}|E_\ell(w)| = r^{-l}\|E\|_r, \quad \sup_{|w|\le r^2}|f_\ell(w)| \le 2r^{-|\ell|-2}\|E\|_r.$$

Hence,

$$|f|_\rho \le \sum_{\ell=1}^\infty \rho^\ell \sup_{|w|\le\rho^2}|f_\ell(w)| \le \sum_{\ell=1}^\infty \rho^\ell 2r^{-|\ell|-2}\|E\|_r = \frac{c_0}{r-\rho}\|E\|_r,$$

$$\sup_{|z|<r}|\phi(z)| \le \|E\|_r.$$

This, in particular, shows that f is holomorphic in any D_ρ for $\rho < r$. Thus we see $f, \phi \in \mathrm{Hol}(D_r)$.

To get the estimates for derivatives, we set $\tau = \frac{r+\rho}{2}$. By the Cauchy estimates, we obtain:

$$|f|_\tau \le \frac{c_1'}{r-\tau}\|E\|_r, \quad |f_z|_\rho \le \frac{|f|_\tau}{\tau-\rho} \le \frac{c'_1\|E\|_r}{(r-\rho)^2}, \quad |f_w|_\rho \le \frac{c_1'\|E\|_r}{(r-\rho)^2}.$$

This completes the proof of the proposition.   □

The following is basic for applying the rapidly convergent power series method to prove Theorem 3.6.

Lemma 3.8: Suppose that $M : w = z\bar{z} + E(z,\bar{z})$ with the s-invariant $s = \infty$. (Namely, suppose that M is formally convergent to the model M_0). Assume that $\mathrm{ord}(E) \ge d$. Then the transformed surface $F(M) : w' = zz'+E'$ obtained above has $\mathrm{ord}(E') \ge 2d - 2$.

Proof of Lemma 3.8: We have the equation:

(3.14) $$\bar{z}f(z,w) + z\overline{f(\bar{z},\bar{w})} = E - |f(z,w)|^2 - E'(z+f, \bar{z}+\overline{f(z,w)}).$$

Apparently, when $\mathrm{ord}(E) = d$, by (3.12), we have $\mathrm{ord}(f) = d-1$ and thus $\mathrm{ord}(|f(z,w)|^2) \ge 2d-2$. Notice that $\mathrm{ord}(f(z,w) - f(z,z\bar{z})) \ge 2d-3$. Since we assumed that $s = \infty$, it must hold that $\mathrm{ord}(\phi(z)) \ge 2d-2$. (Otherwise, $E'(z',\bar{z}') = \mathrm{Re}(b_{s_0}z'^{s_0})+o(|z'^{s_0}|)$ with $2 < s_0 < 2d-2$ and $b_{s_0} \ne 0$.) Therefore it is easy to conclude that $\text{ord}(E') \geq 2d - 2$ by the way f, ϕ were constructed. (See (3.12)). $\qquad\square$

Now let $M' = F(M)$ be as above defined by: $w' = |z'|^2 + E'(z', \overline{z'})$. We will estimate E'. After complexification, namely, after replacing $\overline{z}$ by a new variable ξ, we have

$$(3.15) \qquad \begin{aligned} E'(z', \xi') &= -\xi(f(z, w) - f(z, z\xi)) - z(\overline{f}(\xi, w) \\ &\quad - \overline{f}(\xi, z\xi)) - f(z, w)\overline{f}(\xi, w) + \phi(z) + \overline{\phi}(\xi) \end{aligned}$$

where $z' = z + f(z, w)$, $\xi' = \xi + \overline{f}(\xi, w)$, $w = z\xi + E(z, \xi)$.

Let r' and r be such that $\frac{1}{2} < r' < r < 1$ and choose $\sigma, \rho \in (r', r)$ such that $r - \rho = \rho - \sigma = \sigma - r' = \frac{1}{3}(r - r')$.

Lemma 3.9: Let M be as above with $\text{ord}(E) \geq d$. Then, there exists an absolute constant $1 > \delta > 0$ such that if $\|E\|_r < \delta(r - r')^2$, the above defined mapping $F : (z, w) \to (z', w') = (z + f(z, w), w)$ takes every value in Δ_σ exactly once from Δ_ρ, and takes M into $M' = F(M)$ with $E'(z, \overline{z'})$ holomorphic in $z', \xi' \in \overline{D_{r'}}$ and

$$(3.16) \qquad \|E'\|_{r'} \leq c_2 \|E\|_r \left\{ \frac{\|E\|_r}{(r - r')^2} + \left(\frac{r'}{r}\right)^{\frac{d}{2}} \right\}.$$

Proof of Lemma 3.9: Write $\Psi(z', w') = F^{-1}(z', w') = (\psi(z', w'), w')$. We need to show that for each fixed w with $|w| < \sigma^2$ and z' with $|z'| < \sigma$, we can solve uniquely the equation $z' = z + f(z, w)$ with $|z| < \rho$. For this purpose, we let δ be sufficiently small so that $|f_z|_\tau + |f_w|_\tau < \frac{1}{20}$ and $|f|_\rho \leq \frac{1}{20}(r - r')$ with $\tau = \frac{r+r'}{2}$. Let $z_1 = z'$ and $z_{j+1} = z' - f(z_j, w)$ for $j = 2, \cdots$. By the standard argument on the Picard iteration procedure, one can verify that $|z_j| < \rho$ and $z_j \to z$ with $|z| < \rho$, too. Apparently, z is the solution that we want.

This proves that Ψ biholomorphically maps Δ_σ into its image contained in Δ_ρ. Notice that for $(z, \xi) \in D_\sigma$, $|w(z, \xi)| = |z\xi + E(z, \xi)| \leq |\sigma|^2 + \|E\|_r < \rho^2$ provided that $\|E\|_r < \rho^2 - \sigma^2$, which holds automatically by the way we choose δ above. Hence, we conclude that E' is holomorphic in D_σ. Moreover,

$$\|E'\|_{r'} \leq \|Q\|_\sigma,$$

where

$$(3.17) \qquad \begin{aligned} Q(z, \xi) &= -\xi(f(z, w) - f(z, z\xi)) - \\ &\quad z(\overline{f}(\xi, w) - \overline{f}(\xi, z\xi)) - f(z, w)\overline{f}(\xi, w) + \phi(z) + \overline{\phi}(\xi) \end{aligned}$$

To estimate $\|Q\|_\sigma$, recall that for $(z, \xi) \in D_\sigma$, $|w| \leq \sigma^2 + \|E\|_r < \rho^2$. Hence,

$$|f(z, w) - f(z, z\xi)| \leq \sup_{\Delta_\rho} |f_w| \|E\|_\sigma \leq c_1 (r - \rho)^{-2} \|E\|_r^2,$$

$|\phi(z)| \le \|E\|_r$ for $|z| < r$. Also, by the Schwarz Lemma, $|\phi(z)| \le \left(\frac{\sigma}{r}\right)^d \|E\|_r$ for $|z| < \sigma$. Notice that

$$|f(z,w)\overline{f}(\xi,w)| \le c_1^2 (r-\rho)^{-2}\|E\|_r^2$$

Hence, $\|Q\|_\sigma \le c_2'\{(r-\rho)^{-2}\|E\|_r^2 + \left(\frac{\sigma}{r}\right)^d \|E\|_r\}$. To complete the proof of the lemma, we just need to notice that $r - \rho = \frac{r-r'}{3}$ and thus $(\frac{\sigma}{r})^2 \le \frac{r'}{r}$. $\square$

Proof of Theorem 3.6: We start with $M : w = z\overline{z}+E(z,\overline{z})$ with ord$(E) \ge 3$ and assume that the s-invariant of M is ∞. Choose $\{r_v\}_{v=1}^\infty$ with $r_v = \frac{1}{2}(1+\frac{1}{v+1})$,

$$\rho_v = r_v - \frac{1}{3}(r_v - r_{v+1}), \quad \sigma_v = r_v - \frac{2}{3}(r_v - r_{v+1}).$$

We mentioned that we can a priori make $\epsilon' := \|E\|_{r_1}$ arbitrarily small. Our goal will be proving that when ϵ' is chosen to be sufficiently small, then the $E_v(z,\xi)$ obtained successively will be biholomorphic in D_{r_v} and $\|E_v\|_{r_v} \to 0$ as $v \to \infty$. Moreover, $\Phi_v = \Psi_1 \circ \Psi_2 \circ \cdots \Psi_v$ converges uniformly in $\Delta_{1/2}$. Hence, it follows that $\Phi_v^{-1}(M) = M_v$ converges to $w = z\overline{z}$. Namely, the inverse of the limit of $\{\Phi_v\}$ biholomorphically maps $(M,0)$ into $(M_0,0)$. In details, we explain as follows:

Note that ord$(E_v) \ge d_v = 2^v + 2$ for $v \ge 1$. Set

$$\epsilon_v = (r_v - r_{v+1})^{-2}\|E_v\|_{r_v}.$$

Suppose ϵ_v is smaller than the δ required in Lemma 3.9. Then by (3.16),

$$(3.18) \qquad \epsilon_{v+1} \le \left(\frac{r_v - r_{v+1}}{r_{v+1} - r_{v+2}}\right)^2 c_2 \epsilon_v \left(\epsilon_v + \left(\frac{r_{v+1}}{r_v}\right)^{\frac{d_v}{2}}\right).$$

Hence

$$(3.19) \qquad \epsilon_{v+1} \le c_3 \epsilon_v (\epsilon_v + \lambda_v).$$

Here

$$\lambda_v = \left(1 - \frac{1}{(v+2)^2}\right)^{\frac{d_v}{2}} \to 0.$$

Now, we have the following:

Lemma 3.10: Suppose a positive sequence $\{\epsilon_v\}$ with $\epsilon_1 << 1$ satisfies (3.19). Then when ϵ_1 is sufficiently small, $\epsilon_v \le 2^{-v}$. Moreover, for any $c' < 1$, by making ϵ_1 sufficiently small, one also has $\epsilon_v < c'$

Proof of Lemma 3.10: Notice that $\lambda_v < e^{-(v+2)^2 2^{v-1}} < c_5' e^{-v^2}$. We first choose
$N \gg 1$ and $\epsilon_1 \ll 1$ such that $\lambda_v < 2^{-v}$ for $v \geq N$, $\epsilon_N < 2^{-N} < (4c)^{-1}$.
Then $\epsilon_{N+1} \leq 2^{-N-1}$. By an induction, one sees that $\epsilon_v < 2^{-v}$ for any $v \geq N$.
The rest of the proof is apparent.   □

Hence, once we start with $\epsilon' \ll 1$, then Lemma 3.10 says that Proposition
3.9 can always be applied. We see that $\|E_v\|_{r_v} \leq \epsilon_v \leq 2^{-v} \to 0$. The reader
can easily verify the uniform convergence of $\{\Phi_v\}$ as $v \to \infty$ over $\Delta_{1/2}$.

Now the mapping $\Phi = \lim \Phi_v$ defines a biholomorphic mapping from
$(\mathbf{C}^2, 0)$ to $\mathbf{C}^2, 0)$. Its inverse maps M into the model $w' = |z'|^2$. By Lemma
3.4 or Lemma 3.4', we can also make Φ^{-1} satisfy the normalization in (3.6)
or (3.7), respectively.   □

4 Moser-Webster's Theory on Bishop Surfaces with Non-exceptional Bishop Invariants

Now we turn to real analytic elliptic Bishop surfaces with non-vanishing
Bishop invariant. Let M be defined by

$$(4.1) \qquad\qquad w = q_\lambda(z, \overline{z}) + o(|z|^2),$$

where $q_\lambda(z, \overline{z}) = z\overline{z} + \lambda(z^2 + \overline{z}^2)$ with $0 < \lambda < \frac{1}{2}$. Write $M_\lambda = \{w = q_\lambda(z, \overline{z})\}$.
Then M_λ is still foliated near 0 by closed analytic curves which bound holo-
morphic curves. Hence, a similar argument as in §3 can be used to show that
$\mathrm{Aut}_0(M_\lambda)$ consists of precisely the maps of the form:

$$(4.2) \qquad z' = a(w)z, \ w' = a^2(w)w \ \ \text{with } a = \overline{a}, \ a(0) \neq 0.$$

More generally, we call $0 \in M$ a non-exceptional complex tangent if $0 \in M$ is
a non-degenerate complex tangent with Bishop invariant $\lambda \neq 0, \frac{1}{2}, \infty$ and if
the quadratic equation: $\lambda\gamma^2 - \gamma + \lambda = 0$ has no roots of unity. It is shown in
[MW] that for the general model M_λ with λ non-exceptional, $\mathrm{Aut}_0(M_\lambda)$ also
consists precisely of the maps of the form in (4.2). (See [Corollary 3.5, MW]).

One might also want to use the methods in the previous sections to normalize Bishop surfaces near general non-degenerate Bishop complex tangents.
However, one would find out that even the linear algebra involved for the
linearized equation will immediately become a lot of more complicated. Up
to now, no one seems to have succeeded in obtaining a complete set of invariants in this way. In the paper of Moser-Wester [MW], they reduced the
normalization problem to the normalization problem for a pair of involutions
intertwined by a conjugate holomorphic involution. This reduction enables
them to completely settle the local equivalence problem for elliptic Bishop
surfaces with non-vanishing Bishop invariant. In the following, we present a
quick discussion on the theory of Moser-Webster. The reader is referred to
their original paper [MW] for more details.

4.1 Complexification $\mathcal{M}$ of M and a Pair of Involutions Associated with $\mathcal{M}$

Assume that M is defined by an equation of the form

$$w = z\bar{z} + \lambda(z^2 + \bar{z}^2) + H(z,\bar{z}) \text{ with } H(z,\bar{z}) = o(|z|^2).$$

Replacing $\bar{z}$ by ξ and $\bar{w}$ by η, we obtain a complex surface $\mathcal{M}$ in $\mathbf{C}^4$ near the origin defined by

$$(4.3) \qquad (w,\eta) = \Psi_0(z,\xi) := \begin{cases} w = z\xi + \lambda(z^2 + \xi^2) + H(z,\xi), \\ \eta = z\xi + \lambda(z^2 + \xi^2) + \overline{H}(\xi,z) \end{cases}.$$

Consider the projections π_1 and π_2 from $\mathcal{M}$ to the (z,w) and (ξ,η) spaces, respectively. Then π_j are two-to-one branched covering maps. Write $\hat{\tau}_j$ for the deck transformations of π_j. Namely, for $p, q \in \mathcal{M}$ $\hat{\tau}_j(p) = q$ if and only if $\pi_j(p) = \pi_j(q)$. One sees that $\hat{\tau}_j$ extend to biholomorphic self-maps of $(\mathcal{M}, 0)$. Also, write $\hat{\rho}$ for the conjugate holomorphic self-map of $(\mathcal{M}, 0)$: $\hat{\rho}(z, w, \xi, \eta) = (\bar{\xi}, \bar{\eta}, \bar{z}, \bar{w})$. Then, the following relations are fundamental:

$$(4.4) \qquad \hat{\tau}_j^2 = \hat{\tau}_j, \quad \hat{\rho}^2 = \hat{\rho}, \quad \hat{\tau}_2 = \hat{\rho} \circ \hat{\tau}_1 \circ \hat{\rho},$$

In what follows, we call $(\hat{\tau}_j, \hat{\rho})$ the Moser-Webster triplet. Notice that $\mathcal{M}$ is parameterized by (z, ξ) by (4.3). We can define the following self-maps of $(\mathbf{C}^2, 0)$:

(i): $\tau_1(z, \xi) := (z, \xi')$ if and only if $\pi_1(z, w, \xi, \eta) = \pi_1(z, w, \xi', \eta')$ for a certain $(z, w, \xi', \eta') \in \mathcal{M}$.

(ii): $\tau_2(z, \xi) := (z', \xi)$ if and only if $\pi_1(z, w, \xi, \eta) = \pi_1(z', w', \xi, \eta)$ for a certain $(z', w', \xi, \eta) \in \mathcal{M}$.

(iii): $\rho(z, \xi) := (\bar{\xi}, \bar{z})$

A direct computation shows that τ_2 and τ_1 are given, respectively, by:

$$(4.5) \qquad \begin{cases} z' = -z - \frac{1}{\lambda}\xi + h_1(z, \xi), \\ \xi' = \xi, \end{cases}$$

$$(4.6) \qquad \begin{cases} z' = z, \\ \xi' = -\frac{1}{\lambda}z - \xi + h_2(z, \xi) \end{cases}$$

where $h_j(z, \xi) = o(\|(z, \xi)\|)$.

τ_j are naturally associated to $\hat{\tau}_j$ by (4.3):

$$(4.7) \qquad \tau_j = \Psi_0^{-1} \circ \hat{\tau}_j \circ \Psi_0, \quad \rho = \Psi_0^{-1} \circ \hat{\rho} \circ \Psi_0.$$

The following lemma can be proved by a direct construction:

Lemma 4.1: Bishop surfaces $(M, 0)$ and $(\widetilde{M}, 0)$ with Bishop invariant $\lambda \neq 0, \frac{1}{2}, \infty$ are holomorphic equivalent if and only if there is a biholomorphic map Ψ from $(\mathcal{M}, 0)$ to $(\widetilde{\mathcal{M}}, 0)$ such that $\Psi \circ \hat{\tau}_j = \hat{\tilde{\tau}}_j \circ \Psi$ and $\Psi \circ \hat{\rho} = \hat{\tilde{\rho}} \circ \Psi$.

Suppose that we have a general pair of holomorphic involutions τ_1 and τ_2, together with a conjugate holomorphic involution from $(\mathbf{C}^2, 0)$ to $(\mathbf{C}^2, 0)$. Let $\mathcal{M}$ be the complexification of the Bishop surface in (4.1). Suppose that there is a biholomorphic map Φ from $(\mathbf{C}^2, 0)$ to $(\mathcal{M}, 0)$ such that

$$\Phi \circ \tau_j = \hat{\tau}_j \circ \Phi, \text{ and } \Phi \circ \rho = \hat{\rho} \circ \Phi.$$

Then we say $\{(\mathcal{M}, 0), \hat{\tau}_j, \hat{\rho}\}$ is parameterized by $\{(\mathbf{C}^2, 0), \tau_j, \rho\}$ through Φ. Notice that it then always holds that $\tau_2 = \rho \circ \tau_1 \circ \rho$.

The following is a fundamental fact in the theory of Moser-Webster, whose proof can be reduced to the proof of Lemma 4.1

Proposition 4.2: Let $(M, 0)$ and $(\widetilde{M}, 0)$ be two Bishop surfaces with Bishop invariant $\lambda \neq 0, \frac{1}{2}, \infty$. Suppose that the Moser-Webster triplet of their complexifications are parameterized by $\{\tau_j, \rho\}$ and $\{\tilde{\tau}_j, \tilde{\rho}\}$, respectively. Then $(M, 0)$ is holomorphically equivalent to $(\widetilde{M}, 0)$ if and only if there is a biholomorphic map ψ from $(\mathbf{C}^2, 0)$ to $(\mathbf{C}^2, 0)$ such that

$$\tilde{\tau}_j = \psi \tau_j \psi^{-1}, \quad \tilde{\rho} = \psi \rho \psi^{-1}.$$

4.2 Linear Theory of a Pair of Involutions Intertwined by a Conjugate Holomorphic Involution

Assume that we have two involutions $\tau_j : (\mathbf{C}^2, 0)$ to $(\mathbf{C}^2, 0)$ and an anti-holomorphic involution: $\rho : (\mathbf{C}^2, \mathbf{0}) \to (\mathbf{C}^2, \mathbf{0})$ such that $\tau_2 = \rho \tau_1 \rho$. We always assume that the linear parts T_j of τ_j satisfy the following properties:

$$(4.8) \quad \begin{cases} T_1, \ T_2 \text{ have no common non-trivial eigenvectors;} \\ \det(T_j) + 1 = tr T_j = 0. \end{cases}$$

Notice that this is always the case for the holomorphic involutions obtained from Bishop surfaces with Bishop invariant $\lambda \neq 0, \frac{1}{2}, \infty$. Indeed, for such involutions,

$$(4.9) \qquad T_2 = \begin{pmatrix} -1 & -\frac{1}{\lambda} \\ 0 & 1 \end{pmatrix}, \ T_1 = \begin{pmatrix} 1 & 0 \\ -\frac{1}{\lambda} & -1 \end{pmatrix}.$$

We first present the linear theory for these involutions.

Assume τ_j, ρ are linear. Let $\phi = \tau_1 \circ \tau_2$. By studying the normalization of ϕ, one can find a new coordinates system (see [Lemma 2.2, MW]) (x, y) in which

$$(4.10) \qquad \begin{cases} \tau_1(x, y) = (\gamma y, \gamma^{-1} x), \ \tau_2(x, y) = (\gamma^{-1} y, \gamma x), \\ \phi(x, y) = (\mu x, \mu y^{-1}) \text{ with } \mu = \gamma^2, \mu^2 \neq 1. \end{cases}$$

Also, it holds either

$$(4.11) \qquad \begin{cases} \rho(x,y) = (\overline{y},\overline{x}) \text{ and } \gamma = \overline{\gamma} > 1, \text{ or} \\ \rho(x,y) = (\overline{x},\overline{y}) \text{ and } |\gamma| = 1, \ 0 < \arg(\gamma) < \frac{\pi}{2}. \end{cases}$$

The coordinates system which put τ_j, ρ into the above normalization is unique up to the scaling map, which maps (x,y) to (ax, ay) with $a = \overline{a}$.

We now discuss how to construct a Bishop surface M such that $(\mathcal{M}, \hat{\tau}_j, \hat{\rho})$ is parameterized through a certain biholomorphic map Φ by the above mentioned set of involutions.

First, we let $\xi = b(\gamma x + y)$, $z = a(x + \gamma y)$, $a, b \in \mathbf{C}$. Then z is invariant under the action of τ_1 and ξ is invariant under the action of τ_2. We will also so construct w and η that they are invariant under the action of both τ_1 and τ_2. We need to choose a, b such that ρ will be associated to the mapping $(z, w, \xi, \eta) \to (\overline{\xi}, \overline{\eta}, \overline{z}, \overline{w})$ in the complexification of the surface. Hence, when $\gamma = \overline{\gamma} > 1$, we need to choose $a = \overline{b}$. When $\gamma\overline{\gamma} = 1$ with $0 < arg(\gamma) < \frac{\pi}{2}$, we chose $a\gamma = \overline{b}$. Hence for ρ in (4.11), we have, respectively, the following expressions:

$$(4.12) \qquad \begin{cases} (i): \ \xi = b(\gamma x + y), \ z = \overline{b}(x + \gamma y); \\ (ii): \ \xi = b(\gamma x + y), \ z = \overline{b\lambda}(x + \gamma y). \end{cases}$$

We only consider how to construct Bishop surfaces in Case (ii). The simplest quadratic polynomials that are invariant under the action of both τ_j are cxy with $c \in \mathbf{C}^1$. We then want to choose b so that $q_\lambda(x,y)$ is a multiple of xy. In fact,

$$\begin{aligned} w = q_\lambda(z, \xi) &= z\xi + \lambda(z^2 + \xi^2) \\ &= |b|^2\overline{\gamma}(\gamma x^2 + \gamma^2 xy + xy + \gamma y^2) \\ &\quad + \lambda\left(b^2(\gamma^2 x^2 + 2\gamma xy + y^2) + \overline{b^2\gamma^2}(x^2 + 2\gamma xy + \gamma y^2)\right) \end{aligned}$$

Hence we obtain

$$(4.13) \qquad \begin{cases} |b|^2 + \lambda b^2 \gamma^2 + \overline{b}^2\overline{\gamma^2}\lambda = 0 \\ |b|^2 + \lambda b^2 + \overline{b}^2\lambda = 0. \end{cases}$$

Therefore, we can choose b be such that $|b| = 1$ and $b^2 = -\gamma^{-1}$. For such a choice of b, we have $\lambda = (\gamma + \gamma^{-1})^{-1} > 0$ and $q = z\xi + \lambda(z^2 + \xi^2) = \lambda^{-1}(1 - 4\lambda^2)xy$

Now, it is straightforward to verify that $\{\tau_j, \rho\}$ is a parameterization for the Moser-Webster triplet on the complexification of $w = q_\lambda(z, \overline{z})$ through the map

$$(4.14) \qquad \begin{aligned} \Phi(x,y) &= (z(x,y), w(z,y) = \lambda^{-1}(1 - 4\lambda^2)xy, \xi(x,y), \eta(x,y) \\ &= \overline{w}(x,y) = w(x,y)), \end{aligned}$$

where $z(x,y)$ and $\xi(x,y)$ are given by the second formula in (4.12).

Notice that $\lambda = (\gamma + \gamma^{-1})^{-1}$ or $\lambda\gamma^2 - \gamma + \lambda = 0$. Hence when $\gamma\bar{\gamma} = 1$ with $0 < \arg(\gamma) < \pi/2$, the Bishop invariant of the quadric $\lambda > \frac{1}{2}$. Namely, M has a hyperbolic complex tangent at 0. The reader can verify that in the case of $\gamma = \bar{\gamma}$, the involutions studied above parameterize the Moser-Webster triplet for elliptic Bishop quadrics with $\lambda \neq 0$.

4.3 General Theory on the Involutions and the Moser-Webster Normal Form

We now study the non-linear involutions τ_j and ρ with $\tau_2 = \rho \circ \tau_1 \circ \rho$, whose linear parts satisfy the property in (4.8). For the purpose of studying Bishop surfaces, one can assume that ρ is conjugate linear. By §3, after a linear change of coordinates, we also assume that in (x, y)-coordinates,

$$(4.15) \qquad \tau_j := \begin{cases} x' = \gamma_j y + p_j(x, y) \\ y' = \gamma_j^{-1} x + q_j(x, y) \end{cases}$$

$$(4.15)' \qquad \begin{array}{l} \rho(x, y) = (\bar{y}, \bar{x}) \text{ and } \gamma = \bar{\gamma} > 1; \quad \text{or} \\ \rho(x, y) = (\bar{x}, \bar{y}) \text{ and } |\gamma| = 1, \ \arg(\gamma) \in (0, \pi/2). \end{array}$$

$$(4.16) \qquad \phi = \tau_2 \circ \tau_1 := \begin{cases} x' = \mu x + f(x, y) \\ y' = \mu^{-1} y + g(x, y) \end{cases}$$

where $\gamma_1 = \gamma_2^{-1} = \gamma$, $\mu = \gamma^2$, $\mu^2 \neq 1$, $p_j, q_j, f, g = o(|(x, y)|)$. We will subject to τ_j, ϕ a transformation of the following form:

$$(4.17) \qquad \psi := \begin{cases} x = t + u(t, T) \\ y = T + v(t, T) \end{cases}$$

For any formal power series $p(t, T)$, we can write it as

$$p(t, T) = \sum_{-\infty}^{\infty} p_s(t, T),$$

with $p_s(\tau t, \tau^{-1} T) = \tau^s p_s(t, T)$, for any $\tau \in \mathbf{R}$. We call p_s is of type s. We impose the normalization condition for the transformation in (4.17): $u_1 = v_{-1} = 0$. A fact is that for any ψ as in (4.17), there is a unique factorization: $\psi = \psi_0 \circ \delta$ where ψ_0 is normalized and $\delta(t, T) = (\alpha(tT)t, \beta(tT)T)$ for certain α, β with $\alpha(0) = \beta(0) = 1$.

Theorem 4.3 (Moser-Webster [MW]): Let τ_j, ρ, ϕ be in (4.13)-(4.14) with $\mu = \gamma^2$ not a root of unity. Then there is a unique normalized transformation ψ of the form (4.17) such that

$$\psi^{-1}\tau_1\psi(t,T) = (\Lambda T, \Lambda^{-1}t), \psi^{-1}\tau_2\psi(t,T) = (\Lambda^{-1}T, \Lambda t),$$
$$\psi^{-1}\phi\psi(t,T) = (Mt, MT), \quad \psi^{-1}\rho\psi(t,T) = \rho(t,T)$$

where $\Lambda = \gamma + \sum_{\alpha=1}^{\infty} \Lambda_\alpha(tT) = \gamma + o(1), M = \Lambda^2(tT)$. The most general transformation that makes τ_j into the above normal form is of the form: $\psi \circ \sigma$ with $\sigma(t,T) = (r(tT)t, r(tT)T)$. Here $r(tT) = \bar{r}(tT)$ and $r(0) \neq 0$. Also in these coordinates, Λ satisfies either the property $\Lambda(tT) = \overline{\Lambda}(tT)$ or $\Lambda(tT)\overline{\Lambda}(tT) = 1$, according to the first form or the second form ρ takes in (4.15)', respectively.

Idea of the Proof of Theorem 4.3: The proof is done by an induction argument. Here, we only sketch the the proof how to construct the unique normalized map ψ which puts τ_j, ϕ into their normal form. The reader can find the detailed proof for statements in the theorem, in the original paper of Moser-Webster [§3, MW].

Assume that there is a ψ whose terms of order less than m can be chosen uniquely so that $\psi^{-1}\tau_j\psi$ has the property in the theorem up to order $m-1$. Thus we assume that τ_j take the following form:

$$\tau_j : \begin{cases} x' = \Lambda_j y + p_j + \cdots, \\ y' = \Lambda_j^{-1} x + q_j + \cdots \end{cases}$$

where $\Lambda_j(xy) = \Lambda_j(xy)$ are polynomial of degree $< m-1$, p_j, q_j are holomorphic polynomials of degree $m \geq 2$. The dots denote terms of order at least $m+1$. Using $\tau_j^2 = Id$ and noting that $\Lambda_j(\tau_j) = \Lambda_j + O(m)$, we get

(4.17)'
$$\gamma_j q_j(x,y) + p_j(\gamma_j x, \gamma_j^{-1} y) = 0.$$

It then follows that

$$\phi : \begin{cases} x' = Mx + a + ..., \\ y' = M^{-1}y + b + ..., \end{cases}$$

where $M = \Lambda_1\Lambda_2^{-1}$ and $a = \gamma_1 q_2 + p_1(\gamma_2 y, \gamma_2^{-1} x)$, $b(x,y) = \gamma_1^{-1} p_2(x,y) + q_1(\gamma_2 y, \gamma_2^{-1} x)$.

We want to choose u, v so that $\widetilde{\phi} = \psi^{-1}\phi\psi$ has the form given in Theorem 4.3, modifying terms of order at least $(m+1)$. Then one can see that it forces $\psi^{-1}\tau_j\psi$ also to have the form as in Theorem 4.3 modifying terms of order at least $(m+1)$. Let $\widetilde{\phi}$ be in Theorem 4.3. Since $\psi\widetilde{\phi} = \phi\psi$, we have:

(4.18)
$$\begin{cases} u(\mu t, \mu^{-1}T) - \mu u(t,T) = (a - \widetilde{a})(t,T) \\ v(\mu t, \mu^{-1}T) - \mu^{-1}v(t,T) = (b - \widetilde{b})(t,T). \end{cases}$$

We want to make $\widetilde{a}_s = 0$ for $s \neq 1$ and $\widetilde{b}_s = 0$ for $s \neq -1$. This leads to the equation:

(4.19)
$$\begin{cases} (\mu^s - \mu)u_s = a_s, \ s \neq 1 \\ (\mu^s - \mu^{-1})v_s = b_s, \ s \neq -1, \end{cases}$$

which clearly can be solved by the assumption.

Then, $\tilde{a}_1 = a_1 = A(tT)t$, $\tilde{b}_{-1} = b_{-1} = B(tT)T$, and ψ is uniquely determined up to order m.

We next show that $p_j(x,y) = P_j(xy)y$, $q_j(x,y) = Q_j(xy)x$. By $(4.17)'$, we have

$$\gamma q_2 + p_1\tau_2 = tA, \quad \gamma^{-1}p_2 + q_1\tau_2 = TB.$$

By (4.17), $q_1 = -\gamma^{-1}p_1\tau_1$, $q_2 = -\gamma p_2\tau_2$ up to order m. Therefore, $p_1 - \mu p_2 = \gamma^{-1}yA$, $p_2 - p_1\phi = \gamma_1 xB$. This then leads to $p_1 - \mu p_1\phi = y(\gamma^{-1}A + \mu\gamma B)$, $p_2 - \mu p_2\phi = y(\gamma^{-1}\mu A + \gamma B)$. Since μ is not a root of unity, this implies that p_j are of type $s = -1$. Similarly, we can get q_j are of type $s = +1$.

Returning to τ_j, ϕ, we may write

$$\tau_j : \begin{cases} x' = (\Lambda_j + P_j)y + \ldots \\ T' = (\Lambda_j^{-1} + Q_j)x + \ldots \end{cases}$$

with $\Lambda_2 = \Lambda_1^{-1}$. One can also verifies that $(\Lambda_j + P_j)(\Lambda_j^{-1} + Q_j) = 1 + O(m)$. By induction, we proved the existence of ϕ, which normalizes τ_j and ϕ.

The rest of the proof is similarly done, which we refer the reader to [§3, MW])   □.

The following result of Moser-Webster provides a convergence result when γ does not have norm 1. The proof is based on a majorant argument, motivated by the study of the normalization problem for area preserving mappings in mechanics. (See [SM]). The proof can be found in [§4, MW].

Theorem 4.4: Let $\{\tau_j, \rho\}$ be as in Theorem 4.3. Assume that $|\gamma| \neq 1$. Then the normalization ψ in Theorem 4.3 and the normal forms for τ_j, ϕ are convergent near the origin.

Making using of Theorem 4.2, Theorem 4.3 and a similar way for constructing Bishop surfaces from the involutions as discussed in §4.2, Moser-Webster obtained the following Theorem. (See [§5, MW])

Theorem 4.5 (Moser-Webster): Let $(M, 0)$ be a real analytic Bishop surface with Bishop invariant $\lambda \in (0, 1/2)$. Then there is a holomorphic change of coordinates, such that in the new coordinates, M is represented by an equation of the form:

$$(4.20) \qquad w = z\bar{z} + (\lambda + \epsilon w^s)(z^2 + \bar{z}^2), \quad \epsilon = 0, \pm 1.$$

5 Geometric Method to the Study of Local Equivalence Problems

The method we discussed in the previous sections is fundamentally based on the understanding of the associated power series. Results obtained in such a manner are usually easy to apply; and invariants obtained so are relatively easy to computer. However, it mainly applies to real analytic submanifolds. The convergence issue may also be very difficult to handle in certain cases. In this section, we introduce to the reader a geometric approach for the study of the equivalence problem, initiated from the work of E. Cartan. This method applies to smooth CR generic submanifolds. The invariants are the so-called curvature functions and their covariant derivatives. Notice that the invariants from the power series method are usually embedded into the coefficients of the normal forms. There are many references related to the topics discussed here. We mention [Ga] [CM] [Ch] [CJ] [Ja] [HJY] [HJ2], to name a few.

5.1 Cartan's Theory on the Equivalent Problem

Let $V, \widetilde{V} \subset \mathbf{R}^n$ be open subsets with $p \in V$ and $\widetilde{p} \in \widetilde{V}$. Let $\theta_V = (\theta_V^1, ..., \theta_V^n)^t$ and $\widetilde{\theta}_{\widetilde{V}} = (\widetilde{\theta}_{\widetilde{V}}^1, ..., \widetilde{\theta}_{\widetilde{V}}^n)^t$ be co-frames on V and $\widetilde{V}$, respectively. Let $G \subset GL(n, \mathbf{R})$ be a connected linear subgroup. We would like to understand the following question: When does there exist a diffeomorphism Φ from V to $\widetilde{V}$ with $\Phi(p) = q$ such that $\Phi^*(\widetilde{\theta}_{\widetilde{V}})(p) = \gamma_{V\widetilde{V}}(p)\theta_V$, where $\gamma_{V\widetilde{V}}(p) \in G$ for each p?

To answer the question, we construct its G-co-frame bundle (Y, π, V), where $\forall p \in V$, $\pi^{-1}(p) = \{g \cdot \theta_V(p) : g \in G\}$. (Since we only consider the local problem, we can identify Y as the product manifold $V \times G$.)

Notice that G acts smoothly from the left on Y, which is defined as follows: $\forall C \in G$, and $P = g \cdot \theta_V(p) \in \pi^{-1}(p)$,

$$(C, P) = (Cg) \cdot \theta_V(p) \in \pi^{-1}(p).$$

This action makes Y into a so-called G-structure bundle over V
Now, θ_V can be lifted naturally to globally defined 1-forms $\omega = (\omega_1, \cdots, \omega_n)^t$: $\omega_l|_{g\theta_V(p)} = g\pi_V^*(\theta_V^l(p))$.

Similarly, we can define a G-structure co-frame bundle $(\widetilde{Y}, \widetilde{\pi}, \widetilde{V})$

In what follows, when there is no confusion, we identify the space Y with $V \times G$ through a manner, which should be obvious from the context. For instance, in the following lemma, Y is identified with $V \times G$. through the map $g\theta_V(p) \mapsto (p, g)$.

Lemma 5.1 is simple but important for Cartan's theory.

Lemma 5.1: There exists a diffeomorphism $\Phi : V \to \widetilde{V}$ with $\Phi(p) = \widetilde{p}$ satisfying $\Phi^*(\widetilde{\theta}_{\widetilde{V}}) = \gamma_{V\widetilde{V}}\theta_V$, where $\gamma_{V\widetilde{V}}$ smoothly maps V into G, if and only if there exits a diffeomorphism $\Phi^1 : V \times G \to \widetilde{V} \times G$ such that

$$\Phi^{1*}\widetilde{\omega} = \omega, \quad with \ \Phi^1(P) = \widetilde{P},$$

where $P \in \pi_V^{-1}(p)$ and $\widetilde{P} \in \pi_{\widetilde{V}}^{-1}(\widetilde{p})$.

Proof of Lemma 5.1: We need only to show that the existence of Φ^1 gives the required Φ. (This is because if we know Φ, we can set $\Phi^1(u, S) = (\Phi(u), S\gamma_{V\widetilde{V}}^{-1}(u))$.)

Assume that $\Phi^1 : Y \to \widetilde{Y}$ is a diffeomorphism such that $\Phi^{1*}(\widetilde{\omega}) = \omega$. Write $\Phi^1(u, S) = (\Phi(u, S), T(u, S))$ with $u \in V$, $\Phi(u, S) \in \widetilde{V}$ and $T(u, S) \in G$.

The assumption that $\Phi^*(\widetilde{\omega}) = \omega$ gives that

$$\Phi^{1*}(T\pi_{\widetilde{V}}^*(\widetilde{\theta_{\widetilde{V}}})) = S\pi_V^*(\theta_V), \quad or$$

$$(\pi_{\widetilde{V}} \circ \Phi^1)^*(\widetilde{\theta_{\widetilde{V}}}) = (T(u, S))^{-1}S\pi_V^*(\theta_V), \quad or$$

$$(\Phi(u, S))^*(\widetilde{\theta_{\widetilde{V}}}) = (T(u, S))^{-1}S\theta_V.$$

Since $\{\widetilde{\theta_{\widetilde{V}}}\}$ is a co-frame fro $T^*(\widetilde{V})$ and $\{\theta_V\}$ is a co-frame for T^*V, we conclude that the partial derivatives of $\Phi(u, S)$ with respect to the group variables must be zero. Hence $\Phi(u, S) = \Phi(u)$. In particular, $T(u, S)^{-1}S = \gamma_{V\widetilde{V}}(u)$. The proves the existence of the required map from V to $\widetilde{V}$.   □

Let dim $G = r$. Then dim$(Y) = n + r$. With the forms $\omega_1, ..., \omega_n$, we would like to add r more 1-forms $\omega^{n+1}, ..., \omega^{n+r}$ on $V \times G$ to form a co-frame Ω over Y such that $\Phi^{1*}(\widetilde{\omega}) = \omega$ if and only if $\Phi^{1*}(\widetilde{\Omega}) = \Omega$. If this is the case, we call such an equivalence problem an e-equivalence problem. Whether we can extend ω to Ω to reduce to an e-equivalence problem depends strongly on the property of the group G. Fortunately, for the CR equivalence problem for Levi non-degenerate hypersurfaces, we do have such a reduction which is the content of the Cartan-Chern-Moser theory ([CM]).

Suppose that ω has an extension to Ω such that there is a diffeomorphism Φ from Y to $\widetilde{Y}$ with $\Phi^*(\omega_j) = \omega_j$ $(j \leq n)$ if and only if $\Phi^*(\widetilde{\Omega}) = \Omega$, namely, $\Phi^*(\widetilde{\omega}^j) = \omega^j$ for any $j \in \{1, ..., n + r\}$. The forms $\{\omega^j\}$ for $j \geq n + 1$ are called the connection forms.

Next we introduce Cartan's method for the study of the e-equivalence problem, by introducing a new type of invariant functions from what presented in the above sections.

Let $\Omega = \{\omega_j\}$ be a coframe over a domain $V \subset \mathbf{R}^n$. For any differentiable function γ over V, we define its *covariant partial derivative*:

(5.1)
$$d\gamma = \sum \gamma_{|i}\omega^i.$$

Since $\{\omega_j\}$ is a co-frame, we can uniquely write $d\omega^i = \sum C_{jk}^i\omega^j \wedge \omega^k$ with $C_{jk}^i = -C_{kj}^i$. Let $\widetilde{\Omega} = \{\widetilde{\omega}_j\}$ be a co-frame over another domain $\widetilde{V} \subset \mathbf{R}^n$. Apparently, if there is a Φ with $\Phi^*(\widetilde{\Omega}) = \Omega$, then it mus hold

$$\widetilde{C}^i_{jk|l} \circ \varPhi = Phi^*(\widetilde{C}^i_{jk|l}) = C^i_{jk|l}.$$

Hence $\{C^i_{jk|l}\}$ are the simplest invariant functions attached to the e- equivalence problem for the co-frame $\varOmega$. Now, we can inductively take the covariant derivatives of the obtained invariants to get new invariant functions. More precisely, for each integer s with $s \geq 1$, we define

$$(5.2) \quad \varGamma_s(\varOmega, V) := \left\{ C^i_{jk}, C^i_{jk|l_1}, \cdots, C^i_{jk|l_1 \ldots l_s} \mid 1 \leq i, j, k, l_1, \ldots, l_s \leq n+r \right\},$$

which is written as a lexicographically ordered set. We define

$$(5.3) \qquad k_s(p) := \operatorname{rank}\{d\, \varGamma_s(\varOmega, V)\}(p), \quad p \in V,$$

to be the dimension of the span of the differentials which occur in the ordered set $\varGamma_s(\varOmega, V)$. The *order* of the e-structure $\varOmega$ at $p \in V$ is the smallest $j_0 = j_0(p)$ such that

$$k_{j_0}(p) = k_{j_0+1}(p).$$

In this case, the *rank* of the e-structure $\varOmega$ at p is defined as

$$\rho_0 = \rho_0(p) := k_{j_0}(p).$$

We say that the e-structure $\varOmega$ is *regular* of order j_0 and of rank ρ_0 at $p \in V$ if there exists a neighborhood U_p of p in V such that the order $j_0(q) \equiv$ *constant* and rank $\rho_0(q) \equiv$ *constant*, $\forall q \in U_p$. Then we can find ρ_0- functions $\{g_1, \ldots, g_{\rho_0}\} \subset \varGamma_{j_0}(\varOmega, V)$, and a certain neighborhood U_p of p in Y so that $d\, g_1 \wedge \cdots \wedge d\, g_{\rho_0} \neq 0$,

$$(5.4) \qquad d\, g \wedge d\, g_1 \wedge \cdots \wedge dg_{\rho_0} \equiv 0 \text{ on } U_p, \quad \text{for any } g \in \varGamma_{j_0+1}(\varOmega, V).$$

Notice $0 \leq j_0 \leq n+r-1$. The case $j_0 = 0$ occurs when the functions $C^i_{jk} \equiv$ *constant* for all i, j and k. And the case $j_0 = n+r-1$ occurs if and only if one invariant function is added at each jet level. Notice that $0 \leq \rho_0 \leq n+r-1$. When $\rho_0 = n+r-1$, we say that $\varGamma(\varOmega, V)$ is of the *maximal rank*.

Next, for each $g \in \varGamma_{j_0+1}(\varOmega, Y)$, since $dg \wedge dg_1 \wedge \ldots \wedge dg_{\rho_0} \equiv 0$, we conclude that there is a unique function A_g such that

$$g = A_g(g_1, \ldots, g_{\rho_0})$$

where A_g is defined near a neighborhood of $(g_1(p), \ldots, g_p(p))$ which is called the the *relation function* of g with respect to $\{g_1, \ldots, g_{\rho_0}\}$.

The following fundamental theorem is due to Cartan.

Theorem 5.2 (E. Cartan [Ga]): Let $\varOmega$ and $\widetilde{\varOmega}$ be two smooth regular e-structures of order j_0 and rank ρ_0. Let $g_1, \ldots, g_{\rho_0}$ be as in (5.14). Let $\widetilde{g}_1, \ldots, \widetilde{g}_{\rho_0}$ be such that they have the identical lexicographic indices as for $g_1, \ldots, g_{\rho_0}$. Then the following statements are equivalent:

(i) There exists a C^∞ diffeomorphism $\Phi : (V, p) \to (\widetilde{V}, \widetilde{p})$ with $\Phi^* \widetilde{\Omega} = \Omega$.
(ii) $\widetilde{g}_j(\widetilde{p}) = g_j(p)$ holds for $1 \leq j \leq \rho_0$, and for any function $g \in \Gamma_{j_0+1}(\Omega, V)$, and $\widetilde{g} \in \Gamma_{j_0+1}(\widetilde{\Omega}, \widetilde{V})$ with the same lexicographic order, it holds that $A_g = A_{\widetilde{g}}$ near $(g_1(p), ..., g_p(0))$.

Suppose that Ω is a real analytic co-frame and V is connected. Then, there is a proper real analytic subset E such that any point in $V - E$ is a regular point. Also, from the uniqueness property of real analytic functions, the order and the rank of Ω are all the same in $V - E$. We define the order and the rank of Ω in V to be the order and the rank of Ω at any point in $V - E$.

We call Ω an algebraic co-frame if $\omega^j = \sum h_l^j dx^l$ with h_l^j Nash algebraic smooth functions. We define the algebraic degree of ω^j to be the maximum degree of the algebraic functions h_l^j. Here, we recall that for a Nash algebraic smooth function $h \not\equiv 0$, there is an irreducible polynomial $P(x, X)$ in (x, X) such that $P(x, h) \equiv 0$. Then we define the degree of h to be the degree of the polynomial $P(x, X)$. It is apparent that when Ω is an algebraic co-frame, then any curvature functions and relation functions are algebraic, too. Suppose that the order of Ω is j_0. We set

$$(5.5) \qquad \ell(\Omega) = \max_{g \in \Gamma_{j_0+1}} \deg(g).$$

Then we have the following versions of the Cartan theorem in the analytic category and algebraic category, which are a lot of more convenient to apply:

Theorem 5.3: Let Ω and $\widetilde{\Omega}$ be analytic e- structures at p and $\widetilde{p}$, respectively, with p a regular point of Ω. Then the following are equivalent:
(i). There exists a C^ω diffeomorphism $\Phi : (V, p) \to (\widetilde{V}, \widetilde{p})$ such that

$$\Phi^* \widetilde{\Omega} = \Omega.$$

(ii). $\Gamma_k(\widetilde{\Omega}, V)(\widetilde{p}) = \Gamma_k(\Omega, V)(p)$ holds for all k.
(iii). Suppose that Ω and $\widetilde{\Omega}$ have order j_0 and rank ρ_0 at p and $\widetilde{p}$, respectively. Also assume that $\widetilde{p}$ is a regular point for $\widetilde{\Omega}$. Let $g_1, ..., g_{\rho_0}$ be as above, and let $\widetilde{g}_1, ..., \widetilde{g}_{\rho_0}$ be the corresponding relation functions with the same lexicographic order as for $g_1, ..., g_{\rho_0}$. Then $\widetilde{g}_j(\widetilde{p}) = g_j(p)$ holds for $1 \leq j \leq \rho_0$, and for any function $g \in \Gamma_{j_0+1}(\Omega, V)$, and $\widetilde{g} \in \Gamma_{j_0+1}(\widetilde{\Omega}, \widetilde{V})$ with the same lexicographic order, it holds that $A_g = A_{\widetilde{g}}$ near $(g_1(p), ..., g_p(0))$.

Theorem 5.4([HJ2]): Suppose that Ω and $\widetilde{\Omega}$ are algebraic co-frames with

$$l_0 = \max\{\ell(\Omega), \ell(\widetilde{\Omega})\}.$$

Assume that $p \in V$ is a regular point for Ω. (See (5.5) for the definition of $\ell(\Omega)$ and $\ell(\widetilde{\Omega})$). Let $\widetilde{p} \in \widetilde{V}$. Then the following statements are equivalent:
(i) $\Gamma_{2l_0^3}(\Omega, V)(p) = \Gamma_{2l_0^3}(\widetilde{\Omega}, \widetilde{V})(\widetilde{p})$.

(ii) There is a real analytic diffeomorphism Φ^1 from a neighborhood of $p \in V$ to a neighborhood of $\widetilde{p}$ in $\widetilde{V}$ such that $\Phi^{1*}(\widetilde{\Omega}) = \Omega$.

We mention that both Theorem 5.3 and Theorem 5.4 can be stated in the holomorphic category when Ω, $\widetilde{\Omega}$ are holomorphic or holomorphically algebraic co-frames. For instance, we have the following:

Theorem 5.3′: Let Ω and $\widetilde{\Omega}$ be holomorphic co-frames at $p \in V$ and $\widetilde{p} \in \widetilde{V}$, respectively. Here V (or, $\widetilde{V}$) is a neighborhood of p (or, $\widetilde{p}$, respectively) in $\mathbf{C}^n$. Then the following are equivalent:

(i). There exists a biholomorphic map $\Phi : (V, p) \to (\widetilde{V}, \widetilde{p})$ such that

$$\Phi^* \widetilde{\Omega} = \Omega, \quad \Phi(p) = \widetilde{p}.$$

(ii). $\Gamma_k(\widetilde{\Omega}, V)(\widetilde{p}) = \Gamma_k(\Omega, V)(p)$ holds for all k.

The proof of these results are based on the Frobenius Theorem. We first prove Theorem 5.2. Apparently, we need only to show that $(ii) \Rightarrow (i)$. $((i) \Rightarrow (ii)$ can be seen by the basic fact that if Φ is a C^∞ diffeomorphism from (V, p) to $(\widetilde{V}, \widetilde{p})$ with $\Phi^*(\widetilde{\Omega}) = \Omega$, then $\Gamma_j(\Omega, V) = \Gamma_j(\widetilde{\Omega}, \widetilde{V}) \circ \Phi$.)

Proof of Theorem 5.2: Consider the manifold $M \subset V \times \widetilde{V}$ defined by $g_j(x) = \widetilde{g}_j(\widetilde{x})$ for $(x, \widetilde{x}) \approx (p, \widetilde{p})$. Here $\{g_j\}_{j=1}^{\rho_0}$ and $\{\widetilde{g}_j\}$ are as in the Theorem. M is apparently a smooth manifold of codimension ρ near $(p, \widetilde{p})$, for $dg_1 \wedge ... \wedge dg_{\eta_0}(p) \neq 0$, $d\widetilde{g}_1 \wedge .... \wedge d\widetilde{g}_{\rho_0}(\widetilde{p}) \neq 0$.

Consider the differential ideals Δ: Δ is generated by $\{\pi^*(\Omega) - \widetilde{\pi}^*(\widetilde{\Omega})\}$, where π is the projection from $V \times \widetilde{V}$ to V and $\widetilde{\pi}$ is the projection from $V \times \widetilde{V}$ to $\widetilde{V}$. We first claim that Δ, when restricted to M, is an integral differential system.

Indeed, on $V \times \widetilde{V}$,

$$d(\pi^* \omega^a - \widetilde{\pi}^* \widetilde{\omega}^a) = \sum (C_{jk}^a \circ \pi) \pi^*(\omega^j) \wedge \pi^*(\omega^k) - \sum (\widetilde{C}_{jk}^a \circ \widetilde{\pi}) \widetilde{\pi}^*(\widetilde{\omega}^j) \times \widetilde{\pi}^*(\widetilde{\omega}^k).$$

Since $C_{jk}^a = A_{jk}^a(g_1, ..., g_\rho)$ and $\widetilde{C}_{jk}^a = A_{jk}^a(\widetilde{g}_1, ..., \widetilde{g}_\rho)$, when restricted to M, we see that $C_{jk}^a \circ \pi \equiv \widetilde{C}_{jk}^a \circ \widetilde{\pi}$. Hence on M, we have

$$d(\pi^* \omega^a - \widetilde{\pi}^* \widetilde{\omega}^a) = \sum C_{jk}^a \circ \pi \{\pi^*(\omega^j) \wedge \pi^*(\omega^k) - \widetilde{\pi}^*(\widetilde{\omega}^j) \wedge \pi^*(\widetilde{\omega}^k)\}$$
$$= \sum C_{jk}^a \circ \pi \{\pi^*(\omega^j) \wedge (\pi^* \omega^k - \widetilde{\pi}^* \widetilde{\omega}^k) - \widetilde{\pi}^*(\widetilde{\omega}^k) \wedge (\pi^*(\omega^j) - \widetilde{\pi}^* \widetilde{\omega}^j)\}.$$

Next, we claim that the rank of $\Delta|_M$ is $n - \rho_0$.

Let us write $y_j = g_j$ for $j = 1, ..., \rho_0$ and extend $(y_1, ..., y_{\rho_0})$ to a coordinate system

$$(y_1, ..., y_{\rho_0}, ..., y_n)$$

near $(g_j(p_0), 0, ..., 0)$. Then the regularity assumption at p gives that

$$dy^j = \sum y_{|l}^j(y_1, ..., y_{\rho_0})\omega_V^l \quad for \ j \le \rho_0.$$

Also the matrix $(y_{|j}^a)$ must have rank ρ_0 by the rank assumption of Ω.

By relabelling $\{\omega_j\}$ if necessary, we can assume that $det(y_{|\beta}^\alpha)_{1 \le \alpha, \beta \le \rho_0} \ne 0$. Let $(y_{|\beta}^\alpha)^{-1} = g_{\alpha\beta}$. Then

$$\sum g_{\alpha\beta}dy^\beta = \omega^\alpha + \sum b_{a=\rho_0+1,\cdots,n}\omega^a, \ or$$

$$\omega^\alpha = \sum_{\beta=1}^{\rho_0} g_{\alpha\beta}dy^\beta - \sum_{a=\rho_0+1}^{n} b_{\alpha a}\omega^a.$$

Similarly, we have $\widetilde{g}_{\alpha\beta} = (\widetilde{y}_{|\beta}^\alpha)^{-1}$,

$$\widetilde{\omega}^\alpha = \sum_{\alpha,\beta=1}^{\rho_0} \widetilde{g}_{\alpha\beta}\widetilde{dy^\beta} - \sum_{a=\rho_0+1,\cdots,n} \widetilde{b}_{\alpha a}\widetilde{w}^a.$$

By the assumption in (ii), $g_{\alpha\beta} = \widetilde{g}_{\alpha\beta}$, $b_{\alpha a} = \widetilde{b}_{\alpha a}$ for $y_j = \widetilde{y}_j$ ($j \le \rho_0$). Hence, when restricted to M,

$$\pi^*(\omega^\alpha) = \widetilde{\pi}^*(\omega^\alpha) \ mod_{\rho_0+1 \le a \le n}\{\pi^*(\omega^a) - \widetilde{\pi}^*(\widetilde{\omega}^a)\}.$$

This proves that the rank of Δ, when restricted to M, is bounded by $n - \rho_0$.

We next show that $\{\pi^*(\omega^a) - \widetilde{\pi}^*(\widetilde{\omega}^a)\}_{\rho_0+1 \le a \le n}$, when restricted to M, is linearly independent.

Indeed, since $dy_1 \wedge ... \wedge dy_{\rho_0} \wedge d\omega^{\rho_0+1} \wedge ... \wedge d\omega^n = det(y_\beta^\alpha)\omega_1 \wedge ... \wedge \omega_n \ne 0$,

$$\widetilde{dy}_1 \wedge ... \wedge \widetilde{dy}_{\rho_0} \wedge d\omega^{\rho_0+1} \wedge ... \wedge \widetilde{\omega}^n \ne 0.$$

We see that $\{dy_1, ..., dy_{\rho_0}, d\omega^{\rho_0+1}, ..., d\omega^n\} \ \{\widetilde{dy}_1, ..., \widetilde{dy}_{\rho_0}, \widetilde{d\omega}^{\rho_0+1}, ..., \widetilde{d\omega}^n\}$ are co-frames.

Now, in the $(y, \widetilde{y})$-coordinates, M is defined by $y_j = \widetilde{y}_j$ for $j \le \rho_0$. Hence, it is easy to see that $\{\pi^*(\omega^a)\}_{a \ge \rho_0+1}$ is of rank $n - \rho_0$ when restricted to M near $(p, \widetilde{p})$. Hence the rank of $\pi^*(\omega^a) - \widetilde{\pi}^*(\omega^a)$ is of rank $n - \rho_0$.

Now, Δ induces a foliation in M with each leaf of real dimension $2n - \rho_0 - (n - \rho_0) = n$. Letting $\mathcal{L}$ be a leaf in M, that passes through $(p, \widetilde{p})$. We claim that $\pi_* : T_{(p,\widetilde{p})}\mathcal{L} \to T_p V$ is an isomorphism. For this, we need only to show that π_* is injective.

Suppose $X \in T_{(p,\widetilde{p})}\mathcal{L}$ be such that $\pi_*(X) = 0$. Then

$$0 = < \pi^*(\omega^j) - \widetilde{\pi}^*(\widetilde{\omega}^j), X > = -(\widetilde{\omega}^j, \widetilde{\pi}_*(X))$$

for all j. Since $\{\widetilde{\omega}^j\}$ forms a co-frame in $\widetilde{V}$, we get $\widetilde{\pi}_*(X) = 0$. Thus $X = 0$.

Finally, let $\varPhi$ be such that $\mathcal{L} = \{(x, \varPhi(x)) : x \approx p\}$. Then one sees that $\varPhi$ is precisely the map that we are looking for. $\square$

Next we give the proof of Theorem 5.3:

Proof of Theorem 5.3: Let $(X_1, ..., X_n)$ be the dual frame of $(\omega_1, ..., \omega_n)$. Namely, $\langle \omega_j, X_l \rangle = \delta_j^l$. Let ϵ_0 be sufficient small such that for any constant vector $(a_1, ..., a_n)$ with $\sum_j |a_j|^2 < \epsilon_0^2$. The integral curve $\gamma_a(t)$ with $\gamma_a(0) = p$ of $\sum_j a_j X_j$ is defined for $|t| < 2$. Namely,

$$\frac{\mathrm{d}\gamma_a(t)}{\mathrm{dt}} = \sum_{j=1}^n a_j X_j(\gamma_a(t)), \ \gamma_a(0) = p$$

has a unique solution for $|t| < 2$.

We can similarly define $(\widetilde{X}_1, ..., \widetilde{X}_n)$ and $\widetilde{\gamma_a}(t)$. We then claim that

$$\varGamma_j(\varOmega, V)(\gamma_a(t)) \equiv \varGamma_j(\widetilde{\varOmega}, \widetilde{V})(\widetilde{\gamma}_a(t)) \ \ for \ |t| < 2.$$

To this aim, for $g_j \in \varGamma(\varOmega, V)$ and $\widetilde{g}_j \in \varGamma_j(\widetilde{\varOmega}, \widetilde{V})$ with the same lexicographic order, we first notice that $g_j(\gamma_a(t)) - \widetilde{g}_j(\widetilde{\gamma}_a(t))$ is real analytic for $|t| < 2$. To prove that $g_j(\gamma_a(t)) \equiv \widetilde{g}_j(\widetilde{\gamma}_a(t))$, we need only to verify that $G_j(t) = g_j(\gamma_a(t)) - \widetilde{g}_j(\widetilde{\gamma}_a(t)))$ vanishes to infinite order at 0. In fact,

$$\frac{dG_j(t)}{dt} = \sum_l g_{j|l}(\gamma_a(t))\langle \omega_l, \sum_k a_k X_k \rangle(\gamma_a(t)) - \widetilde{g}_{j|l}(\widetilde{\gamma}_a(t))\langle \widetilde{\omega}_l, \sum_k a_k \widetilde{X}_k \rangle$$
$$= \sum_l \{g_{j|l}(\gamma_a(t))a_l - \widetilde{g_{j|l}}(\widetilde{\gamma}_a(t))a_l\}$$

Hence, it follows that $G_j(0) = G_j'(0)$. By induction and the given hypothesis, we can conclude that $G_j^{(k)}(t) \equiv 0$ for all k.

Now, we can define $M \subset V \times V' := \{(x, \widetilde{x}) : \varGamma(\varOmega, V)(x) = \varGamma(\varOmega, \widetilde{V})(\widetilde{x})\}$. Define Δ the same way as in the proof of Theorem 5.2. Then we can similarly construct the required map $\varPhi$. (In fact, one can choose $\varPhi$ that sends $\gamma_a(1)$ to $\widetilde{\gamma}_a(1)$, when $|a| < \epsilon_0$ varies.) $\square$

For the proof of Theorem 5.4, we refer the reader to [HJ2].

5.2 Segre Family of Real Analytic Hypersurfaces

We now explore how Cartan's method can be adapted to the study of the equivalence problem of real hypersurfaces in $\mathbf{C}^n$. We mainly focus on the real analytic category.

Let M be a real analytic hypersurface in $D \subset \mathbf{C}^n$ with real analytic defining function $r \in C^\omega(D)$. Apparently, for any other local defining function r^* of M, $r^* = s^* r$ with $s^*|_M \neq 0$. Hence we can well define its complexification as the complex submanifold: $\mathcal{M} = \{(z, \xi) \in D \times Conj(D) : r(z, \xi) = 0\}$. $\mathcal{M}$ is a complex submanifold of complex codimension 1 in $\mathbf{C}^n \times \mathbf{C}^n$ near

$M \times conj(M)$. Here for a set $E \subset \mathbf{C}^n$, $Conj(E) := \{\bar{z} \mid z \in E\}$. For each $\xi \sim Conj(M))$, we can define a complex analytic variety $Q_\xi := \{z \in \mathbf{C}^n : r(z, \xi) = 0\}$. We call Q_ξ the *Segre variety* of M with respect to ξ. Notice that $\mathcal{M}$ is foliated by $\{Q_\xi\}$ (In some references, say, in [We1] [Hu4], one defines $Q_\xi := \{z \in D : r(z, \bar{\xi}) = 0\}$ for $\xi \in D$). A fundamental fact for Segre family is its invariant property for holomorphic maps. More precisely, if f is a local holomorphic map from (M, p) to $(\widetilde{M}, \tilde{p})$, then $f(Q_{\bar{\xi}}) \subset \widetilde{Q_{\bar{f}(\bar{\xi})}}$ for any ξ near $\bar{p}$. Here $\widetilde{Q_{\bar{f}(\bar{\xi})}}$ is the Segre variety of $\widetilde{M}$ with respect to $\bar{f}(\bar{\xi})$. In particular, when f is a holomorphic map from (M, p) to $(\widetilde{M}, \tilde{p})$, f induces a holomorphic map $(f(z), \bar{f}(\xi))$ from $(\mathcal{M}, (p, \bar{p}))$ to $(\widetilde{\mathcal{M}}, (\tilde{p}, \bar{\tilde{p}}))$.

We mention that the above simple property for Segre family has been a basic tool to study the analyticity problem for CR mappings between real analytic hypersurfaces, based on ideas from the original paper of Webster. (See [Hu1] [Hu4] for historic discussions and many related references.) Here, we will use it for a different purpose.

In what follows, we assume that $0 \in M$ and we use (z, ξ) for the coordinates of $\mathbf{C}^n \times \mathbf{C}^n$. Also, we can assume, without loss of generality, that M is defined by an equation of the form $r = 2\mathrm{Im}(z_n) + O(|z| + |\mathrm{Re}(z_n)|)$. In what follows, the indices α, β will have range from 1 to $n - 1$. Occasionally, we will write w, η for z_n, ξ_n, respectively. We also use the summation convention: repeated indices imply summation.

On $\mathcal{M}$, there are $(n - 1)$ independent holomorphic one forms

$$(5.6) \qquad \theta^\alpha = dz^\alpha|_\mathcal{M}, \ \theta_\alpha = d\xi_\alpha|_\mathcal{M}, \ \theta = id_z r|_\mathcal{M} = ir_\alpha dz^\alpha|_\mathcal{M} + ir_n dz^n|_\mathcal{M}.$$

$\{\theta, \theta^\alpha, \theta_\alpha\}$ is a co-frame for $\mathcal{M}$, which depends on the choice of the defining functions.

Next, let $(\widetilde{M}, \tilde{p})$ be another real analytic hypersurface near $\tilde{p} = 0$ in $\mathbf{C}^n$ with a defining function $\tilde{r} = 2Im(\tilde{z}_n) + O(|\tilde{z}|)$. Define similarly the co-frame $\{\tilde{\theta}, \tilde{\theta}^\alpha, \tilde{\theta}_\alpha\}$ on $\widetilde{\mathcal{M}}$ near $(\tilde{p}, \bar{\tilde{p}})$.

If there is a biholomorphic map f from $(M, 0)$ to $(\widetilde{M}, 0)$, then we have a holomorphic map $(f(z), \bar{f}(\xi))$ from $(\mathcal{M}, 0)$ to $(\widetilde{\mathcal{M}}, 0)$. We say that $(\mathcal{M}, 0)$ is Segre equivalent to $(\widetilde{\mathcal{M}}, 0)$ if there is a holomorphic map $\Phi = (\Phi_1(z), \Phi_2(\xi))$ from $(\mathbf{C}^{2n}, 0)$ to $(\mathbf{C}^{2n}, 0)$ such that Φ sends each Segre variety Q_ξ of $\mathcal{M}$ near 0 to the Segre variety $Q_{\Phi_2(\xi)}$ of $\widetilde{\mathcal{M}}$. (Apparently, such a map sends $(\mathcal{M}, 0)$ to $(\widetilde{\mathcal{M}}, 0)$). In particular, we see that when $(M, 0)$ is equivalent to $(\widetilde{M}, 0)$, then $(\mathcal{M}, 0)$ is Segre equivalent to $(\widetilde{\mathcal{M}}, 0)$. We mention that even if $M, \widetilde{M}$ are strongly pseudoconvex, Faran constructed in [Fa] examples showing that the converse of the above statement fails. However, see Remark 5.7.

Lemma 5.5: $(\mathcal{M}, 0)$ is equivalent to $(\widetilde{\mathcal{M}}, 0)$ if and only if there is a holomorphic map $\Phi = (\Phi_1, \Phi_2) = (\phi_j, \psi_k)$ from $(\mathcal{M}, 0)$ to $(\widetilde{\mathcal{M}}, 0)$ such that

$$(5.7) \quad \begin{cases} \Phi^*(\widetilde{\theta}) = u\theta, \\ \Phi^*(\widetilde{\theta^\alpha}) = u^\alpha\theta + u^\alpha_\beta\theta^\beta, \\ \Phi^*(\widetilde{\theta}_\alpha) = u_\alpha\theta + v_\alpha\theta_\beta. \end{cases}$$

where $u, u^\alpha, u^\alpha_\beta, v_\beta$ are holomorphic near 0 and the holomorphic 1-forms are defined as in (5.6).

Proof of Lemma 5.5: Suppose the existence of the Segre isomorphism $\Phi = (\Phi_1(z), \Phi_2(\xi))$. Notice that $\widetilde{r}(\Phi)$ is also a defining function for $\mathcal{M}$ near 0. Hence $\widetilde{r}(\Phi) = r(z, \xi)S(z, \xi)$ near $\mathcal{M}$ with $S(z, \xi) \neq 0$ near $\mathcal{S}$.

Since $\widetilde{\theta} = i\partial\widetilde{r}$ on $\mathcal{M}$, we have

$$\Phi^*(\widetilde{\theta}) = i\partial\widetilde{r}(\Phi_1(z), \Phi_2(\xi)) = iS(z, \xi)\partial r(z, \xi) = iS(z, \xi)\theta.$$

$$\begin{aligned} \Phi^*(\widetilde{\theta^\alpha}) &= \Phi^*(d\widetilde{z}_\alpha) = d\phi_\alpha(z) = \frac{\partial\phi_\alpha}{\partial z_\beta}dz_\beta + \frac{\partial\phi_\alpha}{\partial z_n}dz_n \\ &= \frac{\partial\phi_\alpha}{\partial z_\beta}dz_\beta + \frac{\partial\phi_\alpha}{\partial z_n}\left(-i\frac{\theta}{r^n} - \frac{r^\beta}{r_n}dz_\beta\right) \\ &= \left(\frac{\partial\phi_\alpha}{\partial z_\beta} - \frac{\partial\phi_\alpha}{\partial z_n}\frac{r_\beta}{r_n}\right)\theta^\beta - i\frac{\partial\phi_\alpha}{\partial z_n}\frac{1}{r_n}\theta. \end{aligned}$$

Similarly, we can verify the last equality in (5.7). This proves the first part of the lemma. Similarly, if $\Phi = (\Phi_1, \Phi_2)$ is a holomorphic map satisfying (5.7). Then $\frac{\partial}{\partial\xi_\alpha}\Phi_1 = 0$ and $\frac{\partial\Phi_2}{\partial z_\alpha} = 0$. Hence

$$\Phi_1 = \Phi_1(z_1, ..., z_{n-1}, z_n, \xi^n) \quad and \quad \Phi_2 = \Phi_2(\xi_1, ..., \xi_{n-1}, z_n, \xi^n).$$

Since $\mathcal{M}$ can be parameterized either by (z_α, ξ) or (z, ξ_α), Φ_1 and Φ_2 can be completely expressed as holomorphic functions in z or ξ, respectively. Also, it is obvious that Φ preserves the Segre varieties. □

Now let $(\mathcal{M}, P)$ be as before with holomorphic co-frame $\{\theta, \theta^\alpha, \theta_\alpha\}$. Then we can form a G-structure co-frame bundle $\mathcal{Y}$ over $\mathcal{M}$, where G consists of invertible matrices of the form

$$\begin{pmatrix} u & 0 & 0 \\ u^\alpha & u^\beta_\alpha & 0 \\ v_\alpha & 0 & v^\beta_\alpha \end{pmatrix}$$

To solve the Segre equivalence problem by using Cartan's method, the key step is to find the co-frame on $\mathcal{Y}$, through which the G-equivalence can be reduced to the $\{e\}$-equivalence problem.

Assume that M is strongly pseudoconvex at 0 and $\mathcal{M}$ is defined by

$$r = z_n - \rho(z^\alpha, \xi_\alpha, \xi^n) = z_n - \xi^n + o(|z^\alpha| + |\rho^\alpha|).$$

It is easy to see that $(\rho^\beta_\alpha)|_0$ is precisely the Levi-form of M at 0. Hence, $\det(\rho^\beta_\alpha) \neq 0$ near 0.

Following Chern [Ch] and Chern-Ji [CJ], we choose a co-frame over $\mathcal{M}$ of the following form:

(5.8)
$$\begin{cases} \theta = i(dz^n + r_\alpha dz^\alpha) \\ \theta^\alpha = dz^\alpha, \\ \theta_\alpha = i\frac{r_\alpha^n}{r^n}\theta - \left(r_\alpha^\beta - \frac{r_\alpha^n r^\beta}{r^n}\right)d\xi_\beta, \end{cases}$$

where and in what follows, we write $r_\alpha = \frac{\partial r}{\partial z_\alpha}$, $r^\beta = \frac{\partial r}{\partial \xi_\beta}$, $r_\alpha^\beta = \frac{\partial^2 r}{\partial z_\alpha \partial \xi_\beta}$, etc..

For forms in (5.8), we have:

(5.9)
$$d\theta = i\theta^\alpha \wedge \theta_\alpha.$$

Indeed, notice that

$$\begin{aligned} dr_\alpha &= r_{\alpha\beta}dz^\beta + r_\alpha^n d\xi_n + r_\alpha^\beta d\xi^\beta \\ &= r_{\alpha\beta}dz^\beta + \frac{r_\alpha^n}{r^n}(-i\theta - r^\beta d\xi^\beta) + r_\alpha^\beta d\xi_\beta \\ &= r_{\alpha\beta}dz^\beta - \theta_\alpha. \end{aligned}$$

Hence $\theta_\alpha = -dr_\alpha + r_{\alpha\beta}dz^\beta$ and $d\theta = idr_\alpha \wedge dz^\alpha = i\theta^\alpha \wedge \theta_\alpha$.

Now, by the Levi non-degeneracy of M at 0,

$$\begin{cases} z^\alpha = z^\alpha \\ z_n = z_n \\ \rho_\alpha = \rho_\alpha(z^\alpha, \xi_\alpha, \xi_n) \end{cases}$$

can be used to uniquely solve for $(z^\alpha, \xi_\alpha, z_n)$ by the data $(z^\alpha, \rho_\alpha, z^n)$. Hence, we can use $(z^\alpha, \rho_\alpha, z^n)$ for the coordinates of $\mathcal{M}$. In the $(z^\alpha, z_n, \rho_\beta)$ coordinates, we have the following formula:

(5.10)
$$\begin{cases} \theta = i(dz^n - \rho_\alpha dz^\alpha), \\ \theta^\alpha = dz^\alpha, \\ \theta_\alpha = d\rho_\alpha - \rho_{\alpha\beta}dz^\beta, \quad d\theta = i\theta^\alpha \wedge \theta_\alpha. \end{cases}$$

Here $\rho_{\alpha\beta}$ are holomorphic functions in (z, ρ_α).

Next, let $(\widetilde{\mathcal{M}}, 0)$ be the complexification of another real analytic hypersurface $(\widetilde{M}, 0)$. We also choose the same type of the co-frame $(\widetilde{\theta}, \widetilde{\theta}^\alpha, \widetilde{\theta}_\alpha)$ on $(\widetilde{\mathcal{M}}, 0)$ as in (5.10). Now, suppose that Φ is a Segre isomorphism from $(\mathcal{M}, 0)$ to $(\widetilde{\mathcal{M}}, 0)$, then

$$\begin{cases} \Phi^*(\widetilde{\theta}) = u\theta, \\ \Phi^*(\widetilde{\theta}^\alpha) = u_\beta^\alpha \theta^\beta + u^\alpha \theta, \\ \Phi^*(\widetilde{\theta}_\alpha) = v_\alpha^\beta \theta_\beta + v_\alpha \theta. \end{cases}$$

with $d\widetilde{\theta} = i\widetilde{\theta}^\alpha \wedge \widetilde{\theta}_\alpha$ and $u, u_\beta^\alpha, v_\alpha^\beta, u^\alpha, v_\alpha$ holomorphic near the origin.

Hence $du \wedge \theta + ud\theta = i(u_\beta^\alpha \theta^\beta + u^\alpha \theta) \wedge (v_\alpha^\beta \theta_\beta + v_\alpha \theta)$, from which we get the following

$$\begin{cases} \delta_k^l u = u_l^\alpha v_\alpha^k, \\ du = iu_\beta^\alpha v_\alpha \theta^\beta - iu^\alpha v_\alpha^\beta \theta_\beta + t\theta. \end{cases}$$

Next, we consider the $\mathbf{C}^* := \mathbf{C} \setminus \{0\}$ structure bundle $\mathcal{E}_0 = \mathcal{M} \times \mathbf{C}^*$ on $\mathcal{M}$, which can be identified with the $\mathbf{C}^*$-fiber bundle whose fiber $\pi^{-1}(P)$ over $P \in \mathcal{M}$ is precisely $\{u\theta\}$ with $u \in \mathbf{C}^*$. Then $\omega = u\theta$ is a tautological global holomorphic 1-form on $\mathcal{E}_0$. Notice that

$$d\omega = ud\theta + du \wedge \theta = iu\theta^\alpha \wedge \theta_\alpha + \omega \wedge (-\frac{du}{u}).$$

Define co-frame

$$\begin{cases} \omega^\alpha = u^\alpha \theta + u_\beta^\alpha \theta^\beta, \\ \omega_\alpha = v_\alpha \theta + v_\alpha^\beta \theta_\beta, \end{cases}$$

where $u_\beta^\gamma v_\kappa^\beta = \delta_\kappa^\gamma u$. Then

$$d\omega = i\omega^\alpha \wedge \omega_\alpha + \omega \wedge \left(-\frac{du}{u} - i\frac{u^\alpha}{u} v_\alpha^\beta \theta_\beta + iu_\beta^\alpha v_\alpha \theta^\beta \right).$$

Let $\phi = -\frac{du}{u} - i\frac{u^\alpha}{u} v_\alpha^\beta \theta_\beta + iu_\beta^\alpha v_\alpha \theta^\beta + t\omega$. Then, the above motivates us to consider co-frames of the following form:

(5.11)
$$\begin{cases} \omega = ud\theta, \\ \omega^\alpha = u^\alpha \theta + u_\beta^\alpha \theta^\beta, \\ \omega_\alpha = v_\alpha \theta + v_\alpha^\beta \theta_\beta, \\ \phi = -\frac{du}{u} - i\frac{u^\alpha}{u} v_\alpha^\beta \theta_\beta + iu_\beta^\alpha v_\alpha \theta^\beta + t\theta \\ \delta_k^l u = u_\alpha^l v_k^\alpha. \end{cases}$$

A basic property for the above co-frames is the relation:

(5.12)
$$d\omega = i\omega^\alpha \wedge \omega_\alpha + \omega \wedge \phi.$$

Choose a special co-frame:

$$\begin{cases} \omega^0 = ud\theta, \\ \omega^{0\alpha} = u\theta^\alpha, \\ \omega_\alpha^0 = \theta_\alpha, \\ \phi^0 = -\frac{du}{u}. \end{cases}$$

Then, we have

$$\begin{cases} \omega = \omega^0, \\ \omega^\alpha = \frac{u^\alpha}{u}\omega^0 + \frac{u_\beta^\alpha}{u}\omega^{0\beta} \\ \omega_\alpha = \frac{v_\alpha}{u}\omega^0 + v_\alpha^\beta \omega_\alpha^0, \\ \phi = \phi^0 - i\frac{u^\alpha}{u} v_\alpha^\beta \omega_\beta^0 + iu_\beta^\alpha \frac{v_\alpha}{u^2}\omega^{0\alpha} + t\omega^0. \end{cases}$$

Hence the space of the co-frames in (5.11) form a G_1-structure bundle $\mathcal{Y}$ over $\mathcal{M}$, where G_1 consists of matrices of the following form:

$$\begin{pmatrix} 1 & 0 & 0 & 0 \\ \frac{u^\alpha}{u} & \frac{u^\alpha_\beta}{u} & 0 & 0 \\ \frac{v_\alpha}{u} & 0 & v^\beta_\alpha & 0 \\ t & i\frac{v^\alpha}{u^2}u^\beta_\alpha & -i\frac{u_\alpha}{u}v^\alpha_\beta & 1 \end{pmatrix}$$

with $u^l_\alpha v^\alpha_k = \delta^l_k u$. Or

$$\begin{pmatrix} 1 & 0 & 0 & 0 \\ u^\alpha & u^\alpha_\beta & 0 & 0 \\ v_\alpha & 0 & v^\beta_\alpha & 0 \\ t & iv^\alpha u^\beta_\alpha & -iu_\alpha v^\alpha_\beta & 1 \end{pmatrix}$$

with $u^l_\alpha v^\alpha_k = \delta^l_k$.

Now, the Segre family $(\mathcal{M}, 0)$ and $(\widetilde{\mathcal{M}}, 0)$ are equivalent if and only if there is a holomorphic map F from $\mathcal{E}_0$ to $\widetilde{\mathcal{E}}_0$, sending a certain point in the fiber over 0 to a certain point in the fiber of 0, such that

$$F^* \begin{pmatrix} \widetilde{\omega} \\ \widetilde{\omega}^\alpha \\ \widetilde{\omega}_\alpha \\ \widetilde{\phi} \end{pmatrix} = \gamma_F \begin{pmatrix} \omega \\ \omega^\alpha \\ \omega_\alpha \\ \phi \end{pmatrix}$$

with γ_F valued in G_1. Indeed, this assertion follows directly from the holomorphic version of Lemma 5.1.

Now, we consider the G_1-structure bundle $\mathcal{Y}$ over $\mathcal{E}_0$ and lift the above co-frames to globally defined forms over $\mathcal{Y}$. To be able to use the Cartan theorem, one needs to further complete these forms into a certain co-frame over $\mathcal{Y}$ so that the G_1-equivalence problem is to be reduced to an e-equivalence problem over $\mathcal{Y}$. This completion is done in the paper of Chern-Moser and Chern, which we state as follows:

Theorem 5.6 (Chern [Ch], Chern-Moser [CH]): Let $(M, 0)$ be a strongly pseudoconvex real analytic hypersurface at 0 with $(\mathcal{M}, 0)$ its Segre family. From the holomorphic forms $\omega, \omega^\alpha, \omega_\alpha, \phi$ in (5.11)-(5,12), after lifting them up to $\mathcal{Y}$ (which we still denote by the same letters), one can construct holomorphic 1-forms $\phi^\alpha_\beta, \phi^\alpha, \phi_\beta, \psi$ on $\mathcal{Y}$ such that

$$\Omega := \{\Omega^j, \ 1 \le j \le (n+2)^2 - 1\}$$
$$:= \{\omega, \omega^\alpha, \omega_\beta, \phi, \phi^\alpha_\beta, \phi^\alpha, \phi_\beta, \psi\} \text{ forms an e-structure on } \mathcal{Y}$$

and these 1-forms are uniquely determined by the following structure equations

$$d\omega = i\omega^\alpha \wedge \omega_\alpha + \omega \wedge \phi$$
$$d\omega^\alpha = \omega^\beta \wedge \phi^\alpha_\beta + \omega \wedge \phi^\alpha$$
$$d\omega_\alpha = \phi^\beta_\alpha \wedge \omega_\beta + \omega_\alpha \wedge \phi + \omega \wedge \phi_\alpha$$

$$d\phi = i\omega^\alpha \wedge \phi_\alpha + i\phi^\alpha \wedge \omega_\alpha + \omega \wedge \psi$$
$$d\phi_\alpha^\beta = \phi_\alpha^\gamma \wedge \phi_\gamma^\beta + i\omega_\alpha \wedge \phi^\beta - i\phi_\alpha \wedge \omega^\beta - i\delta_\alpha^\beta(\phi_\sigma \wedge \omega^\sigma) - \tfrac{1}{2}\delta_\alpha^\beta \psi \wedge \omega + \Phi_\alpha^\beta$$
$$d\phi^\alpha = \phi \wedge \phi^\alpha + \phi^\beta \wedge \phi_\beta^\alpha - \tfrac{1}{2}\psi \wedge \omega^\alpha + \Phi^\alpha$$
$$d\phi_\alpha = \phi_\alpha^\beta \wedge \phi_\beta - \tfrac{1}{2}\psi \wedge \omega_\alpha + \Phi_\alpha$$
$$d\psi = \phi \wedge \psi + 2i\phi^\alpha \wedge \phi_\alpha + \Psi$$

where $\Phi_\alpha^\beta = S_{\alpha\rho}^{\beta\sigma}\omega^\rho \wedge \omega_\sigma + R_{\alpha\gamma}^\beta \omega \wedge \omega^\gamma + T_\alpha^{\beta\gamma}\omega \wedge \omega_\gamma$

$\Phi^\alpha = T_\beta^{\alpha\gamma}\omega^\beta \wedge \omega_\gamma - \tfrac{i}{2}Q_\beta^\alpha \omega \wedge \omega^\beta + L^{\alpha\beta}\omega \wedge \omega_\beta$

$\Phi_\alpha = R_{\alpha\gamma}^\beta \omega^\gamma \wedge \omega_\beta + P_{\alpha\beta}\omega \wedge \omega^\beta - \tfrac{i}{2}Q_\alpha^\beta \omega \wedge \omega_\beta$

$\Psi = Q_\alpha^\beta \omega^\alpha \wedge \omega_\beta + H_\alpha \omega \wedge \omega^\alpha + K^\alpha \omega \wedge \omega_\alpha$

and $S_{\alpha\rho}^{\beta\sigma} = S_{\rho\alpha}^{\beta\sigma} = S_{\alpha\rho}^{\sigma\beta}$, $R_{\alpha\gamma}^\beta = R_{\gamma\alpha}^\beta$, $T_\beta^{\alpha\gamma} = T_\beta^{\gamma\alpha}$, $L^{\alpha\beta} = L^{\beta\alpha}$, $P_{\alpha\beta} = P_{\beta\alpha}$, $S_{\alpha\sigma}^{\beta\sigma} = R_{\alpha\beta}^\beta = T_\alpha^{\alpha\beta} = Q_\alpha^\alpha = 0$.

Remark 5.7 Since the Segre isomorphism does not induce the equivalence of the underlying hypersurfaces as demonstrated by Faran in [Fa], the existence of the e-equivalence map Ψ from $(\mathcal{Y}, 0)$ to $(\widetilde{\mathcal{Y}}, 0)$ does not induce automatically the biholomorphic equivalence of $(M, 0)$ with $(\widetilde{M}, 0)$. However, if an element $P \in \mathcal{Y}$ with a certain reality condition is mapped to $\widetilde{P}$ with a certain reality property, then we do have the holomorphic equivalence of $(M, 0)$ with $(\widetilde{M}, 0)$. We will briefly discuss this in the following subsection.

5.3 Cartan-Chern-Moser Theory for Germs of Strongly Pseudoconvex Hypersurfaces

The materials in §5.2 can be directly used to study the equivalence problem for strongly pseudoconvex (or Levi non-degenerate) hypersurfaces. Here we give a quick account on this matter. The reader is referred to [CM] for more details.

Let $(M, 0)$ be the germ of a smooth strongly pseudoconvex hypersurface, defined by $r = 0$. Here, we assume that $\frac{\partial r}{\partial z_n}(0) \neq 0$. As before, let $\theta = i\partial r$ and $\theta^\alpha = dz_\alpha$. We have a co-frame $\{\theta, \theta^\alpha, \overline{\theta^\alpha}\}$ on M. Let
(5.13)

$$G := \left\{ \begin{pmatrix} u & 0 & 0 \\ u^\alpha & u_\beta^\alpha & 0 \\ \overline{v^\alpha} & 0 & \overline{u_\beta^\alpha} \end{pmatrix} \middle| u \in \mathbf{R}, \ u_\beta^\alpha, u^\alpha \in \mathbf{C}, \ u > 0, \ det(u_\beta^\alpha) \neq 0, \right\}$$

be the connected linear subgroup of $G(2n - 1, \mathbf{C})$. $M \times G$ is a G-space. Similarly, let $(\widetilde{M}, 0)$ be another strongly pseudoconvex real hypersurface with a similar co-frame $\{\widetilde{\theta}, \widetilde{\theta}^\alpha, \overline{\widetilde{\theta}^\alpha}\}$.

It can be verified that there exists a smooth CR mapping $\Phi(z)$ such that $\Phi(M) \subset \widetilde{M}$ if and only if there is a C^∞ diffeomorphism $\Phi : M \to \widetilde{M}$ satisfying

$$(5.14) \qquad \Phi^* \begin{pmatrix} \widetilde{\theta} \\ \widetilde{\theta}^\alpha \\ \widetilde{\theta}_\alpha \end{pmatrix} = \begin{pmatrix} u & 0 & 0 \\ u^\alpha & u_\beta^\alpha & 0 \\ \overline{u^\alpha} & 0 & \overline{u_\beta^\alpha} \end{pmatrix} \begin{pmatrix} \theta \\ \theta^\alpha \\ \theta_\alpha \end{pmatrix} = (\gamma_\beta^\alpha) \begin{pmatrix} \theta \\ \theta^\alpha \\ \theta_\alpha \end{pmatrix}$$

where the $(2n-1) \times (2n-1)$ matrix (γ_β^α) defines a smooth mapping from M into G. By Lemma 5.1, there exists a CR isomorphism $\Phi : M \to \widetilde{M}$ if and only if there exists a smooth diffeomorphism $\Phi^1 : M \times G \to \widetilde{M} \times G$ such that

$$(5.15) \qquad \Phi^{1*}\widetilde{\omega} = \omega, \ \Phi^{1*}\widetilde{\omega}^\alpha = \omega^\alpha, \ \Phi^{1*}\overline{\widetilde{\omega}_\alpha} = \overline{\omega_\alpha},$$

where ω, ω^α are similarly defined as in Lemma 5.1.

Define

$$E = M \times \{\omega = u\theta : \omega = \overline{\omega}, \ u > 0\}$$

Choose $\theta^\alpha := u^\alpha \theta + u_\beta^\alpha dz^\beta$ for some smooth functions u^α, u_β^α so that $d\theta = i\theta^\alpha \wedge \overline{\theta^\alpha}, \ mod(\theta)$. We obtain a co-frame $(\omega, \theta^\alpha, \overline{\theta^\alpha}, \phi_0)$ on E, where $d\omega = iu\theta^\alpha \wedge \overline{\theta^\alpha} + \omega \wedge \phi_0$. Let G_1 be as before. Let $\widetilde{E}$ be the associated bundle over $\widetilde{M}$ with the corresponding co-frame $\{\widetilde{\omega}, \widetilde{\theta^\alpha}, \overline{\widetilde{\theta^\alpha}}, \widetilde{\phi}_0\}$.

Let $\Phi : M \to \widetilde{M}$ be a CR isomorphism. It is easy to verify that Φ induces a unique smooth diffeomorphism, still denoted as Φ, from E to $\widetilde{E}$ satisfying

$$\Phi^* \begin{pmatrix} \widetilde{\theta} \\ \widetilde{\theta^\alpha} \\ \widetilde{\theta}_\alpha \\ \widetilde{\phi} \end{pmatrix} = \begin{pmatrix} 1 & 0 & 0 & 0 \\ u^\alpha & u_\beta^\alpha & 0 & 0 \\ \overline{u^\alpha} & 0 & \overline{u_\beta^\alpha} & 0 \\ s & iu^\alpha u_\beta^\alpha & -i\overline{u^\alpha}\overline{u_\beta^\alpha} & 1 \end{pmatrix} \begin{pmatrix} \theta \\ \theta^\alpha \\ \theta_\alpha \\ \phi \end{pmatrix} = (\gamma_\beta^\alpha) \begin{pmatrix} \theta \\ \theta^\alpha \\ \theta_\alpha \\ \phi \end{pmatrix}$$

where the $(2n+2) \times (2n+2)$ matrix (γ_β^α) defines a smooth mapping from E into G_1^r, $\theta_\alpha = \overline{\theta^\alpha}$,etc. ($G_1^r$ consists of the matrices of the above form). By Lemma 5.1, the existence of a CR isomorphism $\Phi : M \to \widetilde{M}$ is equivalent to the existence of a smooth diffeomorphism $\Phi^1 : Y := E \times G_1^r \to \widetilde{Y} := \widetilde{E} \times G_1^r$ such that

$$\Phi^{1*}\widetilde{\omega} = \omega, \ \Phi^{1*}\widetilde{\omega}^\alpha = \omega^\alpha, \ \Phi^{1*}\widetilde{\omega}_\alpha = \omega_\alpha, \Phi^{1*}\widetilde{\phi} = \phi.$$

(Y, π, E) is called the *CR-structure bundle over* M.

The fundamental theorem proved by Cartan-Chern-Moser [CM] asserts that from $\omega, \omega^\alpha, \omega_\alpha, \phi$, one can construct 1-forms $\phi_\beta^\alpha, \phi^\alpha, \overline{\phi^\alpha}, \psi$ on Y, with $\omega = \overline{\omega}, \ \phi = \overline{\phi}, \ \psi = \overline{\psi}$, such that

$$\Omega := \{\Omega^j, \ 1 \le j \le (n+2)^2 - 1\} := \{\omega, \omega^\alpha, \overline{\omega^\alpha}, \phi, \phi_\beta^\alpha, \phi^\alpha, \overline{\phi^\alpha}, \psi\}$$

forms an e-structure on Y, and they are uniquely determined by certain structure equations. These structure equations are precisely the restriction of those in Theorem 5.6 from $\mathcal{Y}$ to Y, together with several other reality conditions (see [Theorem 5.5, pp 151, Fa] [BS]), if we assume that M is real analytic.

We let θ, θ_α be again as defined in §5.2. Since we have the embedding $M \to \mathcal{M}$, by mapping $z \to (z, \overline{z})$, we can regard the bundles E, Y as the subbundles of $\mathcal{E}, \mathcal{Y}$, respectively, as follows (cf.[Fa, (5.9)][BS]): Let

$$E^* := \{(z, \overline{z}, u\theta) \mid z \in M, \ , \overline{u\theta} = u\theta, (u\theta)(T) > 0\} \subset \mathcal{E},$$

over M. Here T is a certain real tangent vector field transversal to $T^{(1,0)}M +$ $T^{(0,1)}M$. On E^*, we see $\omega^* = \overline{\omega^*} := u\theta$. Let Y^* be the collection of the frames in $\mathcal{Y}$ restricted to E^* such that $\omega_\alpha^* = \overline{\omega^{*\alpha}}$, $\phi^* = \overline{\phi^*}$ over E^*. Since $\omega^* = \overline{\omega^*}$, $\omega_\alpha^* = \overline{\omega^{*\alpha}}$ and $\phi^* = \overline{\phi^*}$ hold on Y^*, one can check that the structure equations defining Ω^* over Y^* are the same ones defining Ω on Y. Hence, Y and Y^* are G_1^r-isomorphic. Identify E and Y with E^* and Y^*, respectively. Then the restriction of a function $g \in \Gamma(\Omega, \mathcal{Y})$ on Y equals to the lexicographically corresponding function $g|_Y \in \Gamma(\Omega|_Y, Y)$.

Finally, we mention that the reality condition mentioned in Remark 5.7 is precisely the condition that $P, \widetilde{P}$ are in Y^* or $\widetilde{Y}^*$, respectively. This explains the statement in Remark 3.7.

References

[Alx]   H. Alexander, Proper holomorphic maps in $\mathbf{C}^n$, *Ind. Univ. Math. Journal* (26), 137–146, 1977.

[BER1] S. Baouendi, P. Ebenfelt and L. Rothschild, Real submanifolds in complex spaces and their mappings, Princeton Mathematical Series (47), Princeton, New Jersey, 1999

[BER2] M.S. Baouendi, P. Ebenfelt and L.P. Rothschild, Local geometric properties of real submanifolds in complex spaces, Bull. AMS, 37(2000), 309-336.

[BJT]   S. Baouendi, H. Jacobowitz and F. Treves, On the analyticity of CR mappings, *Ann. of Math.*(122), 365-400,1985

[BRZ]   S. Baouendi, L. Rothschild and D. Zaitsev, Equivalences of real submanifolds in complex space, *Jour. Diff. Geom.* 59 (2001), no. 2, 301–351.

[BG]   E. Bedford and B. Gaveau, Envelopes of holomorphy of certain 2-spheres in $\mathbf{C}^2$, *Amer. J. Math.* (105), 975-1009, 1983.

[EK]   E. Bedford and W. Klingenberg, On the envelopes of holomorphy of a 2-sphere in $\mathbf{C}^2$, *Journal of AMS*(4), 623-655, 1991.

[Bis]   E. Bishop, Differentiable manifolds in complex Euclidean space, *Duke Math. J.* (32), 1-21, 1965.

[BS1]   D. Burns and S. Shnider, Projective connections in CR geometry, Manuscripta Math., 33(1980), 1-26.

[BSW]   D. Burns, S. Shnider and R. Wells, On deformations of strictly pseudoconvex domains, *Invent. Math.* (46), 237-253, 1978.

[Ch]   S.-S. Chern, On the projective structure of a real hypersurface in C_{n+1}, Math. Scand. 36(1975), 74-82.

[CJ]   S.-S. Chern and S. Ji, Projective geometry and Riemann's mapping problem, Math Ann. 302(1995), 581-600.

[CM]   S. S. Chern and J. K. Moser, Real hypersurfaces in complex manifolds *Acta Math.* (133), 219-271, 1974

[DF]   K. Diederich and E. Fornaess, Pseudoconvex domains with real analytic boundaries, *Ann. Math.* (107), 371-384, 1978.

[Da]   J. P. D'Angelo, Several Complex Variables and the Geometry of Real Hypersurfaces, CRC Press, Boca Raton, 1993

[EHZ1] P. Ebenfelt, X. Huang and D. Zaitsev, The equivalence problem and rigidity for hypersurfaces embedded into hyperquadrics, *Amer. Jour. of Math.*, 2002, to appear.

[EHZ2] P. Ebenfelt, X. Huang and D. Zaitsev, Rigidity of CR-immersions into spheres, preprint, 2002.

[Ga]   R. Gardner, The method of equivalence and its applications, CBMS-NSF, regional conference series in applied mathematics, 1989.

[Gon]   X. Gong, On the convergence of normalizations of real analytic surfaces near hyperbolic complex tangents, *Comment. Math. Helv.* 69 (1994), no. 4, 549–574.

[Fa]   J. Faran, Segre families and real hypersurfaces, *Invent. Math.* 60, 135-172, 1980.

[Fe]   C. Fefferman, The Bergman kernel and biholomorphic mappings of pseudoconvex domains, *Invent. Math.* 26, 1-65, 1974.

[GK]   R. Greene and S. Krantz, Deformation of complex structures, estimates for $\bar{\partial}$-equation, and stability of the Bergman kernel, *Advanced in Math.* (43), 1-86, 1982.

[HS]   Zheng-Xu He and O. Schramm, Fixed points, Koebe uniformization and circle packings, *Ann. of Math.* (137),369-406, 1993.

[Ho]   L. Hörmander, Introduction to Complex Analysis in Several Variables, North Holland, Amsterdam, 1973.

[Hu1]   X. Huang, On some problems in several complex variables and Cauchy-Riemann Geometry, Proceedings of ICCM (edited by L. Yang and S. T. Yau), AMS/IP Stud. Adv. Math. 20, 383-396, 2001.

[Hu2]   X. Huang, On a linearity problem of proper holomorphic mappings between balls in complex spaces of different dimensions, *Jour of Diff. Geom.* (51), 13-33, 1999

[Hu3]   X. Huang, On an n-manifold in $\mathbf{C}^n$ near an elliptic complex tangent *J. Amer. Math. Soc.* (11), 669–692, 1998.

[Hu4]   X. Huang, Schwarz reflection principle in complex spaces of dimension two, *Comm. in PDE* (21), 1781-1828, 1996.

[HJ1]   X. Huang and S. Ji, Global extension of local holomorphic maps and the Riemann mapping theorem for algebraic domains, *Math. Res. Lett.* (5), 247–260, 1998.

[HJ2]   X. Huang and S. Ji, Cartan-Chern-Moser theory on real algebraic hypersurfaces and applications, *Annales. de L'Inst. Fourier* (52), 1793-1831, 2002.

[HJY]   X. Huang, S. Ji, and S.S.T. Yau, An example of real analytic strongly pseudoconvex hypersurface which is not holomorphically equivalent to any algebraic hypersurfaces, *Ark. Mat.*, 39(2001), 75-93.

[HK]   X. Huang and S. Krantz, On a problem of Moser *Duke Math. J.* (78), 213-228, 1995.

[Ja]   H. Jacobowitz, An introduction to CR structures, Mathematical Surveys and Monographs (32), American Mathematical Society, Providence, RI, 1990

[KW1]   C. Kenig and S. Webster, The local hull of holomorphy of a surface in the space of two complex variables, *Invent. Math.* (67), 1-21, 1982.

[KW2]   C. Kenig and S. Webster, On the hull of holomorphy of an n-manifold in $\mathbf{C}^n$, *Annali Scuola Norm. Sup. de Pisa IV* (11), 261-280, 1984

[Kr]   S. Krantz, Function Theory of Several Complex Variables, John Wiley & Sons, New York, 1982.

[Le]   H. Lewy, On the boundary behavior of holomorphic mappings, *Acad. Zaz. Lincei.* (35), 1-8, 977.

[Mos]   J. Moser, Analytic surfaces in $\mathbf{C}^2$ and their local hull of holomorphy, *Annales Aca -demiæFennicae Series A.I. Mathematica* (10), 397-410, 1985.

[MW]   J. Moser and S. Webster, Normal forms for real surfaces in $\mathbf{C}^2$ near complex tangents and hyperbolic surface transformations, *Acta Math.* (150), 255-296, 1983.

[Pi]   S. Pinchuk, On the proper holomorphic mappings of real analytic pseudoconvex domains, *Siberian Math.* (15), 909-917, 1974.

[SM]   C. L. Siegel and J. Moser, Lectures on Celestial Mechanics, Classics in Mathematics, Springer-verlag, 1995

[Vit]   A.G. Vitushkin, Holomorphic mappings and geometry of hypersurfaces, Encyclopaedia of Mathematical Sciences, Vol. 7, Several Complex Variables I, Springer-Verlag, Berlin, 1985, 159-214.

[We1]   S. Webster, The rigidity of C-R hypersurfaces in a sphere. *Indiana Univ. Math. J.*(28), 405–416, 1979.

[We2]   S.M. Webster, On the mapping problem for algebraic real hypersurfaces, *Invent. Math.* (43), 53-68, 1977.

Introduction to a General Theory of Boundary Values

Jean-Pierre Rosay *

Department of Mathematics, University of Wisconsin
Madison WI 53705, U.S.A.
jrosay@math.wisc.edu

These notes are entirely based on my close collaboration with E. L. Stout. Their goal is to invite and to help the reader to read the papers [RS1] [RS2] [RS3] (later referred to collectively as [RS]), in which E. L. Stout and I have tried to develop a general theory of boundary values. Our work led us to develop certain points in the theory of analytic functionals, for which the seminal work is the beautiful paper of Martineau [M].

There is only one point where some originality can be claimed in these notes. The complete proof of the non-linear Paley-Wiener Theorem in section V is substantially different from our original proof in [RS1]. It has the advantage to connect much faster with the classical case of convex carriers. But I personally find it far less enlightening. Since there would be no point in

repeating here the proofs and arguments given in [RS], most proofs are just sketched. Although they can be found in [RS] the reader should rather try to complete them by him/herself.

Instead of discussing general situations, I discuss here special cases such as the theory on the unit circle where Fourier series allow very simple computations. Also, I show what were some preliminary steps in our work (not to be found in [RS]) that may prepare for an easier reading of [RS]. I wish to

draw the attention to the argument that uses subharmonicity at the end of section 2 (proof of Proposition 1). It is a typical argument for which in these notes details are provided only once. This kind of argument is the true basis for our definition of strong boundary value.

* Partly supported by NSF grant.

Contents

1 Introduction – Basic Definitions

1.1 What Should a General Notion of Boundary Value Be?

Let Ω be a bounded domain in $\mathbb{R}^N$, with real analytic boundary $b\Omega$. Let $P(x, D)$ be a linear partial differential operator defined near $b\Omega$ with real analytic coefficients. Assume that $b\Omega$ is noncharacteristic (it is always the case if the operator is elliptic). Finally let u be a solution to $P(x, D)u = 0$ defined on Ω, near $b\Omega$. There is absolutely no restriction made on the growth of u when approaching $b\Omega$. For simplicity we will always assume that u is continuous but it is not essential. Then u has a boundary value along $b\Omega$. The boundary value is an analytic functional. It can be defined in the following way. Let ρ be a real analytic defining function for $b\Omega$ ($\rho < 0$ on Ω and $\nabla\rho \neq 0$ on $b\Omega$). For any "test function" φ real analytic near $b\Omega$, for $\epsilon < 0$ $|\epsilon|$ small enough set

$$\lambda_\epsilon(\varphi) = \int_{\{\rho=\epsilon\}} u\varphi d\sigma_\epsilon$$

where $d\sigma_\epsilon$ is (say) area measure on $\{\rho = \epsilon\}$. Then $\epsilon \mapsto \lambda_\epsilon(\varphi)$ extends holomorphically to a neighborhood of 0 in $\mathbb{C}$. The functional $\varphi \mapsto \lambda_0(\varphi) = \lim_{\epsilon\to 0^-} \lambda_\epsilon(\varphi)$ is the boundary value of u.

This is what we consider to be the classical theory (going back at least to [LM]), see the discussion at the end of section 9.5 in [H]. The fact that not only is there a limit as $\epsilon \to 0$, but that there is a holomorphic extension has been less noticed but it is absolutely crucial.

Now the general question that we raise is "when can one reasonably define the boundary value of a function?"

Taking Ω and u as above, consider now the function $u^\# = uh$, where h is a real analytic function defined near $b\Omega$. Then obviously $\lambda_\epsilon^\#(\varphi) = \int_{\{\rho=\epsilon\}} u^\# \varphi d\sigma_\epsilon$ extends holomorphically at 0, as a function of ϵ. Indeed $\lambda_\epsilon^\#(\varphi) = \lambda_\epsilon(h\varphi)$.

It is clear that a satisfactory theory of boundary value should include the case of functions such as the function $u^\#$ that has been just considered. But of which PDE would $u^\#$ be a solution? In particular, but not only for that reason we felt the need of going beyond the classical theory.

In the discussion that precedes, whether the choice of ρ does matter is an unpleasant but unavoidable question.

1.2 Definition of Strong Boundary Value (Global Case)

Let Ω be as before (a bounded domain in $\mathbb{R}^N$ with real analytic boundary) and u be a function defined on Ω near $b\Omega$. Let ρ be a real analytic defining function for $b\Omega$, as in (1).

Definition. *u has a strong boundary value along $b\Omega$ if and only if for every real analytic function φ defined near $b\Omega$ the map*

$$\epsilon \longmapsto \int_{\rho=\epsilon} u\varphi d\sigma_\epsilon \quad (\text{ defined for } \epsilon < 0, |\epsilon| \text{ small})$$

extends holomorphically to a neighborhood of 0 *in* $\mathbb{C}$.

It is true but not obvious that this holomorphic extendibility does not depend on the choice of ρ.

It is a consequence of the Cauchy Kovalevsky theorem that solutions of PDE have strong boundary values along noncharacteristic boundaries. The proof follows the line of Proof 2 in 2.

1.3 Remarks on Smooth (Not Real Analytic) Boundaries

Instead of having Ω with real analytic boundaries, consider the case of domains with $\mathcal{C}^\infty$ boundaries. There is an easy theory of boundary values for solutions of PDE with smooth coefficients along noncharacteristic boundaries. But one has then to restrict to the case of functions with "polynomial growth" $\left(|u(x)| \leq \frac{C}{\text{dist}(x,b\Omega)^k}\right)$. Then, for every smooth test function φ, the map $\epsilon \to \int_{\rho=\epsilon} u\varphi d\sigma_\epsilon$ extends smoothly at $\epsilon = 0$. The limit defines a distribution that is the boundary value of u. However we have no idea of what could a general theory be. It would be tempting to give a definition similar to the one above, replacing holomorphic extendibility at 0 by smoothness at 0. But this appears to be clearly not satisfactory, since it then depends on the choice of the defining function ρ (see [RS3]).

1.4 Analytic Functionals

We will always denote by $\mathcal{O}(\mathbb{C}^N)$ the space of (entire) holomorphic functions on $\mathbb{C}^N$. Let K be a compact set in $\mathbb{C}^N$. An *analytic functional* ψ *carried by* K is a linear form on $\mathcal{O}(\mathbb{C}^N)$ such that for every neighborhood V of K there exists $C_V > 0$ such that for every $f \in \mathcal{O}(\mathbb{C}^N)$

$$|\psi(f)| \leq C_V \sup_V |f|.$$

K is called a carrier, ψ is called an analytic functional (on $\mathbb{C}^N$).

Example. In $\mathbb{C}$, $f \mapsto \sum_{n=0}^\infty \frac{f^{(n)}(0)}{n!} a_n^n$ defines an analytic functional if and only if $|a_n|$ is a bounded sequence. This functional is carried by $\{0\}$ if and only if $a_n \to 0$ as $n \to \infty$.

1.5 Analytic Functional as Boundary Values

The functional λ_0 in (1) defines an analytic functional on $\mathbb{C}^N$, carried by $b\Omega \subset \mathbb{R}^N \subset \mathbb{C}^N$. Resume the notations of (1). If $\varphi \equiv 0$ on $b\Omega$ we can write $\varphi = \rho\tilde{\varphi}$, and we get

$$\lambda_\epsilon(\varphi) = \int_{\rho=\epsilon} u\rho\tilde{\varphi}d\sigma_\epsilon = \epsilon \int u\tilde{\varphi}d\sigma_\epsilon.$$

It immediately follows that $\lambda_0(\varphi) = 0$. Hence $\lambda_0(\varphi)$ depends only on the restriction of φ to the boundary of Ω. It would therefore be better to consider φ as an analytic functional not on $\mathbb{C}^N$, but on some complexification of the real analytic manifold $b\Omega$ (so a complex manifold of dimension $N - 1$). For simplicity we will not get in that discussion in these notes.

1.6 Some Basic Properties of Analytic Functionals

Carriers – Martineau's Theorem

Carriers are of course not unique and the intersection of two carriers is not a carrier in general. Indeed in $\mathbb{C}$, $f \mapsto f(0)$ is an analytic functional carried by any curve around 0. In $\mathbb{C}^N (N > 1)$ even the intersection of two convex carriers may fail to be a carrier.

Consider in $\mathbb{C}^2$ the analytic functional ψ defined by

$$\psi(f) = \int_{|z|=2} f(z, \frac{1}{z})dz = \int_{|z|=1/2} f(z, \frac{1}{z})dz.$$

ψ is carried by $\{|z| \leq 2\} \times \{w \leq \frac{1}{2}\}$ and also by $\{|z| \leq \frac{1}{2}\} \times \{|w| \leq 2\}$, but is not carried by $\{|z| \leq \frac{1}{2}\} \times \{|w| \leq \frac{1}{2}\}$. Indeed there is a sequence of polynomial that converges uniformly to $\frac{1}{z}$ on the circle $|z| = 2$ $w = \frac{1}{z}$ and to 0 on the polydisc $|z| \leq \frac{1}{2}, |w| \leq \frac{1}{2}$.

The following positive result is very useful and is due to Martineau.

Remind that a compact set $K \subset \mathbb{C}^N$ is said to be *polynomially convex* if and only if for every $z \in \mathbb{C}^N \setminus K$ there exists $f \in \mathcal{O}(\mathbb{C}^N)$ (or a polynomial f) such that $|f(z)| > \operatorname{Sup}_K |f|$.

Theorem 1 (Martineau). *Let ψ be an analytic functional on $\mathbb{C}^N$. Let K_1 and K_2 be compact sets in $\mathbb{C}^N$. If ψ is carried by K_1 and by K_2 and if $K_1 \cup K_2$ is polynomially convex, then ψ is carried by $K_1 \cap K_2$.*

Sketch of Proof. Since $K_1 \cup K_2$ is Runge, ψ extends by continuity to a linear functional on the space $\mathcal{O}(K_1 \cup K_2)$ of germs of holomorphic functions defined on a neighborhood of $K_1 \cup K_2$, for which (in the appropriate sense) K_1 and K_2 both are carriers (see 1.6 below). Given $f \in \mathcal{O}(K_1 \cup K_2)$, if f is small near $K_1 \cap K_2$ f can be decomposed into a sum $f = f_1 + f_2$ with $f_j \in \mathcal{O}(K_1 \cup K_2)$, f_j small near K_j. This is the standard Cousin problem. When completing the proof the reader should pay attention to the fact that polynomial convexity and not only the existence of pseudoconvex neighborhoods is to be used!

A special case, of great importance, where Martineau's theorem applies is when K_1 and $K_2 \subset \mathbb{R}^N$. For that case there is an interesting, very different approach in [H] (Theorem 9.1.6).

Local Analytic Functionals

Local analytic functionals were introduced by Martineau. A *local analytic functional carried by a compact set* $K \subset \mathbb{C}^N$ is a linear functional on $\mathcal{O}(K)$, the space of germs of holomorphic functions defined on a neighborhood of K (same notation as in 1.6, with the following continuity property: for every neighborhood V of K there exists a constant $C_V > 0$ such that

$$|\psi(f)| \leq C_V \mathrm{Sup}_V |f|$$

for every $f \in \mathcal{O}(V)$ (the space of holomorphic functions on V).

Every local analytic functional defines an analytic functional. If K is a polynomially convex set, any analytic functional ψ_0 carried by K extends to a local analytic functional ψ carried by K. It is rather straightforward by density. For $f \in \mathcal{O}(K)$, set $\psi(f) = \lim \psi_0(P_j)$, where (P_j) is any sequence of polynomials converging uniformly to f on some neighborhood of K. But, when checking the continuity condition for local analytic functionals, the reader will have to use the following fact that is an easy consequence of the Oka-Weil theorem. Given a neighborhood V of K, there exists a neighborhood W of K such that for any $f \in \mathcal{O}(V)$ there exists a sequence of polynomials converging uniformly to f on W. An important point is that W does not depend on f and that kind of property will play a crucial role later (see 3).

If K is not polynomially convex the extension of any analytic functional on $\mathbb{C}^N$ carried by K to a local analytic functional is still possible but it is not an elementary result and it will be discussed in 3.

1.7 Hyperfunctions

The Notion of Functional (Analytic Functional or Distribution, etc.) Carried by a Set, Defined Modulo Similar Functionals Carried by the Boundary of that Set

The notion is extremely natural and it is good to notice that it already appears in elementary distribution theory.

Let ψ be the distribution on $\mathbb{R}$ defined by

$$\psi(\varphi) = vp \int_{-1}^{+1} \frac{\varphi(x)}{x} dx = \lim_{\epsilon \to 0} \int_{-1}^{-\epsilon} + \int_{\epsilon}^{1} \frac{\varphi(x)}{x} dx,$$

for $\varphi \in \mathcal{C}_0^\infty(\mathbb{R})$.

We want to write ψ as a sum $\psi = \psi_1 + \psi_2$ with ψ_1, resp ψ_2, a distribution carried by $[0, \infty)$, resp. $(-\infty, 0]$. Of course one cannot take $\psi(\varphi) = \int_0^1 \frac{\varphi(x)}{x} dx$ which does not make sense if $\varphi(0) \neq 0$. One can instead choose $\psi^+(\varphi) = \int_0^1 \frac{\varphi(x)-\varphi(0)}{x} dx$ and $\psi^- = \psi - \psi^+$. But there are other choices. We could as well take $\psi^+(\varphi) = \int_0^1 \frac{\varphi(x)-(\varphi(0)+x\varphi'(0))}{x} dx$ and $\psi^- = \psi - \psi^+$, although in this example one may find the second choice less "natural".

Both choices differ only by a distribution carried by $\{0\}$ ($\varphi \mapsto \varphi'(0)$), i.e. by a distribution carried by the boundary of $[0, \infty)$. The problem of restricting distributions or analytic functionals to subsets unavoidably leads to the notion of functional carried by a set defined modulo functional carried by its boundary.

The problem of extension also leads to this notion. For example consider the linear form $\varphi \mapsto T(\varphi) = \int_0^\infty \frac{\varphi(x)}{x^4} dx$ defined for $\varphi \in \mathcal{C}_0^\infty((0, +\infty))$. This is the distribution on $(0, +\infty)$ defined by the function $\frac{1}{x^4}$. Now, we want to extend it to a distribution on $\mathbb{R}$, i.e. we wish to define $T(\varphi)$ for $\varphi \in \mathcal{C}_0^\infty(\mathbb{R})$.

We can do it by setting

$$T(\varphi) = \int_0^a \frac{\varphi(x) - \sum_{n=0}^K \frac{\varphi^{(n)}(0)}{n!} x^n}{x^4} dx + \int_a^\infty \frac{\varphi(x)}{x^4} dx,$$

and $K \geq 3$ for some $a > 0$. Again there are some arbitrary choices (a and K), corresponding to distributions carried by $\{0\}$.

A similar construction will allow to extend the distribution on $(0, \infty)$ defined by any continuous function u on $(0, +\infty)$ with polynomial growth at 0 ($|u(x)| \leq \frac{C}{x^p}$, for some p and for $x \in (0, 1]$) to a distribution on $\mathbb{R}$.

For simplicity let us now assume that $u(x) = 0$ if $x > 1$. If one no longer has any restriction on the growth of u when approaching 0 one can still let u act on test functions which now are going to be the restriction to $\mathbb{R}$ of entire functions on $\mathbb{C}$ by setting for every $f \in \mathcal{O}(\mathbb{C})$

$$T(f) = \int_0^1 u(x)[\varphi(x) - \sum_{n=0}^{K(x)} \frac{\varphi^{(n)}(0)}{n!} x^n] dx.$$

Taking $K(x) \equiv +\infty$ would be of no interest ($T = 0!$). But one takes $K(x)$ locally finite on $(0, 1]$. Think of $K(x)$ tending fast enough to $+\infty$ as x approaches 0 so that

$$|\varphi(x) - \sum_{n=0}^{K(x)} \frac{\varphi^{(n)}(0)}{n!} x^n| \leq \frac{1}{|u(x)|}.$$

The object just constructed (T) is an analytic functional, not intrinsically defined, but defined only modulo analytic functionals carried by $\{0\}$ - i.e. a hyperfunction.

The above extension can be seen as an extremely naive introduction to flabbiness.

Hyperfunctions

There are two schools of thought about hyperfunctions and one should never forget that the groundbreaking work has been done by Sato and his followers.

A less algebraic approach that we follow was given by Martineau. Again *in these notes* we will consider only the case of $\mathbb{R}^N$ in $\mathbb{C}^N$, and we will not go beyond the following.

Let K be a compact set in $\mathbb{R}^N$, a hyperfunction on K will be for us an analytic functional on $\mathbb{C}^N$, carried by K, defined modulo analytic functionals carried by the boundary of K (in $\mathbb{R}^N$).

7.3 Hyperfunctions as boundary values.

Here we shall not try to explain the general theory, we just wish to indicate how it is already clear that hyperfunctions come up as boundary values. At this time we will therefore partly keep the previous global setting.

Let u be a holomorphic function defined on the unit disc. It has a boundary value which is an analytic functional ψ, on the unit circle. If we intend to speak about the boundary value along an arc instead of along the unit circle, we have to restrict ψ to that arc. As discussed already in the case of distribution, that will lead to a analytic functional carried by the arc, defined modulo analytic functional carried by its boundary (the two endpoints).

1.8 Limits

From the previous paragraph, it is clear that the question of which set carries a given analytic functional is a crucial one. But serious difficulties arise when taking limits, which is unavoidable when considering boundary values. The following trivial example is enough to make that point. In $\mathbb{C}$ consider the analytic functional ψ_K defined by:

$$\psi_K(\varphi) = \sum_{m=0}^{K} \frac{\varphi^{(n)}(0)}{n!}, \quad \text{for} \quad \varphi \in \mathcal{O}(\mathbb{C}).$$

Then ψ_K is carried by $\{0\}$. For any $\varphi \in \mathcal{O}(\mathbb{C})$ $\psi_K(\varphi) \to \varphi(1)$ as $K \to \infty$. But the limit functional $\varphi \to \varphi(1)$ is carried by $\{1\}$, not by $\{0\}$.

The difficulty of controlling carriers when taking limits should not be under estimated. Overcoming it required us to introduce the notion of strong boundary value and to develop a nonlinear Paley Wiener Theory.

2 Theory of Boundary Values on the Unit Disc

Using polar coordinates we switch to the setting of periodic functions. So we shall consider continuous functions $(t, \theta) \mapsto u(t, \theta)$ defined for $t < 0$ t close to 0, and $\theta \in \mathbb{R}/2\pi\mathbb{Z}$.

2.1 Functions $u(t, \theta)$ That Have Strong Boundary Values (Along $t = 0$)

These are the functions such that for every real analytic function $\varphi(t, \theta)$, defined for $t \in \mathbb{R}$ $|t|$ small and $\theta \in \mathbb{R}/2\pi\mathbb{Z}$, the function

$$t \mapsto I(t) = \int_0^{2\pi} u(t, \theta)\varphi(t, \theta)d\theta,$$

that is defined for $t < 0$ $|t|$ small, extends holomorphically to a neighborhood of 0 in $\mathbb{C}$.

Write $u(t, \theta) = \sum_{n=-\infty}^{+\infty} a_n(t)e^{in\theta}$.

Exercise. u has strong boundary value if and only if

(i) There exists $\rho > 0$ such that all the Fourier coefficients $a_n(t)$ extend holomorphically to $\{t \in \mathbb{C}, |t| < \rho\}$, and
(ii) for every $\epsilon > 0$ there exist $C_\epsilon > 0$ and $\rho_\epsilon > 0 (\rho_\epsilon < \rho)$ such that for any $t \in \mathbb{C}$, satisfying $|t| < \rho_\epsilon$, $|a_n(t)| \leq C_\epsilon e^{\epsilon|n|}$.

If we also decompose φ in Fourier series $\varphi(t, \theta) = \sum \varphi_n(t)e^{in\theta}$. The real analyticity of φ corresponds to some exponential decay of the Fourier coefficients, which is exactly what is needed together with (ii) for evaluating

$$\int_0^{2\pi} u(t, \theta)\varphi(t, \theta)d\theta = 2\pi \sum_{-\infty}^{+\infty} a_n(t)\varphi_{-n}(t).$$

2.2 Boundary Values of Holomorphic Functions on the Unit Disc

Let v be a holomorphic function on the unit disc in $\mathbb{C}$, and set

$$u(t, \theta) = v(e^{t+i\theta}) \qquad (t < 0).$$

We claim that u has strong boundary value along $t = 0$.

We shall present two proofs. The first one is immediate by Fourier series. The second one is given because it generalizes easily to the general case of solutions of PDE.

Proof 1. Write $v(z) = \sum_{n=0}^{\infty} v_n z^n$. Since the radius of convergence of the series is at least 1, for every $\epsilon > 0$ we have an estimate

$$|v_n| \leq C_\epsilon e^{\frac{\epsilon n}{2}}.$$

We have $u(t, \theta) = \sum_{n=0}^{\infty} a_n(t)e^{in\theta}$, with $a_n(t) = v_n e^{nt}$. Therefore for $|t| \leq \frac{\epsilon}{2}$, $|a_n(t)| \leq C_\epsilon e^{\epsilon|n|}$, as desired.

Proof 2. We have to study the integral

$$I(t) = \int_0^{2\pi} u(t, \theta)\varphi(t, \theta)d\theta$$

defined for small negative t, when φ is real analytic (and 2π periodic in θ).

For $t < 0$ ($|t|$ small) let $\varphi_t(s, \theta)$ be a function such that

(1) $\varphi_t(t, \theta) = \varphi(t, \theta)$
(2) φ_t is holomorphic as a function of $s + i\theta$.

So φ_t is obtained by restriction of $(s, \theta) \to \varphi(s, \theta)$ to the line $s = t$ and then by holomorphic extension (in a general situation the Cauchy Kovalevsky Theorem will be used). Provided that $|t|$ is small enough φ_t will be defined in a strip $-2\tau_0 < s < 2\tau_0$, for some fixed τ_0. Using path independence:

$$I(t) = \int_0^{2\pi} u(t, \theta)\varphi_t(t, \theta)d\theta = \int_0^{2\pi} u(\tau_0, \theta)\varphi_t(\tau_0, \theta)d\theta.$$

The dependence on t is now easy to discuss by using the right hand side in the last formula.

2.3 Independence on the Defining Function

In the definition of boundary values (1 above) we considered integrals $I(t) = \int_0^{2\pi} u(t, \theta)\varphi(t, \theta)d\theta$ and we let t tend to 0.

So in some sense we used the approximation of the boundary $t = 0$ by the level sets of the "defining function" t, and by integrating on those.

What happens if we approximate $t = 0$ by other curves? More specifically let $R(t, \theta)$ be a real analytic function defined by t close to 0 and $\theta \in \mathbb{R}/2\pi\mathbb{Z}$ Assume that

$$R(0, \theta) \equiv 0 \quad \text{and} \frac{\partial R}{\partial t}(0, \theta) > 0.$$

Replace the defining function t by the defining function R, i.e. instead of $I(t)$ consider

$$J(t) = \int_0^{2\pi} u(R(t, \theta), \theta)\varphi(R(t, \theta), \theta)d\theta.$$

The question is whether, assuming that u has strong boundary value, J will extend holomorphically at $t = 0$. The question may be unpleasant but it is impossible to avoid it, especially if one has in mind to work later on manifolds.

The answer is positive.

Claim. J also extends holomorphically at $t = 0$.

For small negative t we have

$$u(t, \theta) = \sum_{n=-\infty}^{+\infty} a_n(t)e^{in\theta}.$$

Then $J(t) = \sum_{n=-\infty}^{+\infty} J_n(t)$ with

$$J_n(t) = \int_0^{2\pi} a_n(R(t,\theta))\varphi(R(t,\theta),\theta)e^{in\theta}\,d\theta.$$

Since u has strong boundary value, it is clear that the functions J_n extend holomorphically to a fixed neighborhood of 0 in t. We simply have to get estimates in order that the series $\sum J_n(t)$ be convergent.

In order to estimate $J_n(t)$ integrate k times by parts:

$$J_n(t) = (-1)^k \int_0^{2\pi} \frac{\partial^k}{\partial\theta^k}[a_n(R(t,\theta))\varphi(R(t,\theta),\theta)]\frac{e^{in\theta}}{(in)^k}\,d\theta.$$

For $|t|$ small $\theta \mapsto a_n(R(t,\theta))\varphi(R(t,\theta),\theta)$ defines a holomorphic function defined for $\theta \in \mathbb{C}, |\mathrm{Im}\,\theta| \leq \alpha$ (for some $\alpha > 0$). Shrinking α, $\varphi(R(t,\theta),\theta)$ will be bounded. For $a_n(R(t,\theta))$ we have good bounds $|a_n| \leq Ce^{\epsilon|n|}$ provided that $R(t,\theta)$ is small enough depending on ϵ, say $|R(t,\theta)| \leq \rho_\epsilon$, with $\rho_\epsilon > 0$.

But notice that for any θ such that $|\mathrm{Im}\,\theta| < \alpha$ we will have $|R(t,\theta)| \leq \rho_\epsilon$ provided that $|t|$ is small enough, since $R(0,\theta) \equiv 0$. Consequently, and this is a crucial point, provided that $|t|$ is small enough, depending on ϵ, we have

$$|a_n(R(t,\theta))\varphi(R(t,\theta),\theta)| \leq Ce^{\epsilon|n|}$$

for all $\theta \in \mathbb{C}$, with $|\mathrm{Im}\,\theta| \leq \alpha$ (α *not depending on* ϵ). We can now use the Cauchy estimates for $\frac{\partial^k}{\partial\theta^k}$ and for $|t|$ small

$$|J_n(t)| \leq C\frac{k!\alpha^{-k}e^{\epsilon|n|}}{n^k}.$$

By Stirling formula ($k! \simeq k^k e^{-k}\sqrt{2\pi k}$), and changing C

$$|J_n(t)| \leq C\frac{(\frac{k}{\alpha})^k e^{-k}\sqrt{k}e^{\epsilon|n|}}{n^k}$$

It is time to choose how many times one wishes to integrate by parts, take $k \simeq \alpha|n|$ (k an integer). Then one gets

$$|J_n(t)| \leq Ce^{-k}\sqrt{k}e^{\epsilon|n|}.$$

That gives us exponential decay of $J_n(t)$ provided that we took $\epsilon < \alpha$. This establishes the convergence of the series $\sum J_n(t)$ for t in a neighborhood of 0 in $\mathbb{C}$.

2.4 The Role of Subharmonicity (Illustrated Here by Discussing the Independence on the Space of Test Functions)

Let $u(t,\theta)$ be a continuous function defined for small negative t and $\theta \in \mathbb{R}/2\pi\mathbb{Z}$.

Proposition 1. *The following are equivalent*

(1) u has strong boundary value (along $t = 0$)
(2) For every holomorphic entire function $(z, w) \mapsto \Phi(z, w)$

defined on $\mathbb{C}^2$ and 2π periodic in w the function

$$I_\Phi(t) = \int_0^{2\pi} u(t, \theta)\Phi(t, \theta)d\theta$$

(defined for small negative t) extends holomorphically at 0.

Of course (1) $\Rightarrow$ (2). The point in (2) $\Rightarrow$ (1) is that we do not need to test on all real analytic functions defined on a neighborhood of $\{0\} \times \mathbb{R}/2\pi\mathbb{Z}$, it is enough to test on the global holomorphic functions (here on $\mathbb{C} \times (\mathbb{R}/2\pi\mathbb{Z} + i\mathbb{R})$).

Sketch of proof. Assume (2). Set $\Omega = \mathbb{C} \times (\mathbb{R}/2\pi\mathbb{Z} + i\mathbb{R})$. So the hypothesis is that for every $\Phi \in \mathcal{O}(\Omega)$ I_Φ extends holomorphically at $t = 0$. By a Baire category argument, all the I_Φ's extend holomorphically to a same neighborhood of 0 in $\mathbb{C}$.

Together with the Banach Steinhaus theorem, it implies that there exist a compact set $K \subset \Omega$, $\alpha > 0$ and $C > 0$ such that for every $\Phi \in \mathcal{O}(\Omega)$:

$$|I_\Phi(t)| \leq C \mathrm{Sup}_K |\Phi|, \quad \text{for} \quad t \in \mathbb{C}, |t| \leq \alpha.$$

Again for $t < 0$ $|t|$ small, write

$$u(t, \theta) = \sum_{n=\infty}^{+\infty} a_n(t)e^{in\theta}.$$

Applying the above inequality to $\Phi(t, \theta) = e^{-in\theta}$, one sees that each Fourier coefficient $a_u(t)$ extends holomorphically as a function of t to $\{|t| < \alpha\}$ with an estimate

$$|a_n(t)| \leq Ae^{B|n|} \quad \text{for some} \quad A \text{ and } B > 0.$$

On the other hand, for t negative and small $a_n(t) = \int_0^{2\pi} u(t, \theta)e^{-in\theta}\frac{d\theta}{2\pi}$. So (after shrinking α if needed) if $-\alpha < -\beta < 0$, for any $t \in [-\alpha, -\beta]$ we have

$$|a_n(t)| \leq C_{\alpha\beta}, \quad \text{for some constant} \quad C_{\alpha\beta}.$$

Let us summarize. We have two inequalities

$$\begin{cases} |a_n(t)| \leq Ae^{B|n|} & \text{for } |t| < \alpha \ (t \in \mathbb{C}) \\ |a_n(t)| \leq C_{\alpha\beta} & \text{for } -\alpha \leq t \leq -\beta \ . \end{cases}$$

Now, it is simply a matter of using the fact that $a_n(t)$ is a holomorphic function of t and that hence $\mathrm{Log}|a_n|$ is subharmonic.

Let $D_{\alpha\beta}$ the disc with a slit defined by

$$D_{\alpha\beta} = \{s \in \mathbb{C} : |s| < \alpha, \ s \notin [-\alpha, -\beta]\}.$$

If $t \in D_{\alpha\beta}$, there is a so called harmonic measure μ_t representing t on the boundary of $D_{\alpha\beta}$ (the slit treated as usual with an upper slit S^+ and a lower slit S^-). By subharmonicity one has

$$\mathrm{Log}|a_n(t)| \leq \int_{S^+ \cup S^-} \mathrm{Log}|a_n(s)| d\mu_t(s) + \int_{|s|=\alpha} \mathrm{Log}|a_n(s)| d\mu_t(s).$$

Hence

$$\mathrm{Log}|a_n(t)| \leq \mathrm{Log} C_{\alpha\beta} + \mathrm{Log} A + B\mu_t\{|s| = \alpha\}|n|.$$

The key point is that provided we take β close enough to 0, and $|t|$ small enough, the harmonic measure of the circle $|s| = \alpha$, in the boundary of the domain $D_{\alpha\beta}$, is arbitrarily small. Hence for any $\epsilon > 0$ we have an estimate

$$\mathrm{Log}|a_n(t)| \leq Constant + \epsilon|n|,$$

i.e. $|a_n(t)| \leq C e^{\epsilon|n|}$, valid for $t \in \mathbb{C}$ with $|t|$ small enough (depending on ϵ). So according to (1) u has a strong boundary value.

The argument that will not be repeated in these notes in crucial in several parts of our theory. It consists in blending a rather weak estimate ($|a_n(t)| \leq Ae^{B|n|}$) with a stronger one ($|a_n(t)| \leq C$) holding say only on an interval, to get an estimate better than the first one due to subharmonicity.

3 The Hahn Banach Theorem in the Theory of Analytic Functionals

There are two important problems of Hahn Banach type in the theory of analytic functionals, and they are related.

(A) The extension of an analytic functional carried by a compact set $K \subset \mathbb{C}^N$ to a local analytic functional (see 1.6) i.e. the extension of an analytic functional that is defined on $\mathcal{O}(\mathbb{C}^N)$ to the space of germs of holomorphic functions near K, with the appropriate notion of continuity.

(B) The decomposition of an analytic functional carried by the union of two compact sets K_1 and K_2 into the sum of an analytic functional carried by K_1 and one carried by K_2. In $\mathbb{C}^N$ this is always possible but the result is nontrivial.

That this is a problem of Hahn Banach type is easily understood by considering the similar problem for measures. Decompose a measure μ carried by $K_1 \cup K_2$ into a sum $\mu = \mu_1 + \mu_2$, μ_j carried by K_j. The measure μ defines a linear form on $\mathcal{C}(K_1 \cup K_2)$. Consider, in an obvious way, $\mathcal{C}(K_1 \cup K_2)$ as a subspace of $\mathcal{C}(K_1) \oplus \mathcal{C}(K_2)$. Extending the linear functional to $\mathcal{C}(K_1) \oplus \mathcal{C}(K_2)$ gives us μ_1 and μ_2. But in both problems (A) and (B) there is a serious topological question to be faced with. Although we have not phrased the continuity condition that analytic functional must satisfy in those terms, we are clearly meeting the problem of inductive limits and of the Hahn Banach Theorem for inductive limits. The reader with a taste for abstract functional analysis should rather read [M].

3.1 A Hahn Banach Theorem

If one is not willing to read about bornology, etc., the following theorem will suffice.

Theorem 2. *Let E be a complex vector space, and for each j, let E_j be a vector subspace of E that is normed with a norm $\| \cdot \|_j$. Assume that $E_1 \subset \cdot \subset E_n \subset \cdot \subset \cup_j E_j = E$ and that for each j, $E_j \subset E_{j+1}$ is a compact embedding. For each j, let F_j be a closed subspace of E_j such that, for each j, $F_j = F_{j+1} \cap E_j$. Put $F = \cup_j F_j$. If $\psi : F \to \mathbb{C}$ is a $\mathbb{C}$-linear map such that for each j there exists a constant C_j with $|\psi(x)| \leq C_j \|x\|_j$ for every $x \in F_j$, then there exists a $\mathbb{C}$-linear map $\tilde{\psi} : E \to \mathbb{C}$ such that for every j $\psi|_{f_j} = \tilde{\psi}|_{f_j}$ and such that there are constants $\tilde{C}_j$ such that for each $y \in E_j$, $|\tilde{\psi}(y)| \leq \tilde{C}_j \|y\|_j$.*

The following diagram, in which $\hookrightarrow$ denotes non isometric embeddings, while $\subset$ denotes (isometric) inclusions, may help:

$$
\begin{array}{ccccccc}
E_1 & \hookrightarrow & E_2 & \hookrightarrow & \hookrightarrow & E_n & \hookrightarrow & \cup E_j = E \\
\cup & & \cup & & & \cup & & \searrow^{\tilde{\psi}} \\
F_1 & \hookrightarrow & F_2 & \hookrightarrow & \hookrightarrow & F_n & \hookrightarrow & \cup F_j = F & \xrightarrow{\psi} & \mathbb{C}
\end{array}
$$

3.2 Some Comments

There are two crucial hypotheses in the theorem: the map $E_j \to E_{j+1}$ should be a compact map, the subspace F_j should be a closed subspace of E_j. In [RS1] counterexamples are given to illustrate the need of these hypotheses. In the theory of analytic functionals restricting holomorphic functions to relatively compact subsets leads naturally to compact maps, but closedness of F_j will be source of trouble.

Here is an extremely elementary proposition and an example which already give some idea of how proofs can work, and how the hypotheses come to play.

Proposition 2. *Let the following be given*

$$E_1 \hookrightarrow E_2$$

$$\cup \qquad \cup$$

$$F_1 \hookrightarrow F_2 \xrightarrow{\psi} \mathbb{C},$$

with: E_1 and E_2 Banach spaces, $E_1 \hookrightarrow E_2$ a continuous map from E_1 into E_2, F_1 and F_2 closed subspaces of E_1 and E_2, $F_1 = F_2 \cap E$, ψ a continuous linear form. If the closed unit ball of E_1 is a compact subset of E_2, then for every $\epsilon > 0$ there exists a continuous linear from $\tilde{\psi}$ on E_2 such that

$$\mathrm{Sup}_{\substack{x \in E_1 \\ \|x\|_1 \leq 1}} |\tilde{\psi}(x)| \leq \mathrm{Sup}_{\substack{x \in F_1 \\ \|x\|_1 \leq 1}} |\psi(x)| + \epsilon,$$

where $\| \ \|_1$ denotes E_1 norm.

Proof. Without loss of generality we assume that $\mathrm{Sup}_{\substack{x \in F_1 \\ \|x\|_1 \leq 1}} |\psi(x)| = 1$. Let $S = \{x \in F_2, \ \psi(x) = 1 + \epsilon\}$ and let $\overline{B} = \{x \in E_1, \|x\|_1 \leq 1\}$. In E_2, S is a closed affine subspace and $\overline{B}$ is a convex compact set, and $\overline{B} \cap S = \phi$. For $\delta > 0$ small enough the δ neighborhood of $\overline{B}$ in E_2 is an open convex set that still does not intersect S. So by the standard geometric Hahn Banach Theorem there exists an affine hyperplane $\tilde{S}$ containing S and not intersecting this neighborhood of $\overline{B}$, and hence not intersecting $\overline{B}$, nor a neighborhood of 0 in E_2. Let $\tilde{\psi}$ be the unique linear form on E_2 such that $\tilde{\psi} \equiv 1 + \epsilon$ on $\tilde{S}$. Then $\tilde{\psi}$ is a continuous linear form on E_2 whose restriction to F_2 is ψ and the operator norm of $\tilde{\psi}$ on E_1 is at most $1 + \epsilon$.

In Proposition 2 it would not be enough to assume that $E_1 \hookrightarrow E_2$ is a compact embedding (i.e. that the closed unit ball of E_1 is only relatively compact in E_2). This is illustrated by the following example which on one hand explains why the proof of Theorem 2 simplifies if one makes the stronger hypothesis that the closed unit ball of E_j is a compact set in E_{j+1}. (Note that it is necessarily the case if the E_j's are reflexive Banach spaces), and which on the other hand leads to counter examples when one tries to weaken the hypotheses of Theorem 2.

Example. Let E_1 be the space of continuous functions on the closed unit disc in $\mathbb{C}$ whose restriction to the open unit disc is holomorphic, with norm $\|f\|_1 = \mathrm{Sup}_{|z| \leq 1} |f(z)|$. For E_2 take the corresponding space with the disc of radius $1/2$ instead of 1 and norm $\|f\|_2 = \mathrm{Sup}_{|z| \leq 1/2} |f(z)|$. Finally let g be a bounded holomorphic function on the unit disc with $\mathrm{Sup}_{|z| < 1} |g(z)| = 1$ but $g \notin E_1$, (i.e. g does not extend continuously to the closed unit disc). Take $F_2 = \mathbb{C}g$ (a one dimensional subspace). Accordingly $F_1 = F_2 \cap E_1 = \{0\}$. Define ψ by imposing $\psi(g) = 1$. Although $F_1 = \{0\}$, if $\tilde{\psi}$ is any continuous linear extension of ψ to E_2 we have

$$\mathrm{Sup}_{\substack{f \in E_1 \\ \|f\|_1 \leq 1}} |\tilde{\psi}(f)| \geq 1.$$

Indeed there exists a sequence of polynomials P_j satisfying $|P_j| < 1$ on the unit disc and $P_j \to g$ uniformly on compact sets in the unit disc, therefore in $\|\quad\|_2$ norm. So $\tilde{\psi}(P_j) \to \tilde{\psi}(g) = 1$.

3.3 The Notion of Good Compact Set

Given an open set U in $\mathbb{C}^N$ and a compact set $K \subset U$, for applying the Hahn Banach theorem stated in (1), the crucial question arises whether the subspace of $\mathcal{O}(U)$ that consists of all functions $f \in \mathcal{O}(U)$ that can be approximated by polynomials on some neighborhood of K is a closed subspace of $\mathcal{O}(U)$. Compact sets K such that for all $U \supset K$ the answer is positive are called *good compact sets*. The practical definition (not Martineau's definition but equivalent to it) is

Definition. *A compact set $K \subset \mathbb{C}^N$ is called a good compact set if for every neighborhood U of K there exists a (smaller) neighborhood V of K such that: if $f \in \mathcal{O}(U)$ and f is uniformly approximable by polynomials on some neighborhood of K then f is uniformly approximable by polynomials on V.*

It happens that

Theorem 3. *Every compact set in $\mathbb{C}^N$ is a good compact set.*

The proof is rather easy if one assumes some mild hypothesis on the polynomial hull of K (see [M] Theorem 1.1' in Chapter 1) but the general case (so far) relies on a deep theory of Bishop ([RS2]).

Theorem 2 allows one to solve both problems (A) and (B) in $\mathbb{C}^N$, or more generally in arbitrary Stein manifolds.

Now, let K be a polynomially convex set in $\mathbb{C}^N$ (remember that for such a set goodness is a very easy result), and let ψ be an analytic functional carried by K. If $K = K_1 \cup K_2$ we can write $\psi = \psi_1 + \psi_2$, with ψ_j carried by K_j. By Theorem 1, ψ_1 is unique modulo analytic functionals carried by $K_1 \cap K_2$ and that allows to define a restriction of ψ to K_1 (modulo etc. ...).

3.4 The Case of Non-Stein Manifolds

Let $\mathcal{M}$ be a complex manifold. We will not repeat the definitions given in $\mathbb{C}^N$. Simply replace $\mathcal{O}(\mathbb{C}^N)$ by $\mathcal{O}(\mathcal{M})$.

There are complex manifolds that contain "bad" compact sets (even a point can be bad). That gives counter examples to (A) and (B). In such examples the complex manifold $\mathcal{M}$ must have an interesting space of global holomorphic functions: not trivial but not too rich (non Stein). Examples are given in [RS1] pp 28–31. Here I show some preliminary steps that led to the examples:

(i) an abstract example, seemingly of little relevance

(ii) how this example can be made more concrete

(i) "A bad algebra".

We want to find a subalgebra of $\mathcal{O}(\mathbb{C}^3)$ with the property that for every $n > 0$ there exists an entire function $f \in \mathcal{O}(\mathbb{C}^3)$ such that:

(a) f is approximable by A on some neighborhood of 0.

(b) but f is not approximable by A in any neighborhood of the point $(0,0,\frac{1}{n})$ (so in any neighborhood of 0 containing that point).

So contrary to the requirement made in the definition of good compact sets, the set on which f can be approximated is not independent of f.

In $\mathbb{C}^3_{z,w,t}$, let A be the algebra generated by the polynomials

$$w, t, (t-1)zw, (t-1)(t-\frac{1}{2})z^2w, \ldots, (t-1)(t-\frac{1}{2})\ldots(t-\frac{1}{n})z^nw, \ldots$$

The function $z^n w$ is approximable in the region $\{|t| < \frac{1}{n}\}$, since in that region $(t-1)(t-\frac{1}{2})\ldots((t-\frac{1}{n})$ is invertible (its inverse is the limit of polynomials in t). However in no neighborhood of $(0,0,\frac{1}{n})$ is $z^n w$ approximable. More: $z^n w$ is not even approximable in any neighborhood of $(0,0,\frac{1}{n})$ in $\mathbb{C}^2_{z,w} \times \{0\}$, since the algebra of restrictions of A to $t = \frac{1}{n}$ is generated by monomials $z^p w^q$ with $p < nq$ (the other generators have $(t-\frac{1}{n})$ as a factor).

(ii) We are not interested in an abstract algebra A, but in an algebra A that would be the algebra of all holomorphic functions on some complex manifold. The following is a first step in order to turn the abstract example (i) into a concrete one.

Find a 2 dimensional complex manifold containing $\mathbb{C}^2_{z,w}$ as a dense open subset and such that $w \in \mathcal{O}(\mathcal{M})$ and $z^k w \in \mathcal{O}(\mathcal{M})$ if and only if $k \leq k_0$ (for some given integer k_0). In particular $z \notin \mathcal{O}(\mathcal{M})$. Of course we mean that there is no $f \in \mathcal{O}(\mathcal{M})$ whose restriction to $\mathbb{C}^2$ is the function z, and similarly above. This is rather easy to do, roughly said, by adding a point of "compactification" at infinity for z to each level set of $z^{k_0}w$ (some kind of blow up is involved).

Precisely.

Let $\mathcal{M}$ be the union of $\mathbb{C}^2_{z,w}$ and $\mathbb{C}^2_{\zeta,\eta}$ with the identification of $(z,w) \in \mathbb{C}^2$ for $z \neq 0$ with (ζ,η) for $\zeta = \frac{1}{z}$ and $\eta = z^{k_0}w$. Then the function $z^k w$ in the $\mathbb{C}^2_{z,w}$ chart corresponds to the function $\zeta^{k_0-k}\eta$ in the $\mathbb{C}^2_{\zeta,\eta}$ chart. So it extends to a global holomorphic function on $\mathcal{M}$ if and only if $k \leq k_0$.

4 Spectral Theory

I will mostly refer the reader to our paper [RS3]. In that paper, we prove in full generality, for compact boundaries, results analogous to the results explained in 2.3 and 2.4. There, we needed to replace Fourier series that allowed immediate computation in 2. Consider a relatively compact domain Ω in $\mathbb{R}^d$, with smooth (later real analytic) boundary. On its boundary $b\Omega$ equipped with some measure consider a positive elliptic linear partial differential operator P (e.g. $\mathbb{1} + \Delta$, when Δ is the Laplace Beltrami operator for some Riemannian metric). One then replaces the orthonormal basis of $L^2(\mathbb{R}/2\pi\mathbb{Z})$ by an orthonormal basis of $L^2(b\Omega)$ made of the eigenvalues of P. Following what has been explained in 2, one has to do repeated integrations by parts ($\int (Pu)v = \int u(Pv)$) of various terms and one should add them up.

Two questions arise:

(A)estimate $P^{(k)}\varphi$ for φ real analytic (the Cauchy estimate was used in 2 for $\frac{\partial^k \varphi}{\partial \theta^k}$)

(B)roughly said: how many terms we have to add up.

The second question is related to the asymptotics of the eigenvalues of P for which there is an extensive literature. An extremely elementary result suffices (see the appendix in [RS3]).

For the first question, the following argument gives a good hint of how little one needs to do.

Assume that we know only the following: for any f holomorphic on the disk $\{z \in \mathbb{C}, |z| \leq \rho\}$, $|f'(0)| \leq \frac{1}{\rho}\text{Sup}_{|z|<\rho}|f(z)|$. We want to estimate $f^{(k)}(0)$ for a function f holomorphic on the unit disk and satisfying $|f| \leq 1$. By our assumption $|f'(z)| \leq \frac{1}{k}$ for $|z| \leq 1 - \frac{1}{k}$. By the same, $|f''(z)| \leq \frac{1}{k^2}$ for $|z| \leq 1 - \frac{2}{k}$, $|f'''(z)| \leq \frac{1}{k^3}$ for $|z| \leq 1 - \frac{3}{k}$, etc. Finally $|f^{(k)}(0)| \leq \frac{1}{k^k}$, which is not so far off the Cauchy estimate with $\frac{1}{k!}$ ($k! \simeq k^k e^{-k}\sqrt{2\pi k}$). This strategy allows one to easily estimate $P^k\varphi$, and the resulting non sharp estimate happened to be sufficient for our purpose.

5 Non-linear Paley Wiener Theory and Local Theory of Boundary Values

5.1 The Paley Wiener Theory

The Paley Wiener theory is a theory that characterizes carriers of analytic functionals. The characterization of the support of functions or distributions on $\mathbb{R}^N$ can be considered as a special case.

A distribution on $\mathbb{R}^N$, with compact support, has its support included in a compact set $K \subset \mathbb{R}^N$ if and only if the analytic functional that this distribution naturally defines is carried by K (Proposition 4.10 in [RS1]).

(1) The classical Paley Wiener theory gives a characterization of convex carriers. The following version for analytic functionals (usually stated by using the indicatrix of K) is due to Martineau.

Theorem 4. Let K be a compact convex set in $\mathbb{C}^N$, and let ψ be an analytic functional on $\mathbb{C}^N$. The following are equivalent:

(1) ψ is carried by K

(2) For any neighborhood V of K, there exists a constant C_V such that for every linear function L on $\mathbb{C}^N$:

$$|\psi(e^L)| \leq C_V \mathrm{Sup}_V |e^L|.$$

Of course (1) $\Rightarrow$ (2) trivially. Indeed to say that K is a carrier is to say that an inequality $|\psi(\varphi)| \leq C_V \mathrm{Sup}_V |\varphi|$ holds for all $\varphi \in \mathcal{O}(\mathbb{C}^N)$. The theorem says that it is enough to consider the case of $\varphi = e^L$ for L linear. It clearly has to be the dual result to a result usually unnoticed about the density, in a precise appropriate sense, of the exponential functions e^L. This was the path followed in [RS1], leading to a totally new approach to the classical Paley Wiener theory. Here we will not follow that path, but we will simply reduce the case of non convex carriers to the case of convex carriers.

In fact we will need only the following special case of the classical theorem (for which the elementary Lemma 4.2 in [RS1] gives a transparent explanation).

Corollary *An analytic functional ψ on $\mathbb{C}^N$ is carried by $\overline{\Delta}^N$ if and only if for any $\delta > 0$ there exists C_δ such that for every $\lambda_1, \ldots, \lambda_N \in \mathbb{C}$*

$$|\psi(e^{\lambda_1 z_1 + \cdots + \lambda_N z_N})| \leq C_\delta e^{(1+\delta)\sum |\lambda_j|}$$

where $e^{\lambda_1 z_1 + \cdots + \lambda_N z_N}$ designates the function $(z_1, \ldots, z_N) \mapsto e^{\lambda_1 z_1 + \cdots + \lambda_N z_N}$.
Our first result is:

Theorem 5. *Let $P_1, \ldots, P_R \in \mathcal{O}(\mathbb{C}^N)$ and let*

$$K = \{z \in \mathbb{C}^N; |z_j| \leq 1 \ j = 1, \ldots, N, \ |P_r(z)| \leq 1, r = 1, \ldots, R\}.$$

Let ψ be an analytic functional on $\mathbb{C}^w$. The following are equivalent:

(1) ψ is carried by K.

(2) For every $\delta > 0$ there exists $C_\delta > 0$ such that for every $\lambda_1, \ldots, \lambda_{N+R} \in \mathbb{C}$.

$$|\psi(e^F)| \leq C_\delta e^{(1+\delta)\sum_{j=1}^{N+R} |\lambda_j|},$$

if

$$F(z) = \sum_{j=1}^{N} \lambda_j z_j + \sum_{r=1}^{R} \lambda_{N+r} P_r(z).$$

Again (1) $\Rightarrow$ (2) is trivial. If $\delta > 0$ is given, for an appropriate small neighborhood V of K, we have

$$\mathrm{Sup}_V |e^F| \leq \mathrm{Sup}_V e^{|F|} \leq e^{(1+\delta)\sum |\lambda_j|};$$

and if ψ is carried by K we have $|\psi(e^F)| \leq C_V \mathrm{Sup}_V |e^F|$. Now we prove (2) $\Rightarrow$ (1).

Consider the analytic functional $\tilde{\psi}$ defined on $\mathbb{C}^{N+R}$ in the following way: Let $\tau : \mathbb{C}^N \to \mathbb{C}^{N+R}$ be defined by

$$\tau(z_1,\ldots,z_N) = (z_1,\ldots,z_N, P_1(z),\ldots,P_R(z)).$$

For $f \in \mathcal{O}(\mathbb{C}^{N+R})$ set

$$\tilde{\psi}(f) = \psi(f \circ \tau).$$

Since ψ is carried by some big ball $B \subset \mathbb{C}^N$, it is immediate that $\tilde{\psi}$ is carried by $\tau(B) \subset \tau(\mathbb{C}^N) \subset \mathbb{C}^{N+R}$. But $\tilde{\psi}$ is also carried by the unit polydisc $\overline{\Delta}_{N+R}$ in $\mathbb{C}^{N+R}$, as we will now show. In $\mathbb{C}^{N+R}$ we denote the coordinates by (z,w) with $z = (z_1,\ldots,z_N)$ $w = (w_1,\ldots,w_R)$. For any $(\lambda_1,\ldots,\lambda_{N+R}) \in \mathbb{C}^{N+R}$, by definition of $\tilde{\psi}$ we have

$$\tilde{\psi}[e^{(\sum_{j=1}^N \lambda_j z_j + \sum_{r=1}^R \lambda_{N+r} w_r)}] = \psi(e^F),$$

with

$$F(z,w) = \sum_{j=1}^N \lambda_j z_j + \sum_{r=1}^R \lambda_{n+r} P_r(z).$$

By condition (2) we get

$$|\tilde{\psi}[e^{(\sum_{j=1}^w \lambda_j z_j + \sum_{r+1}^R \lambda_{N+r} w_r)}]| \leq C_\delta e^{(1+\delta) \sum |\lambda_j|}.$$

By the above corollary it implies that $\tilde{\psi}$ is carried by $\overline{\Delta}_{N+R}$ as claimed.

Next, we observe that the union of $\overline{\Delta}_{N+R}$ and $\tau(B)$ is polynomially convex. If $(z^0, w^0) \in (\mathbb{C}^N \times \mathbb{C}^R) - [\overline{\Delta}_{N+R} \cup \tau(B)]$, there are two cases to consider.

If $w^0 = (w_1^0,\ldots,w_R^0) \notin \tau(\mathbb{C}^N)$, there exists $r \in \{1,\ldots,R\}$ such that $w_r^0 \neq P_r(z)$ and there exists $j \in \{1,\ldots,N\}$ such that $|z_j^0| > 1$. Set $f_p(z,w) = [w_r - P_r(z)]z_j^p$. Then $f_p \equiv 0$ on $\tau(\mathbb{C}^N)$ and for p large enough $|f_p(z^0, w^0)| > \mathrm{Sup}_{\overline{\Delta}_{N+R} \cup \tau(B)}|f_p|$. If $w^0 \in \tau(\mathbb{C}^N)$ then $z^0 \notin B$, set $f(z,w) = \sum_{j=1}^N z_j \overline{z}_j^0$. Then $|f(z^0, w^0)| > \mathrm{Sup}_{\overline{\Delta}_{N+R} \cup \tau(B)}\|f|$.

Since $\overline{\Delta}_{N+R} \cup \tau(B)$ is polynomially convex we can apply Theorem 1 (on the intersection of carriers). Hence $\tilde{\psi}$ is carried by $\tau(\mathbb{C}^N) \cap \overline{\Delta}_{N+R}$.

Let V be a neighborhood of K in $\mathbb{C}^N$ and set $\hat{V} = \{(z,w) \in \mathbb{C}^{N+R}; z \in V\}$. Then $\hat{V}$ is a neighborhood of $\tau(\mathbb{C}^N) \cap \overline{\Delta}_{N+R}$ in $\mathbb{C}^{N+R}$. Let Π be the projection map from $\mathbb{C}^{N+R}$ onto $\mathbb{C}^N$, $\Pi(z,w) = z$. For some constant $C > 0$, for any $f \in \mathcal{O}(\mathbb{C}^N)$:

$$|\psi(f)| = |\tilde{\psi}(f \circ \Pi)| \leq C \, \mathrm{Sup}_{\hat{V}}|f \circ \Pi| = C \, \mathrm{Sup}_V|f|.$$

This shows that ψ is carried by K.

Comment. The proof that has just been given seems so easy that one may wonder whether some deep point has been hidden. It seems to me that for that, one has to look at the proof of the Martineau Theorem on the intersection of carriers (crucially used). There one has to solve a Cousin problem with bounds and this seems to somewhat correspond in the proofs given in [RS1] to the proof of a "Oka extension result" with bounds [RS1] Lemma 4.4. Everything else in both proofs is rather "mechanical".

Theorem 6. *Let K be a compact set in $\mathbb{C}^N$. Let ψ be an analytic functional. The following are equivalent:*

(1) ψ is carried by K

(2) For every neighborhood V of K, and every $d \in \mathbb{N}$ there exists a constant $C_{V,d}$ such that for every polynomial P of degree $\leq d$

$$|\psi(e^P)| \leq C_{V,d} \operatorname{Sup}_V |e^P|.$$

(3) For every neighborhood V of K and every $d \in \mathbb{N}$ there exists a constant $A_{V,d}$ such that for every polynomial P of degree $\leq d$

$$|\psi(e^P)| \leq A_{V,d} \operatorname{Sup}_V e^{|P|}.$$

Proof. Trivially $(1) \Rightarrow (2) \Rightarrow (3)$. We have to prove that $(3) \Rightarrow (1)$. We can assume that K is polynomially convex since obviously K is a carrier if and only if the polynomial hull of K is a carrier, and since we can replace V by its polynomial hull (we mean the union of the polynomial hulls of the compact sets included in V) also. We can of course assume that $K \subset \overline{\Delta}_N$.

Let W be a neighborhood of K. There exist $R \in \mathbb{N}$ and polynomials $P_1, \ldots, P_R$ such that

$$K \subset \{z \in \mathbb{C}^N; |z_j| \leq 1 \, j = 1, \ldots, N, \ |P_r(z)| \leq 1 \, r = 1, \ldots, R\} \subset W.$$

Fix d so that each P_r is of degree at most d. Set $K_1 = \{z \in \mathbb{C}^N; |z_j| \leq 1, |P_r(z)| \leq 1\}$. Let $\delta > 0$. Fix a neighborhood V of K_1 on which $|z_j| \leq 1 + \delta$ and $|P_r(z)| \leq 1 + \delta$ $(r = 1, ..., R)$. If $\lambda_1, \ldots, \lambda_{N+R} \in \mathbb{C}^{N+R}$ and

$$F(z) = \sum_{j=1}^N \lambda_j z_j + \sum_{r=1}^R \lambda_{N+r} P_r(z),$$

by (3), we have:

$$|\psi(e^F)| \leq A_{V,d} \operatorname{Sup}_V e^{|F|} \leq A_{V,d} e^{(1+\delta) \sum |\lambda_j|}.$$

Hence by Theorem 5 ψ is carried by K_1, and therefore by K since $K_1 \subset W$ and W was an arbitrary neighborhood of K.

5.2 Application

Definition. For t negative and small let ψ_t be an analytic functional on $\mathbb{C}^N$. If for every $f \in \mathcal{O}(\mathbb{C}^N)$ $t \mapsto \psi_t(f)$ extends holomorphically to a neighborhood of 0 in $\mathbb{C}$, we say that (ψ_t) is an *analytic family* of analytic functionals.

A first application of the non linear Paley Wiener theory is:

Theorem 7. *Let (ψ_t) be an analytic family of analytic functionals. Let K be a compact set in $\mathbb{C}^N$. If for each small negative t ψ_t is carried by K then $f \mapsto \lim_{t \to 0_-} \psi_t(f)$ defines an analytic functional carried by K.*

So having a hypothesis sufficiently strong allows one to control the carrier of a limit, which is one of the main difficulties.

Quick sketch of proof. General functional analysis arguments allow to define ψ_t for $|t| < \rho$ (for some $\rho > 0$), and all the ψ_t's are carried by a same compact set H in $\mathbb{C}^N$. For $-\rho < t < 0$ we know more, ψ_t is carried by K and we wish to show that ψ_0 is carried by K itself. Using the Paley Wiener theory one has to control $\psi_0(e^P)$ in terms of the supremum of $e^{|P|}$ on some arbitrary neighborhood of K for all polynomials of fixed degree. This can be done by using the subharmonicity of $t \mapsto \mathrm{Log}|\psi_t(e^P)|$, and by taking advantage that one is working with a very small set of functions (e^P for P of fixed degree). The following totally elementary Lemma [RS1] page 54 is to be used:

Lemma. *Let $d \in \mathbb{N}$, and V_0, V_1, V be non empty relatively compact open subsets in $\mathbb{C}^N$, with $V_1 \subset\subset V_0 \subset\subset V$. There exists $\epsilon > 0$ such that for all polynomial of degree $\leq d$ on $\mathbb{C}^N$:*

$$(1 - \epsilon)\, \mathrm{Sup}_{V_1}|P| + \epsilon\, \mathrm{Sup}_V|P| \leq \mathrm{Sup}_{V_0}|P|.$$

Extensions of Theorem 7 are needed and are given in [RS1].

5.3 Application to a Local Theory of Boundary Values

Our general definition [RS1] (section 5) is rather cumbersome (but should one blame nature?). So are also the proofs of basic properties (such as restrictions). Here I will just try to sketch what our general definition is when applied to the case of holomorphic function defined in the upper half plane (in $\mathbb{C}$).

So, we start with a holomorphic function u defined on the intersection of the upper half plane $\prod^+ = \{z \in \mathbb{C}, \mathrm{Im}\, z > 0\}$ and of a neighborhood of the interval $[-1, +1]$. We wish to define the boundary values along $[-1, +1]$. It will be an analytic functional on $\mathbb{C}$ carried by $[-1, 1]$ defined modulo analytic functionals carried by $\{-1\} \cup \{1\}$. There are several ways of proceeding and the most elementary one may be the way explained in [RS1] example 4 page 21.

A slight variant of it is by making sense of "$\int_{-1}^{+1} u(x)\varphi(x)dx$" for $\varphi \in \mathcal{O}(\mathbb{C})$ by setting it equal to "$\int_{-1}^{-1+i\epsilon}$" $+ \int_{-1+i\epsilon}^{1+i\epsilon} +$ "$\int_{1+i\epsilon}^{1}$" $u(z)dz$, and by making sense of "$\int_{-1}^{-1+i\epsilon}$" and "$\int_{1+i\epsilon}^{1}$" as in [RS1] example 2 page 17 (elementary version of flabbiness). See also 1.7 in 1 above.

Let us rather follow the general approach.

At this point we wish to change notations and $\mathbb{C}$ (used before) should be replaced by $\mathbb{R}^2$ and $\mathbb{R}^2 \subset \mathbb{C}^2$.

So let us start again:

Let u be a function defined on the set $\{(x_1, x_2) \in \mathbb{R}^2; -2 < x_1 < +2, 0 < x_2 < 1\}$, and satisfying the equation $\frac{\partial u}{\partial x_1} + i\frac{\partial u}{\partial x_2} = 0$. There is no restriction on the growth of $u(x, y)$ as x approaches 0. We intend to give a meaning to "the boundary value of u along $[-1, 1]$" ($[-1, 1] \times \{0\}$).

Let $\varphi \in \mathcal{O}(\mathbb{C}^2)$. For $t \in (0, 1)$ set

$$I_\varphi(t) = \int_{-1}^{+1} u(x_1, t)\varphi(x_1, t)dx_1.$$

There is of course no reason that $I(t)$ has a limit as $t \to 0$. For example there exists u such that $u(x_1, \frac{1}{2n})$ tends uniformly to 0 as $n \to \infty$, while $u(x_1, \frac{1}{2n+1})$ tends to 1.

So we should *not* try to define an analytic functional by setting $\psi(\varphi) = \lim_{t\to 0} I_\varphi(t)$.

Instead we fix an arbitrarily small neighborhood V of $\{(-1, 0), (1, 0)\}$ in $\mathbb{C}^2$. For $t > 0$ let $\varphi_t \in \mathcal{O}(\mathbb{C}^2)$ be defined by:

$$\begin{cases} \varphi_t(x_1, t) = \varphi(x_1, t) \\ \frac{\partial \varphi_t}{\partial z_1} + i\frac{\partial \varphi_t}{\partial z_2} = 0 \end{cases}.$$

The point here is that the restriction of φ_t to $\mathbb{R}^2$ satisfies $\frac{\partial \varphi_t}{\partial x_1} + i\frac{\partial \varphi_t}{\partial x_2} = 0$. And up to a $-$ sign $\frac{\partial}{\partial x_1} + i\frac{\partial}{\partial x_2}$ is the transpose of the operator $\frac{\partial}{\partial x_1} + i\frac{\partial}{\partial x_2}$ itself of which u is a solution. In general such a situation is reached by application of the Cauchy Kovalevsky Theorem and it does not lead to entire functions, but to functions defined on a large enough domain.

In our case we can write φ_t explicitly: $\varphi_t(z_1, z_2) = \varphi(z_1 + iz_2 - it, t)$.

Now fix $t_0 \in (0, 1)$. For $t \notin (0, t_0)$ we defined $I_\varphi(t) = \int_{-1}^{+1} u(x, t)\varphi(x, t)dx$. So $I_\varphi(t) = \int_{-1}^{+1} u(x, t)\varphi_t(x, t)dx$, since $\varphi_t(\cdot, t) \equiv \varphi(\cdot, t)$. By the Cauchy formula

$$I_\varphi(t) = \int_{-1}^{+1} u(x, t_0)\varphi_t(x, t_0)dx + E_t(\varphi).$$

where

$$E_t(\varphi) = i\int_t^{t_0} u(-1, s)\varphi_t(-1, s)ds + i\int_{t_0}^t u(1, s)\varphi_t(1, s)ds.$$

The following is to be noticed

(i) $t \mapsto I_\varphi(t) - E_t(\varphi)$ extends holomorphically at $t = 0$.
(ii) Provided that t_0 has been taken small enough E_t is an analytic functional carried by $\overline{V}$ the closure of the neighborhood V.

Now forget entirely the above construction and consider any other family of analytic functionals F_t each carried by $\overline{V}$, with similarly: $t \mapsto I_\varphi(t) - F_t(\varphi)$ extending holomorphically at $t = 0$.

Note that the above conditions make absolutely no reference to the equation that u satisfies.

We have now two analytic functionals to consider:

$$\psi(\varphi) = \lim_{t \to 0} [I_\varphi(t) - E_t(\varphi)]$$

and

$$\psi^\#(\varphi) = \lim_{t \to 0} [I_\varphi(t) - F_t(\varphi)].$$

We have $\psi - \psi^\#(\varphi) = \lim_{t \to 0} [E_t - F_t(\varphi)]$, $t \in (0, t_0)$. On the right hand side we have an analytic family of analytic functionals which, for $t \in (0, t_0)$, are carried by $\overline{V}$. By Theorem 7 $\psi - \psi^\#$ is also carried by $\overline{V}$.

It is also clear by the same reason that ψ and $\psi^\#$ are carried by $[-1, 1] \times \{0\} \cup \overline{V} \subset \mathbb{C}^2$. To make things more precise take

$$V = \{(z_1, z_2) \in \mathbb{C}, \min(|z_1 - 1|, |z_1 + 1|) < \epsilon, |z_2| < \epsilon\} (\epsilon > 0, \text{small}).$$

We can decompose ψ as a sum $\psi = \psi_1 + \psi_2$ with ψ_1 carried by $[-1 + \epsilon, 1 - \epsilon] \times \{0\}$ and ψ_2 carried by $\overline{V}$. Similarly $\psi^\# = \psi_1^\# + \psi_2^\#$. Applying Martineau's theorem to the polynomially convex set $[-1, 1] \times \{0\} \cup \overline{V} \subset \mathbb{C}^2$, we see that $\psi_1 - \psi_1^\#$ is an analytic functional carried by $\{(-1+\epsilon), (1-\epsilon)\} \times \{0\}$. So starting from u, we got an analytic functional ψ_1 or $\psi_1^\#$ carried by $[-1+\epsilon, 1-\epsilon] \times \{0\}$, unique modulo analytic functionals carried by the endpoints. This is exactly the kind of object that one is looking for. There is still some distance to cover:

(a) Shrink ϵ and in the limit get an analytic functional carried by $[-1, 1] \times \{0\}$. Here taking advantage of the situation this step can obviously be avoided (replace $[-1, 1]$ by $[-1 + \epsilon, 1 - \epsilon]$), but in general this step will simply involve flabbiness.
(b) Finally one has to observe that the analytic functionals that one gets are (up to the arbitrary term carried by the endpoints) analytic functionals in $\mathbb{C}^2$ which come from analytic functional in $\mathbb{C}$ ($= \mathbb{C} \times \{0\} \subset \mathbb{C}^2$). But after a final remark we shall stop this discussion here referring the reader to [RS1] for details and for a systematic discussion.

Remark. The theory of analytic functionals carried by compact sets in $\mathbb{R}^N \subset \mathbb{C}^N$ offers some specificity and things simplify as clearly demonstrated in [H] Chapter 9. However our theory required us to consider analytic functionals carried by sets not in $\mathbb{R}^N$ ($\overline{V}$ above), even if the final object to be found in a hyperfunction that is given by an analytic functional carried by some set in $\mathbb{R}^N$, unique modulo analytic functionals carried by the boundary of that set in $\mathbb{R}^N$.

The following is just a minimal list of references for these notes. Many more references are given in [RS].

References

[H]   L. V. Hörmander, "The Analysis of Linear Partial Differential Operators", Grund. Math. Wis. 256 I Springer Verlag 1983.

[LM]   J. L. Lions and E. Magenes, "Problèmes aux limites non homogènes (VII)", Ann. Mat. Pura Appl. 63 (1963), 201–224.

[M]   A. Martineau, "Sur les fonctionelles analytiques et la transformation de Fourier Borel", J. An. Math. XI (1963) 1–164.

[RS1]   J-P. Rosay and E. L. Stout, "Strong boundary values, analytic functionals, and non linear Paley-Wiener theory", Memoirs of the AMS, 725 (2001).

[RS2]   J-P. Rosay and E. L. Stout, "An approximation theorem related to good compact sets in the sense of Martineau", Ann. Inst. Fourier 50 (2000) 677–687, and corrigendum to appear.

[RS3]   J-P. Rosay and E. L. Stout, "Strong boundary values: Independence of the defining function and spaces of test functions", To appear in Ann. Sc. Norm. Sup. Pisa.

Extremal Discs
and the Geometry of CR Manifolds

Alexander Tumanov

Department of Mathematics, University of Illinois,
Urbana, IL 61801
tumanov@math.uiuc.edu

The theory of extremal discs introduced by Lempert in 1881 is a far-reaching generalization of the classical Riemann mapping theorem. It has found many applications including the regularity of biholomorphic mappings, Green's function for Monge-Ampère equations, moduli space and normal forms of domains, polynomial hulls, deformations and embeddibility of CR structures.

In these lectures, we describe recent results on the theory of extremal discs for CR manifolds and give an application to the regularity of CR maps. We do not assume any knowledge of the theory of extremal discs. Some background on CR manifolds would be useful but is not required.

The author would like to thank CIME and the organizers of the session in Martina Franca in July 2002 for the pleasure to lecture there.

Contents

1 Extremal Discs for Convex Domains

In his celebrated paper, Lempert [L1] introduced extremal analytic discs for a strictly convex domain $D \subset \mathbf{C}^N$.

Let $\Delta = \{\zeta \in \mathbf{C} : |\zeta| < 1\}$ be the unit disc in complex plane. An *analytic disc* f in D is a holomorphic mapping $f : \Delta \to D$. An analytic disc f in D is called *extremal* if $|f'(0)| > |g'(0)|$ for every other analytic disc g in D with the same "center" and direction at 0 as f, that is $g(0) = f(0)$ and $g'(0) = \sigma f'(0)$, $\sigma > 0$.

Extremal discs are important holomorphic invariants of D. Lempert [L1] proved that for every $p \in D$ and every $v \in \mathbf{C}^N$ there exists a unique extremal disc f in D such that $f(0) = p$ and $f'(0) = \sigma v$, where $\sigma > 0$.

Let $N = 1$. Then according to Koebe's proof of Riemann's mapping theorem, an extremal disc f in a simply connected domain D gives a conformal mapping of Δ onto D. If D has a smooth boundary, then f extends smoothly to the circle $b\Delta$ and $f(b\Delta) = bD$.

For $N > 1$ Lempert [L1] proved that extremal discs f in a strictly convex domain D are smooth up to $b\Delta$ and $f(b\Delta) \subset bD$. Furthermore for given $p \in D$ the images of all extremal discs f in D such that $f(0) = p$ form a foliation of $D \setminus \{p\}$. Lempert also proved that the extremal discs coincide with the complex geodesics of the Kobayashi metric and minimize the Kobayashi distance between points. Slodkowsky [Sl] found simpler proofs of some of the results of [L1]. We refer to [Bl] [BD] [L2] [Se] for applications and further results.

We extend some of the above results to the situation in which the hypersurface bD is replaced by a real manifold of higher codimension $\mathbf{C}^N$.

2 Real Manifolds in Complex Space

Let M be a smooth real manifold in $\mathbf{C}^N$ of real codimension k. By "smooth" we usually mean C^∞ unless smoothness is specified. Recall that M is *generic* if $T_p(M) + JT_p(M) = T_p(\mathbf{C}^N)$, $p \in M$, where $T(M)$ denotes the tangent bundle to M, and J is the operator of multiplication by the imaginary unit in $T(\mathbf{C}^N)$. If M is generic, then locally M can be defined as $\rho(z) = 0$, where $\rho = (\rho_1, \ldots, \rho_k)$ is a smooth real vector function such that $\partial \rho_1 \wedge \ldots \wedge \partial \rho_k \neq 0$.

Recall the *complex tangent space* $T_p^c(M)$ of M at $p \in M$ is defined as $T_p^c(M) = T_p(M) \cap JT_p(M)$. If M is generic, then M is a CR manifold, which means that $\dim_{\mathbf{C}} T_p^c(M)$ is independent of p, and $T^c(M)$ forms a bundle. Recall the space $T_p^{(1,0)}(M) \subset T_p(M) \otimes \mathbf{C}$ of complex $(1,0)$-vectors is defined as $T_p^{(1,0)}(M) = \{X \in T_p(M) \otimes \mathbf{C} : X = \sum a_j \, \partial/\partial z_j\}$.

Recall that M is *totally real* at $p \in M$ if $T_p^c(M) = 0$. Recall $M \subset \mathbf{C}^N$ is *maximally real* if M is totally real and $\dim M = N$.

From now on M will denote a generic manifold in $\mathbf{C}^N$. The CR dimension $\mathrm{CRdim}(M)$ of M is equal to $\dim_{\mathbf{C}} T_p^c(M) = \dim_{\mathbf{C}} T_p^{(1,0)}(M)$. If $\mathrm{CRdim}(M) = n$, then $N = n + k$.

Let $T^*(\mathbf{C}^N)$ be the real cotangent bundle of $\mathbf{C}^N$. Since every $(1,0)$ form is uniquely determined by its real part, we represent $T^*(\mathbf{C}^N)$ as the space of $(1,0)$ forms on $\mathbf{C}^N$. Then $T^*(\mathbf{C}^N)$ is a complex manifold. Let $N^*(M) \subset T^*(\mathbf{C}^N)$ be the real conormal bundle of $M \subset \mathbf{C}^N$. Using the representation of $T^*(\mathbf{C}^N)$ by $(1,0)$ forms, we define the fiber $N_p^*(M)$ at $p \in M$ as

$$N_p^*(M) = \{\phi \in T_p^*(\mathbf{C}^N) : \mathrm{Re}\phi|_{T_p(M)} = 0\}.$$

The angle brackets $\langle,\rangle$ denote the natural pairing between vectors and covectors. If $a, b \in \mathbf{C}^m$, then we put $\langle a, b \rangle = \sum a_j b_j$.

In a fixed coordinate system, we will identify $\phi = \sum \phi_j \, dz_j \in T^*(\mathbf{C}^N)$ with the vector $\phi = (\phi_1, \ldots, \phi_N) \in \mathbf{C}^N$. Then for $\phi \in N_p^*(M)$, the vector ϕ is orthogonal to M in the real sense, that is $\mathrm{Re}\langle \phi, X \rangle = 0$ for all $X \in T_p(M)$.

The forms $\partial\rho_j$, $(j = 1, \ldots, k)$, define a basis of $N_p^*(M)$, so every $\phi \in N_p^*(M)$ can be written as $\phi = \sum c_j \partial\rho_j$, $c_j \in \mathbf{R}$.

Using the matrix notation, we write ϕ and its coordinate representation respectively in the form $\phi = c\partial\rho(p)$ and $\phi = c\rho_z(p)$, where $c \in \mathbf{R}^k$ is considered a row vector, and ρ_z is the $k \times N$ matrix with the entries $(\rho_z)_{ij} = \partial\rho_i/\partial z_j$.

For every $\phi \in N_p^*(M)$ we define the Levi form $L(p, \phi)$ of M at $p \in M$ in the conormal direction $\phi = c\partial\rho$ as

$$L(p, \phi)(X, \bar{Y}) = -c\partial\bar{\partial}\rho(X, \bar{Y}),$$

where $X, Y \in T_p^{1,0}(M)$. The form $L(p, \phi)$ is a hermitian form on $T_p^{1,0}(M)$. This definition is independent of the defining function because, by Cartan's formula, $c\partial\bar{\partial}\rho(X, \bar{Y}) = \frac{1}{2}\langle \phi, [\xi, \bar{\eta}] \rangle$, where ξ, η are $(1,0)$ extensions of X, Y to a neighborhood of p in M.

Proposition 2.1. $N^*(M)$ is totally real at $\phi \in N_p^*(M)$ if and only if $L(p, \phi)$ is non-degenerate.

In the hypersurface case, this fact is essentially due to Webster (1978). For a proof in higher codimension, see, e.g. [T2].

The forms $L(p, \phi)$ can be regarded as components of the $N_p(M)$-valued Levi form $L(p)$, where $N(M) = T(\mathbf{C}^N)|_M/T(M)$ is the normal bundle of $M \subset \mathbf{C}^N$. Indeed, $L(p)(X, \bar{X}) \in N_p(M)$ is such an element that

$$\mathrm{Re}\langle \phi, L(p)(X, \bar{X}) \rangle = L(p, \phi)(X, \bar{X}) \quad \text{for all} \quad \phi \in N_p^*(M).$$

The Levi cone $\Gamma_p \subset N_p(M)$ is defined as the convex span of the values of the Levi form $L(p)$, that is

$$\Gamma_p = \mathrm{Conv}\{L(p)(X, \bar{X}) : X \in T_p^{1,0}(M)\}.$$

The dual Levi cone Γ_p^* is defined as

$$\Gamma_p^* = \{\phi \in N_p^*(M) : L(p,\phi) > 0\},$$

where $L(p,\phi) > 0$ means that the form $L(p,\phi)$ is positive definite.

Definition 2.2.

(i) We say that M is *strictly pseudoconvex* at p if $\Gamma_p^* \neq \emptyset$. We say that M is strictly pseudoconvex if it holds at every $p \in M$.

(ii) We say that the Levi form $L(p)$ is *generating* if Γ_p has nonempty interior.

The condition in (i) holds iff locally near p, M is contained in a strictly pseudoconvex hypersurface. It implies that Γ_p does not contain an entire line. The condition in (ii) means that $L(p,\phi) \neq 0$ as $\phi \neq 0$. A strictly pseudoconvex manifold with generating Levi form is an analogue of a strictly pseudoconvex hypersurface.

Changing notations, we introduce the coordinates $Z = (z,w) \in \mathbf{C}^N$, $z = x + iy \in \mathbf{C}^k$, $w \in \mathbf{C}^n$, so that the defining function of M can be chosen in the form $\rho = x - h(y,w)$, where $h = (h_1, \ldots, h_k)$ is a smooth real vector function with $h(0) = 0$, $dh(0) = 0$. Furthermore, we can choose the coordinates in such a way that each term in the Taylor expansion of h contains both w and $\bar{w}$ variables (see, e.g., [BER] [Bo]). Then the equations of M take the form

$$x_j = h_j(y,w) = \langle A_j w, \bar{w} \rangle + O(|y|^3 + |w|^3), \qquad 1 \leq j \leq k, \qquad (2.1)$$

where A_j are hermitian matrices. Using the coordinates, $N_0(M)$ and $N_0^*(M)$ are identified with $\mathbf{R}^k \subset \mathbf{C}^k$, $T_0^{1,0}(M)$ is identified with $\mathbf{C}^n$ and for $\phi = \sum c_j dz_j \in N_0^*(M)$, the Levi form $L(0,\phi)$ has the matrix $\sum c_j A_j$. Thus the matrices A_j are the components of the Levi form $L(0)$ of M at 0.

The manifold M of the form (2.1) is strictly pseudoconvex at 0 if and only if there exists $c \in \mathbf{R}^k$ such that $\sum c_j A_j > 0$. It has a generating Levi form at 0 if and only if the matrices $A_1, \ldots, A_k$ are linearly independent.

A strictly pseudoconvex hypersurface is locally strictly convex after a suitable change of coordinates. We give an analogue of this statement in higher codimension. For $p \in M$ we introduce the following realization of Γ_p^* in $\mathbf{R}^k$.

$$C_p = \{c \in \mathbf{R}^k : c\partial\rho \in \Gamma_p^*\} \subset \mathbf{R}^k.$$

Proposition 2.3. *Let M be a strictly pseudoconvex manifold with generating Levi form. For every $o \in M$, there exists a local holomorphic coordinate system with origin at o such that for every cone C' finer than C_o (that is $\bar{C}' \setminus \{0\} \subset C_o$) there is a neighborhood $U \subset M$ of o such that for every $p \in U$, $q \in U$, $p \neq q$, and $c \in C'$, $|c| = 1$, we have $\mathrm{Re}\langle \xi, p - q \rangle > 0$, where $\xi = c\partial\rho(q)$.*

The proof is similar to the one in the hypersurface case. See [T2] for the details.

3 Extremal Discs and Stationary Discs

Let M be a smooth generic manifold in $\mathbf{C}^N$.

An *analytic disc* in $\mathbf{C}^N$ is a mapping $f : \Delta \to \mathbf{C}^N$ holomorphic in the unit disc Δ. We say that f is *attached* to M if f is continuous in the closed disc $\bar{\Delta}$ and $f(b\Delta) \subset M$.

There is no natural domain associated with a manifold of higher codimension, therefore in the definition of extremal discs we restrict to discs attached to M.

Definition 3.1. An analytic disc $f : \bar{\Delta} \to \mathbf{C}^N$ attached to M is called *extremal* if there exists $a \in \mathbf{C}^N$ such that for every analytic disc $g : \bar{\Delta} \to \mathbf{C}^N$ attached to M such that $g \neq f$ and $g(0) = f(0)$ the following inequality holds:

$$\mathrm{Re}\langle \bar{a}, g'(0) - f'(0)\rangle > 0. \tag{3.1}$$

Let f be extremal. Then for every analytic disc $g : \bar{\Delta} \to \mathbf{C}^N$ attached to M such that

$$g \neq f, \quad g(0) = f(0), \quad g'(0) = \sigma f'(0), \quad \sigma > 0 \tag{3.2}$$

we have $|f'(0)| > |g'(0)|$. Indeed, first consider the disc g of the form $g(\zeta) = f(\zeta^2)$. Then $g'(0) = 0$, and by (3.1) we have

$$\mathrm{Re}\langle \bar{a}, -f'(0)\rangle > 0. \tag{3.3}$$

For a general disc g satisfying (3.2), (3.1) takes the form $\mathrm{Re}\langle \bar{a}, -(1-\sigma)f'(0)\rangle > 0$. Then by (3.3) $\sigma < 1$ and $|f'(0)| > |g'(0)|$.

For a strictly convex hypersurface, Definition 3.1 is equivalent to the fact that $|f'(0)| > |g'(0)|$ for all discs g satisfying (3.2), which is the original definition by Lempert [L1]. In higher codimension, the author does not know under what circumstances they are equivalent.

Definition 3.2. An analytic disc f attached to M is called *stationary* if there exists a nonzero continuous holomorphic mapping $f^* : \bar{\Delta} \setminus \{0\} \to T^*(\mathbf{C}^N)$, such that ζf^* is holomorphic in Δ and $f^*(\zeta) \in N^*_{f(\zeta)}M$ for all $\zeta \in b\Delta$. In other words, f^* is a punctured analytic disc with a pole of order at most one at zero attached to $N^*(M) \subset T^*(\mathbf{C}^N)$ such that the natural projection sends f^* to f. We call f^* a *lift* of f. In the rest of the paper, the term "lift" is used in this sense.

Example 3.3. Let $M = S^{2N-1}$ be the unit sphere in $\mathbf{C}^N$. Let $f(\zeta) = a + b\zeta$ be an analytic disc attached to M, where $a, b \in \mathbf{C}^N$. Then $|a|^2 + |b|^2 = 1$, $\langle a, \bar{b}\rangle = 0$. Then $f^*(\zeta) = -\bar{a} - \bar{b}\zeta^{-1}$ (as well as any real multiple of it) is a lift of f. All stationary discs for M have this form or differ from it by an automorphism of Δ. We will see that for $b \neq 0$ the disc f is extremal.

We call a disc f *defective* if it has a nonzero lift f^* holomorphic in the whole unit disc including 0. Defective discs arise in the problem of holomorphic extendibility of CR functions, see e.g. [BER] [T1]. Being defective is an anomaly. For instance, for a strictly convex hypersurface, all defective discs are constant. This is not true for strictly pseudoconvex manifolds of higher codimension.

Example 3.4. Let $M = S^3 \times S^3$, where S^3 is the unit sphere in $\mathbf{C}^2$. Then M is a generic manifold of codimension 2 in $\mathbf{C}^4$. Let $f = (f_1, f_2)$ be an analytic disc attached to M, so each f_j is a disc attached to S^3, $j = 1, 2$. Let $f_1 = p \in S^3$ be constant. Let $\phi \in N_p^*(S^3)$, $\phi \neq 0$. Let f_2 be any disc attached to S^3. Then the disc f is defective because it has the lift $f^* = (\phi, 0)$. Note that the discs of this form fill the set $S^3 \times \mathbf{B}^2$, where $\mathbf{B}^2$ is the unit ball in $\mathbf{C}^2$.

Conjecture 3.5. Defective discs attached to M can cover at most a set of measure zero in $\mathbf{C}^N$.

One can see that a disc f is defective iff f is a critical point of the evaluation map $f \mapsto f(0)$ defined on the set of all small analytic discs attached to M. Hence, if the set of defective discs was "finite dimensional", then the conjecture would follow by the Sard theorem. In general, the question is open.

For a strictly convex hypersurface, Lempert [L1] shows that the extremal and stationary discs are the same. This is not the case in higher codimension.

Example 3.6. Let $M = S^3 \times S^3$, where S^3 is the sphere in $\mathbf{C}^2$. Let $f = (f_1, f_2)$ be an analytic disc attached to M. Let f_1 be an extremal disc for the unit ball $\mathbf{B}^2 \subset \mathbf{C}^2$ and let f_2 be a nonconstant analytic disc attached to S^3 which is not extremal. Then the disc f is stationary. Indeed, it admits the lift $f^* = (f_1^*, 0)$, where f_1^* is a lift of f_1. The disc f is not extremal. Indeed, there exists a disc $g_2 \neq f_2$ attached to S^3 such that $g_2'(0) = f_2'(0)$. Take $g_1 = f_1$. Then for the disc $g = (g_1, g_2)$ we have $g'(0) = f'(0)$, so f is not extremal. Note however, that the disc f is "weakly" extremal, that is for every disc g attached to M with the same value and direction at 0, we have $|f'(0)| \geq |g'(0)|$, because $g'(0) = \sigma f'(0)$, $\sigma \geq 0$, implies $g_1'(0) = \sigma f_1'(0)$, whence $\sigma \leq 1$. The set of such weakly extremal discs f has infinite dimension because f includes an arbitrary component f_2.

We call a lift f^* *supporting* if for all $\zeta \in b\Delta$, $f^*(\zeta)$ defines a (strong) supporting real hyperplane to M at $f(\zeta)$, that is

$$\mathrm{Re}\langle f^*(\zeta), p - f(\zeta)\rangle > 0 \text{ for all } \zeta \in b\Delta \text{ and } p \in M, p \neq f(\zeta). \qquad (3.4)$$

For instance, in Example 3.3, the lift f^* is supporting.

We will show that stationary discs that admit supporting lifts are extremal and that extremal discs are stationary.

We put $\mathrm{Res}\psi = \mathrm{Res}(\psi, 0)$, the residue of ψ at 0.

Proposition 3.7. *Let f be a stationary disc with a supporting lift f^*. Then f is extremal. Furthermore, for every disc g attached to M such that $g(0) = f(0)$ the inequality (3.1) holds, where $\bar{a} = \mathrm{Res}f^*$.*

Proof. The proof is a repetition of the one by Lempert [L1].

Plugging $p = g(\zeta)$ in (3.4) yields $\mathrm{Re}\langle f^*(\zeta), g(\zeta) - f(\zeta)\rangle \geq 0$ for $\zeta \in b\Delta$, and the equality holds only if $f(\zeta) = g(\zeta)$. Since $\zeta \mapsto \zeta f^*(\zeta)$ is holomorphic and $f(0) = g(0)$, then

$$\mathrm{Re}\langle f^*(\zeta), g(\zeta) - f(\zeta)\rangle = \mathrm{Re}\langle \zeta f^*(\zeta), \zeta^{-1}(g(\zeta) - f(\zeta))\rangle$$

is harmonic and positive a. e. on $b\Delta$. Hence, it is positive at 0, and (3.1) holds, where $\bar{a} = \lim_{\zeta \to 0} \zeta f^*(\zeta) = \mathrm{Res} f^*$. The proposition is proved.

4 Coordinate Representation of Stationary Discs

Let M be a generic manifold in $\mathbf{C}^N$. We recall the equation for constructing analytic discs attached to M introduced by Bishop (1965, see e.g. [Bo] [T1]). Let M be defined by a local equation

$$x = h(y, w), \tag{4.1}$$

where $h(0) = 0$ and $dh(0) = 0$. Let $\zeta \mapsto f(\zeta) = (z(\zeta) = x(\zeta) + iy(\zeta), w(\zeta))$ be an analytic disc attached to M. If $\zeta \mapsto w(\zeta)$ and $y_0 = y(0)$ are given, then the function $\zeta \mapsto y(\zeta)$ on the circle $b\Delta$ satisfies the Bishop equation

$$y = Th(y, w) + y_0, \tag{4.2}$$

where T is the Hilbert transform on the unit circle $b\Delta$. The operator T is bounded on the space $C^{k,\alpha}$ of functions with derivatives up to order $k \geq 0$ satisfying a Lipschitz condition with exponent $0 < \alpha < 1$. We often write C^α for $C^{0,\alpha}$.

If h is C^k, where $k \geq 2$, then for sufficiently small $y_0 \in \mathbf{R}^k$ and $\|w\|_{C^\alpha}$ there exist a unique solution y of class C^α and depends C^{k-1} smoothly on y_0 and w (see [Bo]). The solution defines the z component of f by harmonically extending $\zeta \mapsto h(y(\zeta), w(\zeta)) + iy(\zeta)$ from $b\Delta$ to Δ.

Let f be an analytic disc attached to M. We introduce a $k \times k$ matrix function G on $b\Delta$ such that

$$H = G(I + ih_y \circ f) \tag{4.3}$$

extends holomorphically from $b\Delta$ to Δ, where h_y denotes the matrix of partial derivatives and I is the identity matrix. For a small disc f, the matrix G always exists and it is unique up to a (left) constant factor. We define G by the equation

$$G = I - T(Gh_y \circ f). \tag{4.4}$$

For a fixed disc f, we will omit writing "$\circ f$" in most formulas.

We interpret the definition of a stationary disc using the equation of M in a fixed coordinate system. Let $Z = (z, w)$ be a coordinate system as above, and let M be defined by (4.1). Let $\rho = x - h(y, w)$.

Proposition 4.1. *Let M be C^2 smooth. Let $f \in C^{0,\alpha}(\bar{\Delta})$, $0 < \alpha < 1$, be a small disc attached to M. Then the following (i) to (v) hold.*

(i) *Every lift f^* of f holomorphic at 0 has the form $f^*|_{b\Delta} = cG\partial\rho$, where $c \in \mathbf{R}^k$.*

(ii) *f is defective if and only if there exists a nonzero $c \in \mathbf{R}^k$ such that cGh_w extends holomorphically to Δ.*

(iii) *Every lift f^* of f has the form $f^*|_{b\Delta} = \text{Re}(\lambda\zeta + c)G\partial\rho$, where $\lambda \in \mathbf{C}^k$, $c \in \mathbf{R}^k$.*

(iv) *f is stationary if and only if there exist $\lambda \in \mathbf{C}^k$ and $c \in \mathbf{R}^k$ such that $\zeta\text{Re}(\lambda\zeta + c)Gh_w$ extends holomorphically to Δ.*

(v) *In both (i) and (iii) $f^* \in C^{0,\alpha}(b\Delta)$. Moreover, if M is $C^{m,\alpha}$, where $m \geq 2$, and $f \in C^{m-1,\alpha}(\bar{\Delta})$, then $f^* \in C^{m-1,\alpha}(b\Delta)$.*

Proof. We skip (i) and (ii) because they are similar to (iii) and (iv). Let f^* be a lift of f. Then there is a real row vector function r on $b\Delta$ such that

$$f^*|_{b\Delta} = r\partial\rho = r(\rho_z\, dz + \rho_w\, dw).$$

Note $\rho_z = \frac{1}{2}(I + ih_y)$, $\rho_w = -h_w$. The function

$$\zeta r\rho_z = \frac{1}{2}\zeta r(I + ih_y) = \frac{1}{2}\zeta rG^{-1}H$$

extends holomorphically to Δ as z-component of ζf^*. Since $H = G(I + ih_y)$ is a nondegenerate matrix that extends holomorphically to Δ, then ζR, where $R = rG^{-1}$, extends holomorphically to Δ. Since R is real, then the Fourier series of R must have the form $R(\zeta) = \text{Re}(\lambda\zeta + c)$ for some $\lambda \in \mathbf{C}^k$ and $c \in \mathbf{R}^k$. Then $r = \text{Re}(\lambda\zeta + c)G$, and (iii) follows. Now f^* in (iii) defines a lift if and only if the w-component of ζf^* is also holomorphic, and (iv) follows. Further, (v) immediately follows from (i) and (iii). The proposition is now proved.

The extremal discs are stationary, furthermore the following holds.

Proposition 4.2. *Let $f \in C^\alpha(\bar{\Delta})$, $0 < \alpha < 1$, be an extremal disc attached to a manifold M of class C^2 so that (3.1) holds for some $a \in \mathbf{C}^N$. Then f is stationary. Suppose f is not defective. Then there exists a unique lift f^* such that $\text{Res} f^* = \bar{a}$. Moreover, f^* depends continuously on f and a in the C^α norm.*

The proof consists of quite technical calculations in coordinates, see [T2]. It does not help understand the result and we omit it. We would appreciate a natural coordinate-free proof.

A question of interest is the regularity of extremal and stationary discs. Let M be strictly pseudoconvex with generating Levi form. Let f be a stationary disc with lift f^*. Assume $N^*(M)$ is totally real at $f^*(\zeta_0)$ for some

$\zeta_0 \in b\Delta$, which by Proposition 2.1 is equivalent to the Levi form $L(f^*(\zeta_0))$ being nondegenerate. Then by the smooth version of Schwarz reflection principle (see Proposition 9.5 below, in which $n_1 = 1$) f^* whence f is smooth up to $b\Delta$ near ζ_0. Unfortunately, we don't know whether the above assumption should hold for extremal discs. In fact for the stationary disc f in Example 3.6 it does not hold, and f is smooth iff f_2 is smooth. Therefore we cannot guarantee that extremal discs are smooth up to the boundary.

This difficulty persists only for $k > 1$.

Proposition 4.3. *Let $M \subset \mathbf{C}^N$ be a strictly pseudoconvex hypersurface. Then all small stationary discs of class $C^\alpha(\bar\Delta)$, $0 < \alpha < 1$, are actually smooth up to the boundary.*

Proof. Let f be a stationary disc of class C^α with lift f^*. By Proposition 4.1(v), the lift f^* is also C^α. By Proposition 2.1 for $k = 1$, $N^*(M)$ is totally real off the zero-section. It f^* does not vanish on $b\Delta$, then by the smooth reflection principle, f^* whence f is smooth up to $b\Delta$.

Now assume that f^* does vanish on $b\Delta$, but not identically, say $f^*(1) = 0$. By Proposition 4.1 (iii), $f^*|_{b\Delta} = \mathrm{Re}(\lambda\zeta + c)G\partial\rho$, where $\lambda \in \mathbf{C}$, $c \in \mathbf{R}$. Then $f^*(1) = 0$ implies $\mathrm{Re}\lambda + c = 0$. Since f^* is nontrivial, then $\lambda \neq 0$. Then for $|\zeta| = 1$ we have $2\mathrm{Re}(\lambda\zeta + c) = \zeta^{-1}(\zeta - 1)(\lambda\zeta - \bar\lambda)$. By Proposition 4.1 $\zeta\mathrm{Re}(\lambda\zeta + c)Gh_w$ whence $(\zeta - 1)(\lambda\zeta - \bar\lambda)Gh_w$ extends holomorphically from $b\Delta$ to Δ. Then so does Gh_w because $(\zeta - 1)(\lambda\zeta - \bar\lambda)$ vanishes only on $b\Delta$. Then $\tilde{f}^*|_{b\Delta} = G\partial\rho$ is a nonzero lift of f holomorphic at 0, and f is defective.

For a strictly pseudoconvex hypersurface, small defective discs are constant. Indeed, the lift $\tilde{f}^*$ above is an analytic dicsc attached to a totally real manifold $N^*(M)$. By Bishop's equation (4.2) for a totally real manifold, the attached discs are constant. Hence $\tilde{f}^*$ and f are constant, and the proof is complete.

5 Stationary Discs for Quadrics

We first construct stationary discs for a quadratic CR manifold $M \subset \mathbf{C}^N$.

Definition 5.1. Let f be a stationary disc attached to M and let f^* be a lift of f. We call f^* *positive* if $f^*(\zeta) \in \Gamma^*_{f(\zeta)}$ for $\zeta \in b\Delta$.

The positivity condition is independent of the coordinate system. We will describe stationary discs with positive lifts for a quadratic strictly pseudoconvex manifold M with generating Levi form.

Let the quadric $M \subset \mathbf{C}^N$, $N = n + k$, be defined by the equation

$$x_j = h_j(y, w) = \langle A_j w, \bar{w} \rangle, \qquad 1 \le j \le k, \tag{5.1}$$

where the hermitian matrices $A_1, \ldots, A_k$ are linearly independent and a linear combination of A_j-s is positive definite.

Let $\zeta \mapsto f(\zeta) = (z(\zeta), w(\zeta))$ be a stationary disc and let $\zeta \mapsto f^*(\zeta)$ be a lift of f. Note that $G = I$ since $h_y = 0$. Also note that the disc f can be made small by a change of coordinates of the form $(z, w) \mapsto (t^2z, tw)$. Hence, by Proposition 4.1(iii) there exist $\lambda \in \mathbf{C}^k$ and $c \in \mathbf{R}^k$ such that

$$f^*|_{b\Delta} = \mathrm{Re}(\lambda\zeta + c)\partial\rho, \tag{5.2}$$

where $\rho_j = x_j - \langle A_j w, \bar{w}\rangle$. The lift f^* is positive if

$$\sum \mathrm{Re}(\lambda_j\zeta + c_j)A_j > 0 \quad \text{for} \quad \zeta \in b\Delta. \tag{5.3}$$

By a classical factorization theorem (see e.g. [L1]), there exist a holomorphic in Δ nondegenerate in $\bar{\Delta}$ $n \times n$ matrix function Φ such that

$$\sum \mathrm{Re}(\lambda_j\zeta + c_j)A_j = \Phi^*(\zeta)\Phi(\zeta) \quad \text{for} \quad \zeta \in b\Delta. \tag{5.4}$$

The structure of the left hand part implies that Φ is linear in ζ that is $\Phi(\zeta) = \Phi_0(I - \zeta X)$, where Φ_0 and X are constant matrices. The matrix Φ is unique up to a left multiplicative constant, so X is uniquely defined by λ and c. Then (5.4) takes the form

$$2\sum \mathrm{Re}(\lambda_j\zeta + c_j)A_j = (I - \zeta X)^* B(I - \zeta X), \tag{5.5}$$

where $B = 2\Phi_0^*\Phi_0 > 0$.

Since $\Phi(\zeta)$ is nondegenerate and holomorphic in $\bar{\Delta}$, the series $(I - \zeta X)^{-1} = \sum \zeta^m X^m$ converges in $\bar{\Delta}$, which holds if and only if all the eigenvalues of X are in Δ.

Let $P = \sum \lambda_j A_j$ and $Q = \sum c_j A_j$. Then (5.5) is equivalent to the system

$$P = -BX, \qquad 2Q = B + X^*BX. \tag{5.6}$$

Eliminating B, we get a matrix quadratic equation

$$P^*X^2 + 2QX + P = 0. \tag{5.7}$$

The equations (5.6) and (5.7) are not needed for the existence result, but they are useful for practical calculating stationary discs. Here is a description of stationary discs with positive lifts for a quadric.

Proposition 5.2. *Let M be a quadratic manifold given by the equations (5.1). For every $\lambda \in \mathbf{C}^k$ and $c \in \mathbf{R}^k$ satisfying (5.3) and every $w_0, v \in \mathbf{C}^n$, $y_0 \in \mathbf{R}^k$, there exists a unique stationary disc $\zeta \mapsto f(\zeta) = (z(\zeta), w(\zeta))$ with lift f^* given by (5.2) such that $w(0) = w_0$, $w'(0) = v$, $y(0) = y_0$. The disc f is extremal. Moreover, the w-component of f is given by $w(\zeta) = w_0 + \zeta(I - \zeta X)^{-1}v$, where X is part of a unique solution (X, B) of the system (5.6) such that all eigenvalues of X are in Δ. Also, X is a unique solution of the quadratic equation (5.7) with all eigenvalues in Δ.*

Proof. By (5.2) $\zeta\mathrm{Re}(\lambda\zeta + c)h_w = \zeta\sum \mathrm{Re}(\lambda_j\zeta + c_j)\bar{A}_j\bar{w}(\zeta)$ extends holomorphically to Δ. By (5.5) we get that $\bar{\zeta}(I - \zeta X)^*B(I - \zeta X)w(\zeta)$ extends antiholomorphically to Δ. Since $(I - \zeta X)^*B$ is invertible and antiholomorphic, then $\bar{\zeta}(I - \zeta X)w(\zeta)$ also extends antiholomorphically.

Let $w(\zeta) = w(0) + \zeta u(\zeta)$, where u is holomorphic. Then $(I - \zeta X)u(\zeta)$ extends antiholomorphically. Since it is also holomorphic, then it is constant. Hence, $(I - \zeta X)u(\zeta) = u(0) = v$, and we immediately obtain the expression for the w-component of f. The z-component is uniquely defined by $w(\zeta)$ and y_0. Thus the pair (f, f^*) is uniquely defined.

The disc f is extremal. Indeed, by Proposition 2.3, f^* can be made supporting in a suitable system of coordinates.

Finally, we prove that if X is a solution of (5.7) with eigenvalues in Δ, then there exists B such that (5.5) is fulfilled. We put $B = 2Q + P^*X$. Plug the latter in (5.7) to obtain $P = -BX$ and then $2Q = B + X^*B^*X$. To satisfy (5.6) whence (5.5), we only need to show that $B^* = B$. Consider $S = B - B^*$. Then $S = X^*SX$. Then $S = X^{*m}SX^m \to 0$ as $m \to \infty$ since the eigenvalues of X are in Δ. Thus $S = 0$, $B^* = B$. The proposition is now proved.

Proposition 5.3. *Let M be a quadratic manifold given by the equations (4.1). For every $\lambda \in \mathbf{C}^n$ and $c \in \mathbf{R}^k$ satisfying (4.3) and every $w_0, v \in \mathbf{C}^n$, $y_0 \in \mathbf{R}^k$, there exists a unique stationary disc $\zeta \mapsto f(\zeta) = (z(\zeta), w(\zeta))$ with lift f^* given by (4.2) such that $w(1) = w_0, w'(1) = v, y(1) = y_0$. The disc f is extremal. Moreover, the w-component of f is given by $w(\zeta) = w_0 + (\zeta - 1)(I - \zeta X)^{-1}(I - X)v$.*

The proof is similar to that of Proposition 5.2 and we omit it.

6 Existence of Stationary Discs

Theorem 6.1. *Let $M \subset \mathbf{C}^N$ be a smooth strictly pseudoconvex manifold with generating Levi form. Let M be defined by the equations*

$$\rho_j(z, w) = x_j - h_j(y, w) = x_j - \langle A_j w, \bar{w} \rangle + O(|y|^3 + |w|^3) = 0, \qquad 1 \le j \le k.$$

Then for every $\epsilon > 0$ there exists $\delta > 0$ such that for every $\lambda \in \mathbf{C}^k$, $c \in \mathbf{R}^k$, $w_0 \in \mathbf{C}^n$, $y_0 \in \mathbf{R}^k$, $v \in \mathbf{C}^n$ such that

$$\sum \mathrm{Re}(\lambda_j \zeta + c_j)A_j > \epsilon(|\lambda| + |c|)I \qquad (6.1)$$

and $|w_0| < \delta, |y_0| < \delta, |v| < \delta$ there exists a unique stationary disc $\zeta \mapsto f(\zeta) = (z(\zeta), w(\zeta))$ such that $w(0) = w_0$, $w'(0) = v$, $y(0) = y_0$, and there is a lift f^* such that $f^*|_{b\Delta} = \mathrm{Re}(\lambda \zeta + c)G\partial\rho$. The disc f is extremal, and depends smoothly on $\zeta \in \bar{\Delta}$ and all the parameters λ, c, w_0, y_0, v.

Proof. We will convert the condition that the disc f is stationary into an equation so that, together with Bishop's equation (4.2), it will describe small stationary discs. The idea of the proof is that Gh_w is close to that for the corresponding quadric. We then introduce the error term $\Phi(G, y, w) = Aw - Gh_{\bar{w}}(y, w)$, where Aw is the $k \times n$ matrix with the entries $(Aw)_{jl} = \sum_{m=1}^n A_{jlm}w_m$, and $A_{jlm} = (A_j)_{lm}$.

The condition in Proposition 4.1(iv) holds if and only if $\bar{\zeta} \mathrm{Re}(\lambda\zeta + c)Aw +$ ... extends antiholomorphically to Δ where the dots mean the error term involving Φ. By the decomposition (5.5), $\bar{\zeta}(I - \zeta X)^* B(I - \zeta X)w + \ldots$ extends antiholomorphically to Δ.

We put $w(\zeta) = w(0) + \zeta u(\zeta)$, where u is holomorphic. We divide by $(I - \zeta X)^* B$ and get that $(I - \zeta X)u(\zeta) + \ldots$ extends antiholomorphically to Δ. We put

$$K_0\phi = K\phi - (K\phi)(0), \qquad (K\phi)(\zeta_1) = \frac{1}{2\pi i} \int_{|\zeta|=1} \frac{\phi(\zeta)\, d\zeta}{\zeta - \zeta_1},$$

the latter being the standard Cauchy transform in Δ. For $\zeta_1 \in b\Delta$, we understand $(K\phi)(\zeta_1)$ as the inner limiting value for Δ. Obviously, if ϕ is holomorphic, then $K_0\phi = \phi - \phi(0)$. If ϕ is antiholomorphic, then $K_0\phi = 0$.

Using K_0, we rewrite our condition in the form $(I - \zeta X)u(\zeta) = u(0) + K_0(\Psi)$, where $\Psi = 2\bar{\zeta}B^{-1}(I - \zeta X)^{*-1}(\mathrm{Re}(\lambda\zeta + c)\Phi)^T$ is obtained by replacing the dots above by their actual expressions. Hence $w = w_0 + \zeta(I - \zeta X)^{-1}v + \zeta(I - \zeta X)^{-1}K_0(\Psi)$. This is the equation we need. Finally,

$$\begin{aligned}
y &= y_0 + Th(y, w) \\
w &= w_0 + \zeta(I - \zeta X)^{-1}v + \zeta(I - \zeta X)^{-1}K_0(\Psi) \\
G &= I - T(Gh_y(y, w))
\end{aligned} \qquad (6.2)$$

form a system with the unknowns y, w, G and parameters λ, c, v, w_0, y_0.

The existence, uniqueness, and regularity of solutions of such system follow from the implicit function theorem in Banach spaces, see [Bo]. The condition (6.1) ensures that for given ϵ, the expression $(I - \zeta X)^{-1}$ is uniformly bounded in λ, c, and $\zeta \in \bar{\Delta}$.

The disc f obtained by solving the system is extremal. Indeed, (6.1) implies that for small ϵ and δ the values of the lift $f^*(\zeta)$, $\zeta \in b\Delta$, are in the cone finer than the dual Levi cone, and by Proposition 2.3, there is a suitable change of coordinates that makes the lifts f^* supporting simultaneously for all λ and c satisfying (6.1). The theorem is now proved.

Theorem 6.2. *The statement of Theorem 6.1 will be valid if the initial values of f are defined at $1 \in \bar{\Delta}$ instead of 0, that is $w(1) = w_0^-$, $w'(1) = v$, and $y(1) = y_0$, and the lift f^* is defined by $f^*|_{b\Delta} = \mathrm{Re}(\lambda\zeta + c)G^{-1}(1)G\partial\rho$.*

The theorem is derived from Theorem 6.1 by applying the implicit function theorem to the mapping $(\lambda, c, w_0, y_0, v) \mapsto (\lambda G(1), cG(1), w(1), y(1), w'(1))$, see [T2] for the details.

Theorems 6.1 and 6.2 do not describe all small extremal discs even in the case $k = 1$. Indeed, when ϵ shrinks relaxing the restrictions on λ and c, then δ can also shrink restricting the other parameters. Therefore, we cannot even guarantee that all extremal discs are smooth.

The set $\{(f, f^*)\}$ of extremal discs with lifts in Theorems 6.1 and 6.2 depends on $4N$ real parameters. Indeed, by adding the dimensions of λ, c, w_0, y_0, v, we get $2k + k + 2n + k + 2n = 4N$. Surprisingly, this number depends only on the dimension of the ambient space $\mathbf{C}^N$, not on the dimension of M itself. Unfortunately, the parameters do not have a clear geometric meaning. For instance, in Theorem 6.2, w_0 and y_0 define a geometric object $f(1)$, whereas v defines only the w-component of $f'(1)$, which depends on the coordinate system. Also, the expression $\mathrm{Re}\lambda + c$ defines $f^*(1)$, but the significance of each parameter λ and c is unclear.

A relevant question is the dimension of the set of extremal discs. Obviously, it is related to the dimension of the spaces of lifts of individual discs. For instance, if each extremal disc had only one lift up to proportionality, then the space of extremal discs would have dimension $4N - 1$.

Example 6.3. Let $M = S^3 \times S^3 \subset \mathbf{C}^4$. Then every extremal disc $f = (f_1, f_2)$ has 2-dimensional space of lifts of the form $f^* = (c_1 f_1^*, c_2 f_2^*)$, where f_j^* is a fixed nontrivial lift of f_j, and $c_j \in \mathbf{R}$, $j = 1, 2$. The set of extremal discs depends on $16 - 2 = 14$ parameters.

Example 6.4. (Scalari [Sc]) Let M_σ be defined by

$$x_1 = |w_1|^2 + \sigma\mathrm{Re}(w_1^2 \bar{w}_2), \quad x_2 = |w_2|^2.$$

Then M_0 is equivalent to the previous example $S^3 \times S^3$. For $\sigma \neq 0$ all M_σ are equivalent. It turns out that for $\sigma \neq 0$ every extremal disc for M_σ constructed by Theorem 6.1 has only one lift up to proportionality, hence the set of extremal discs depends on 15 parameters. Thus the dimension of this set is not stable under small perturbations.

7 Geometry of the Lifts

We now describe the set in $T^*(\mathbf{C}^N)$ covered by the pairs (f, f^*) where f is a stationary disc with lift f^*. We will prove that the set contains a wedge with edge $N^*(M)$.

We need the following uniqueness property of stationary discs.

Proposition 7.1. *Let f and g be small stationary discs of class $C^{1,\alpha}(\bar{\Delta})$ ($1/2 < \alpha < 1$) attached to M and let f^* and g^* be their supporting lifts of class $C^{1,\alpha}(b\Delta)$ such that $f(1) = g(1)$, $f^*(1) = g^*(1)$, $f'(1) = g'(1)$, $f^{*\prime}(1) = g^{*\prime}(1)$. Assume in addition that f is not defective. Then $f = g$ and $f^* = g^*$.*

Proof. Suppose $f \neq g$. Then the inequalities

$$\mathrm{Re}\langle f^*(\zeta), g(\zeta) - f(\zeta)\rangle > 0$$
$$\mathrm{Re}\langle g^*(\zeta), f(\zeta) - g(\zeta)\rangle > 0$$

hold almost everywhere on $b\Delta$. We add the above inequalities, divide by $|\zeta - 1|^4$ and integrate along the circle. We get

$$\text{Re} \int_0^{2\pi} |\zeta - 1|^{-4} \langle f^*(\zeta) - g^*(\zeta), g(\zeta) - f(\zeta) \rangle d\theta > 0,$$

where $\zeta = e^{i\theta}$. Note that on $b\Delta$ we have $|\zeta - 1|^{-2} = -\zeta(\zeta - 1)^{-2}$, $d\zeta = i\zeta\, d\theta$. Therefore, the above integral after "Re" turns into

$$\int_{b\Delta} \left\langle \frac{\zeta(f^*(\zeta) - g^*(\zeta))}{(\zeta - 1)^2}, \frac{g(\zeta) - f(\zeta)}{(\zeta - 1)^2} \right\rangle d\zeta = 0,$$

since the integrand is holomorphic in Δ. The contradiction proves that $f = g$.

Now the lift $\tilde{f}^*(\zeta) = \zeta(\zeta - 1)^{-2}(f^*(\zeta) - g^*(\zeta))$ is holomorphic at 0. Since f is not defective, then $\tilde{f}^* = 0$, and $f^* = g^*$. The proposition is now proved.

We need the following

Lemma 7.2. Let $F : \mathbf{R}^n \times \mathbf{R}^k \times \mathbf{R} \to \mathbf{R}^n \times \mathbf{R}^k$, $F(x, y, t) = (u, v)$, be a C^2 smooth mapping, such that $F(x, y, 0) = (x, 0)$ and $y \mapsto \frac{\partial v(0, y, 0)}{\partial t}$ is a diffeomorphism at 0, that is $\det \frac{\partial^2 v(0,0,0)}{\partial y \partial t} \neq 0$. Then there exists an open cone $\Gamma \subset \mathbf{R}^k$, and $\epsilon > 0$ such that $\frac{\partial v(0,0,0)}{\partial t} \in \Gamma$ and the image of F contains the wedge $W = \{(u, v) : |u| < \epsilon, |v| < \epsilon, v \in \Gamma\}$

Proof. The mapping F is compared with the first nonzero term in its Taylor expansion, whose image obviously contains the desired wedge. Then a simple topological argument implies that the image of F contains a finer wedge of the same form. The argument is somewhat similar to that in [Bo], section 15.5, Lemma 3. We leave the details to the reader.

Let $\mathcal{E}$ be the set of all pairs (f, f^*) constructed by Theorem 6.2.

Theorem 7.3. Let $(f_0, f_0^*) \in \mathcal{E}$, $(p, \phi) = (f_0(1), f_0^*(1))$. Assume that f is not defective. Then for every $\epsilon > 0$ the set

$$W = \{ (f(1 - t), f^*(1 - t)) : (f, f^*) \in \mathcal{E}, \quad \|f - f_0\| < \epsilon,$$
$$\|f^* - f_0^*\|_{b\Delta} < \epsilon, \quad 0 < t < \epsilon\},$$

covers an open wedge with edge $N^*(M)$ in $T^*(\mathbf{C}^N)$ near (p, ϕ).

Proof. Note that the type of the norm above is unimportant because $\mathcal{E}$ has finite dimension. Without loss of generality, $p = 0$. Consider the map

$$F : (\lambda, c, y_0, w_0, v) \mapsto (f(1 - t), f^*(1 - t)) \in T^*(\mathbf{C}^N),$$

where $(f, f^*) \in \mathcal{E}$, $0 < t < 1$. We apply Lemma 7.2, where the variables x, y, t in the lemma correspond to our variables λ, c, y_0, w_0, v as follows: $x \leftrightarrow (\text{Re}(\lambda) + c - \phi, y_0, w_0)$, $y \leftrightarrow (\lambda, v)$, where ϕ is represented by its coordinates in the basis $\partial \rho_j(0) = dz_j/2$. In the target $T^*(\mathbf{C}^N)$, we introduce the coordinates (u, v) such that $N^*(M)$ is defined by $v = 0$. (This v has nothing in common with the parameter $v = w'(1)$.)

We should check that $y \mapsto \frac{\partial v(0, y, 0)}{\partial t}$ is a diffeomorphism. This reduces to checking that the mapping $\Phi : (v, \lambda) \mapsto (f'(1), f^{*'}(1))$, for fixed $(\text{Re}(\lambda) +$

$c - \phi, y_0, w_0) = 0$, is a diffeomorphism onto an open set in a complement to $T_\phi N^*(M)$ in $T_\phi T^*(\mathbf{C}^N)$.

Since (f, f^*) is attached to $N^*(M)$, then by the Cauchy-Riemann equations, we have $(f'(1), f^{*'}(1)) \in JT_\phi N^*(M)$, where J is the operator of multiplication by the imaginary unit in $T_\phi T^*(\mathbf{C}^N) \simeq \mathbf{C}^N \times \mathbf{C}^N$. Since $N^*(M)$ is maximally real at ϕ, $JT_\phi N^*(M)$ forms a complement to $T_\phi N^*(M)$. Finally, by Proposition 7.1, Φ is injective, so Φ is a diffeomorphism in an open set as a smooth injective mapping of Euclidean spaces of equal dimensions. The theorem now follows from Lemma 7.2.

Theorem 7.3 implies that the directions of the boundary curves of the extremal discs cover an open set in the tangent space.

Corollary 7.4. *Under the hypotheses of Theorem 7.3, for every $\epsilon > 0$, the set*

$$\Sigma' = \{\frac{d}{d\theta}|_{\theta=0} f(e^{i\theta}) : (f, f^*) \in \mathcal{E}, f(1) = p, \|f - f_0\| < \epsilon\} \subset T_p(M)$$

is open.

Proof. According to the proof of Theorem 7.3, the set $\{(f'(1), f^{*'}(1))\}$, where the boundary point $\{(f(1), f^*(1))\}$ is fixed, is an open set in $JT_\phi N^*(M)$. Therefore, by the Cauchy-Rieman conditions, the set $\{\frac{d}{d\theta}|_{\theta=0}(f(e^{i\theta}), f^*(e^{i\theta}))\}$ is open in $T_\phi N^*(M)$. Then its projection Σ' is open in $T_p(M)$ and the corollary follows.

8 Defective Manifolds

The results of the previous section rely on the existence of non-defective discs. Usually this is not a problem, but we need to consider the question carefully.

Definition 8.1. Let $A_1, \ldots, A_k$ be $n \times n$ linearly independent hermitian matrices. We call the tuple (λ, c, v) *defective* if the linear operators $\mathbf{C}^N \to \mathbf{C}^N$ with matrices $A_1, \ldots, A_k$ are linearly dependent on the subspace $S = \text{Span}\{X^m v : m = 0, 1, \ldots\}$, where X is defined by (λ, c) by (5.7). If a tuple is not defective, then all tuples except possibly a proper algebraic set are not defective.

We call the set $(A_1, \ldots, A_k)$ and the corresponding quadric

$$x_j = \langle A_j w, \bar{w} \rangle, \qquad j = 1, \ldots, k \tag{8.1}$$

defective if all tuples (λ, c, v) are defective.

We call a generic manifold M *non-defectve* if for all $p \in M$, the quadric defined by the Levi form $L(p)$ is not defective.

Conjecture 8.2. No quadrics are defective.

According to this conjecture, the subject of this section is an empty set! Defective quadrics, if exist at all, form a proper algebraic set in the set of all strictly pseudoconvex generating quadrics. Indeed, if the matrix X for some λ and c has no multiple eigenvalues, then there is $v \in \mathbf{C}^N$ such that $S = \mathbf{C}^N$, so (λ, c, v) is not defective. This will be the case if the matrix $Q^{-1}P$, where $P = \sum \lambda_j A_j$ and $Q = \sum c_j A_j$, has no multiple eigenvalues for some λ and c. It is easy to see that no quadrics of codimension 2 are defective, but in general Conjecture 8.2 is open.

Lemma 8.3. *The tuple (λ, c, v) is defective for the quadric (8.1) iff the disc f constructed for the quadric (8.1) by Proposition 5.3 with given λ, c, v and $w_0 = 0$, $y_0 = 0$ is defective.*

Proof. By Proposition 5.3, $w(\zeta) = (\zeta - 1)(I - \zeta X)^{-1}u$, where $u = (I - X)v$. Let $S_u = \mathrm{Span}\{X^m u : m = 0, 1, \ldots\} \subset S$. Then $S_u = S$. Indeed, $v - X^m v = \sum_{j=0}^{m-1}(X^j v - X^{j+1} v) = \sum_{j=0}^{m-1} X^j u \in S_u$. By letting $m \to \infty$, we get $v \in S_u$ because $X^m v \to 0$. Hence $v \in S_u$ and $S_u = S$.

By Proposition 4.1 (ii), where $G \equiv I$, the disc f is defective iff there exists nonzero $\mu \in \mathbf{R}^k$ such that μh_w extends holomorphically to Δ. We put $R = \sum \mu_j A_j$. Then

$$\mu h_w = \bar{R}\bar{w} = \overline{R(\zeta - 1)(I - \zeta X)^{-1}u}.$$

By expanding into a Fourier series, μh_w extends holomorphically to Δ iff $RX^m u = 0$ for all $m \geq 0$, which means that the tuple (λ, c, v) is defective and the lemma follows.

Proposition 8.4. *Let $M \subset \mathbf{C}^N$ be a smooth generic strictly pseudoconvex manifold as in Theorem 6.2, and let (λ, c, v) be a non-defective tuple for the quadric defined by the Levi form of M at 0. Then there exists $\delta > 0$ such that for every $0 < t < \delta$, the stationary disc $f^{(t)}$ constructed by Theorem 6.2 with $w(1) = 0$, $y(1) = 0$, $w'(1) = tv$, $f^*|_{b\Delta} = \mathrm{Re}(\lambda\zeta + c)G^{-1}(1)G\partial\rho$ is not defective.*

Proof. We apply Pinchuk's scaling method. Let M be given by the equations

$$x_j = \langle A_j w, \bar{w}\rangle + \chi_j(y, w), \qquad j = 1, \ldots, k,$$

where $\chi_j(y, w) = O(|y|^3 + |w|^3)$. Let $\Phi^{(t)}(z, w) = (t^{-2}z, t^{-1}w)$, $t > 0$. Let $M^{(t)} = \Phi^{(t)}(M)$. Then $M^{(t)}$ has defining equations

$$x_j = \langle A_j w, \bar{w}\rangle + \chi_j^{(t)}(y, w), \qquad j = 1, \ldots, k,$$

where $\chi_j^{(t)}(y, w) = t^{-2}\chi_j(t^2 y, tw) \to 0$ as $t \to 0$ (say, in the C^3 norm), so $M^{(t)}$ approaches the corresponding quadric M_0 given by (8.1).

Let f be the disc constructed for M_0 by Proposition 5.3. By Lemma 8.3, f is not defective. Let $\tilde{f}^{(t)} = \Phi^{(t)} \circ f^{(t)}$. Then $\tilde{f}^{(t)}$ and f are stationary discs attached to $M^{(t)}$ and M_0 with the same values of all parameters. The solution of (6.2) continuously depends on h. Since $M^{(t)}$ approaches M_0, then $\tilde{f}^{(t)} \to f$ as $t \to 0$. The property of not being defective is stable under small perturbations. Hence $\tilde{f}^{(t)}$ is not defective for small t. Since $f^{(t)}$ differs from $\tilde{f}^{(t)}$ by a change of coordinates, then $f^{(t)}$ is not defective and the proposition follows.

9 Regularity of CR Mappings

We apply the extremal discs to the question of the regularity of CR mappings.

Let M_1 and M_2 be C^∞ smooth generic manifolds in $\mathbf{C}^N$, and let $F : M_1 \to M_2$ be a homeomorphism. We call F a CR homeomorphism if both F and F^{-1} are CR mappings, that is the components of F and F^{-1} are CR functions (see e.g. [BER] [Bo] [T1]). We prove the following.

Theorem 9.1. *Let M_1 and M_2 be C^∞ smooth generic strictly pseudoconvex non-defective (see Definition 8.1) generic manifolds in $\mathbf{C}^N$ with generating Levi forms, and let $F : M_1 \to M_2$ be a CR homeomorphism such that both F and F^{-1} satisfy a Lipschitz condition with some exponent $0 < \alpha < 1$. Then F is C^∞ smooth.*

In the hypersurface case, Theorem 9.1 reduces to the following Fefferman's (1974) theorem (see e.g. [F3]).

Theorem 9.2. *Let $D_1, D_2 \subset \mathbf{C}^N$ be C^∞ smoothly bounded strictly pseudoconvex domains and let $F : D_1 \to D_2$ be a biholomorphic mapping. Then F is C^∞ up to bD_1.*

The original proof by Fefferman was quite difficult. Simpler proof were offered by Bell and Ligocka (1980), Nirenberg, Webster and Yang (1980), Lempert [L1], Pinchuk and Khasanov (1987), Forstnerič [F3]. We give another simple proof based on small extremal discs. It is essentially the same as the one by Lempert, but instead of rather difficult global results of [L1], we use the simpler local theory.

Proof. The first simple step, which we omit, consists of showing that F satisfies a Lipschitz condition with exponent $1/2$ in D_1 (Henkin, 1973) therefore F extends to a $C^{1/2}$ homeomorphism $bD_1 \to bD_2$, see [F3].

It is immediate that F maps extremal discs to extremal discs. By Corollary 7.4, the directions of the boundary curves of the extremal discs span all directions in $T(bD_1)$. By Proposition 4.3, the extremal discs are smooth. Therefore F maps a large family of smooth curves to smooth curves. The images are uniformly bounded in the $C^m, m \geq 1$, norms because one can see that the C^m norms of small extremal discs in the hypersurface case are estimated in terms of their $C^\alpha, 0 < \alpha < 1$, norms. Hence F is smooth on bD_1, and the proof is complete.

In higher codimension this proof does not work because we do not know whether all extremal discs are smooth up to the boundary. Forstnerič [F1], [F2] proved the smoothness of CR homeorphisms without the initial Lipschitz regularity but under some additional geometric restrictions on the manifolds or mapping. If the mapping F has initial C^1 regularity, then the C^∞ smoothness easily follows from the smooth reflection principle (Proposition 9.5) applied to the induced mapping $N^*(M_1) \to N^*(M_2)$.

We first show that the extension of F is locally biholomorphic.

Proposition 9.3. *Let M_1 and M_2 be smooth generic manifolds in $\mathbf{C}^N$, and let $F : M_1 \to M_2$ be a CR homeomorphism. Suppose M_1 is minimal (see e.g. [BER] [T1]; "generating Levi form" implies "minimal"). Let D be the interior of the union of all small analytic discs attached to M_1 and let F_1 be the holomorphic extension of F to D. Then the Jacobian determinant of F_1 does not vanish in D.*

Proof. Since M_1 is minimal, then $D \neq \emptyset$. By the Baouendi-Treves approximation theorem (see e.g. [BER] [Bo] [T1]), the mappings F and F^{-1} are limits of sequences of holomorphic polynomials. We define F_1 and F_2 as the limits of these sequences wherever they converge. In particular, F_1 is holomorphic in D and continuous up to M_1.

We claim that $F_2 \circ F_1 = \mathrm{id}$ in D. Indeed, for every analytic disc f_1 attached to M_1, the disc $f_2 = F_1 \circ f_1$ is attached to M_2. Then $F_2 \circ f_2$ is well defined. Then for $\zeta \in b\Delta$, $F_2 \circ F_1 \circ f_1(\zeta) = F^{-1} \circ F \circ f_1(\zeta) = f_1(\zeta)$. Hence $F_2 \circ F_1 \circ f_1 = f_1$. Since D is covered by the discs, then $F_2 \circ F_1 = \mathrm{id}$ in D, so F_1 is injective in D. Since F_1 is holomorphic, the Jacobian cannot vanish. The proof is complete.

The idea of the proof of the main result is that a CR mapping preserves the lifts of the extremal discs.

Proposition 9.4. *Let M_1 and M_2 be smooth generic manifolds in $\mathbf{C}^N$, and let $F : M_1 \to M_2$ be a CR homeomorphism such that both F and F^{-1} are C^α, $0 < \alpha < 1$. Let D and F_1 be the same as in Proposition 9.3. Let f_1 be a small stationary disc attached to M_1 such that $f(\Delta) \subset D$, and let f_1^* be a supporting lift of f_1. Then the disc $f_2 = F_1 \circ f_1$ is also stationary and $f_2^* = f_1^*(F_1' \circ f_1)^{-1}$ (where f_1^* and f_2^* are considered as row vectors) is a lift of f_2.*

We will prove Proposition 9.4 in Section 10.

We will use the following smooth version of the Schwarz reflection principle [PK].

Proposition 9.5. *Let $M_1 \subset \mathbf{C}^{n_1}$ and $M_2 \subset \mathbf{C}^{n_2}$ be $C^{m,\alpha}, m \geq 1, 0 < \alpha < 1$, maximally real manifolds, and let $F : M_1 \to M_2$ be a continuous mapping that holomorphically extends to a wedge W with edge M_1. Then F is $C^{m,\alpha}$ smooth.*

Proof of Theorem 9.1. We will use the notation F_1, F_2 and D introduced in the proof of Proposition 9.3. The mapping F_1 induces the mapping $\mathcal{F}_1 : T^*(D) \to T^*(\mathbf{C}^N)$. Using the identification $T^*(\mathbf{C}^N) \simeq \mathbf{C}^N \times \mathbf{C}^N$, the mapping $\mathcal{F}_1$ is defined as $\mathcal{F}_1(p, \phi) = (F_1(p), \phi F_1'(p)^{-1})$, where ϕ is considered a row vector.

By Proposition 8.4 there exist many non-defective stationary discs attached to M_1. Fix such a disc f_1 with supporting lift f_1^*. Let $f_2 = F_1 \circ f_1$.

By Proposition 9.4, for every stationary disc g_1 close to f_1, and its lift g_1^* close to f_1^*, the map $g_2^* = \mathcal{F}_1 \circ g_1^*$ is a lift of $g_2 = F_1 \circ g_1$. By Theorem 10.5, $g_2^*(\zeta) = a_2 \zeta^{-1} + b_2 + \dots$ is uniquely determined by and continuously depends on (g_2, a_2, b_2). The latter is uniquely determined by (g_1, a_1, b_1), where $g_1^*(\zeta) = a_1 \zeta^{-1} + b_1 + \dots$. The expressions of a_2 and b_2 in terms of a_1 and b_1 only involve F_1 in a neighborhood of $f_1(0)$. Hence for (g_1, g_1^*) close to (f_1, f_1^*), the lift g_2^* is uniformly bounded in the C^α norm on $b\Delta$.

By Theorem 7.3, the lifts g_1^* of g_1 cover an open wedge $W \subset T^*(D)$ with edge $N^*(M_1)$. Hence, $\mathcal{F}_1$ is C^α in W up to the edge $N^*(M_1)$

Let (p, ϕ) be a totally real point of $N^*(M_1)$. If $\mathcal{F}_1(p, \phi)$ also is a totally (maximally) real point of $N^*(M_2)$, then $\mathcal{F}_1$ is smooth at (p, ϕ) by Proposition 9.5, whence F is smooth at p as desired.

The difficulty is that a priori $\mathcal{F}_1(p, \phi)$ is not necessarily a totally real point. However, this difficulty is not essential and we do not address it here. See [T2] for the details.

10 Preservation of Lifts

We prove Proposition 9.4. The proof is based on another extremal property of analytic discs which is more suitable for application to CR mappings than Definition 3.1.

We call p a *real trigonometric polynomial* if it has the form $p(\zeta) = \sum_{j=-m}^{m} a_j \zeta^j$, where $a_{-j} = \bar{a}_j$. We call a real trigonometric polynomial p *positive* if $p(\zeta) > 0$ for $|\zeta| = 1$.

We put $\Delta_r = \{\zeta \in \mathbf{C} : |\zeta| < r\}$; $\Delta = \Delta_1$.

Recall the notation $\mathrm{Res}\phi = \mathrm{Res}(\phi, 0)$, the residue of ϕ at 0.

Definition 10.1. Let f be an analytic disc attached to a generic manifold $M \subset \mathbf{C}^N$. Let $f^* : \Delta \setminus \{0\} \to T^*(\mathbf{C}^N)$ be a holomorphic map with a pole of order at most 1 at 0. We say that the pair (f, f^*) has a *special extremal property* (SEP) if there exists $\delta > 0$ such that for every positive trigonometric polynomial p there exists $C \geq 0$ such that for every analytic disc $g : \Delta \to \mathbf{C}^N$ attached to M such that $\|g - f\|_{C(\bar{\Delta})} < \delta$ we have

$$\mathrm{Re}\,\mathrm{Res}(\zeta^{-1}\langle f^*, g - f \rangle p) + C\|g - f\|^2_{C(\bar{\Delta}_{1/2})} \geq 0. \tag{10.1}$$

The above extremal property is close to the one introduced by Definition 3.1. In particular, we note (Lemma 10.2) that stationary discs with supporting lifts have SEP. Conversely, we prove (Proposition 10.6) that SEP implies that f^* is a lift of f.

In formulating SEP we no longer restrict to the discs g with fixed center $g(0) = f(0)$. This helps prove Proposition 10.6 in case f is defective; see remark after Lemma 10.4. The radius $1/2$ plays no special role here. In Definition 10.1, we could even consider f^* defined only in a neighborhood of 0 and replace $1/2$ by a smaller number. Then SEP would still imply that f^* is a lift of f.

Proof of Proposition 9.4. Since f_1 has a supporting lift f_1^*, then the pair (f_1, f_1^*) has SEP. Then we prove (Lemma 10.3) that the pair (f_2, f_2^*) also has SEP. Then by Proposition 10.6, SEP implies that f_2^* is a lift of f_2 and the proposition follows.

Lemma 10.2. *Let f be a stationary disc attached to a generic manifold $M \subset \mathbf{C}^N$. Let f^* be a supporting lift of f. Then the pair (f, f^*) has SEP with $C = 0$.*

Proof. For every analytic disc g attached to M (not necessarily close to f), we have $\mathrm{Re}\langle f^*, g - f\rangle \geq 0$ on the unit circle $b\Delta$. Multiplying by a positive trigonometric polynomial p and integrating along the circle we immediately get (10.1) with $C = 0$. The lemma is proved.

Lemma 10.3. *Under the assumptions of Proposition 9.4, the pair (f_2, f_2^*) has SEP.*

Proof. For every small disc g_2 attached to M_2, we put $g_1 = F_2 \circ g_2$, where F_2 is the extension of F^{-1} as in Proposition 7.3. If g_2 is close to f_2 in the sup-norm, then for $\zeta \in \bar{\Delta}_{1/2}$ we have

$$|g_1(\zeta) - f_1(\zeta)| \leq C_1|g_2(\zeta) - f_2(\zeta)|,$$

where C_1 is the maximum of $\|F_2'\|$, the norm of the derivative of F_2 in a neighborhood of the compact set $f_2(\bar{\Delta}_{1/2})$. Likewise, for $\zeta \in \bar{\Delta}_{1/2}$ we have

$$
\begin{aligned}
g_2(\zeta) - f_2(\zeta) &= F_1(g_1(\zeta)) - F_1(f_1(\zeta)) = \\
&= F_1'(f_1(\zeta))(g_1(\zeta) - f_1(\zeta)) + R(\zeta)|g_1(\zeta) - f_1(\zeta)|^2,
\end{aligned}
$$

where $|R(\zeta)| \leq C_2$, and C_2 is the maximum of $\|F_1''\|$ in a neighborhood of the compact set $f_1(\bar{\Delta}_{1/2})$. For every positive trigonometric polynomial p, recalling that $f_2^* = f_1^*(F_1' \circ f_1)^{-1}$, we obtain

$$
\begin{aligned}
|\mathrm{Res}(\zeta^{-1}\,&\langle f_2^*, g_2 - f_2\rangle p) - \mathrm{Res}(\zeta^{-1}\langle f_1^*, g_1 - f_1\rangle p)| \\
&= \left|\tfrac{1}{2\pi i}\int_{|\zeta|=1/2}\zeta^{-1}\langle f_2^*(\zeta), R(\zeta)|g_1(\zeta) - f_1(\zeta)|^2\rangle p(\zeta)\,d\zeta\right| \quad (10.2) \\
&\leq C_1^2 C_2\|p f_2^*\|_{C(b\Delta_{1/2})}\|g_2 - f_2\|_{C(\bar{\Delta}_{1/2})}^2
\end{aligned}
$$

Now by Lemma 4, $\mathrm{Re}\,\mathrm{Res}(\zeta^{-1}\langle f_1^*, g_1 - f_1\rangle p) \geq 0$, and (10.2) implies that (f_2, f_2^*) has SEP. The proof is complete.

A (tangential) *infinitesimal perturbation* of an analytic disc f attached to M is a continuous mapping $\dot{f} : \bar{\Delta} \to T(\mathbf{C}^N)$ holomorphic in Δ such that $\dot{f}(\zeta) \in T_{f(\zeta)}(M)$ for $\zeta \in b\Delta$. Infinitesimal perturbations $\dot{f} = (\dot{z}, \dot{w})$ are solutions of the linearized Bishop equation

$$\dot{y} = T(h_y\dot{y} + h_w\dot{w} + h_{\bar{w}}\dot{\bar{w}}) + \dot{y}(0).$$

In particular, it follows that $\dot{y}(0)$ and $\dot{w}(0)$ can be chosen arbitrarily.

Lemma 10.4. *Assume a pair (f, f^*) has SEP, where f is a small analytic disc of class $C^\alpha(\bar{\Delta})$ attached to a generic manifold M. Then for every infinitesimal perturbation $\dot{f}$ of f of class $C^\alpha(\bar{\Delta})$, and every real trigonometric polynomial p, we have*

$$\operatorname{Re}\operatorname{Res}(\zeta^{-1}\langle f^*, \dot{f}\rangle p) = 0. \tag{10.3}$$

Proof. Let $\dot{f} = (\dot{z}, \dot{w})$. By solving Bishop's equation (4.2) with $w(\zeta, t) = w(\zeta) + t\dot{w}(\zeta)$, we construct a one parameter family of discs $\zeta \mapsto g(\zeta, t)$ defined for small $t \in \mathbf{R}$ so that $g(\zeta, 0) = f(\zeta)$ and $\frac{d}{dt}\big|_{t=0} g = \dot{f}$. Plugging g in (10.1) we get

$$\operatorname{Re}\operatorname{Res}(\zeta^{-1}\langle f^*, g - f\rangle p) + O(t^2) \geq 0.$$

Differentiating at $t = 0$, we obtain (10.3) for every positive trigonometric polynomial p. Since positive trigonometric polynomials span the set of all real trigonometric polynomials, then the lemma follows.

Remark. We note that if SEP only held for the discs g with fixed center $g(0) = f(0)$, then in the last proof we would have to realize an infinitesimal perturbation $\dot{f}$ with $\dot{f}(0) = 0$ by a family $\zeta \mapsto g(\zeta, t)$ with fixed center $g(0, t) = f(0)$. However, it turns out to be a problem if f is defective.

We observe that for $p \equiv 1$, the condition (10.3) takes the form

$$\operatorname{Re}(\langle \bar{a}, \dot{f}'(0)\rangle + \langle \bar{b}, \dot{f}(0)\rangle) = 0, \tag{10.4}$$

where $\bar{a}$ and $\bar{b}$ are the first two Laurent coefficients of f^*, that is

$$f^*(\zeta) = \bar{a}\zeta^{-1} + \bar{b} + \dots \tag{10.5}$$

The following theorem and its proof are similar to those of Proposition 4.2. The advantage of the new result is that it holds even if the disc f is defective.

Theorem 10.5. *Let f be a small analytic disc of class $C^\alpha(\bar{\Delta})$ attached to a generic manifold $M \subset \mathbf{C}^N$. Assume $a, b \in \mathbf{C}^N$ are such that (10.4) holds for every infinitesimal perturbation $\dot{f}$ of f of class $C^\alpha(\bar{\Delta})$. Then there exists a unique lift f^* of f of the form (10.5). Moreover, the correspondence $(f, a, b) \mapsto f^*$ is continuous in the $C^\alpha(b\Delta)$ norm.*

The proof is quite technical and we omit it for the same reason as for Proposition 4.2. See [T3] for the proof.

Proposition 10.6. *Assume a pair (f, f^*) has SEP, where f is a small analytic disc of class $C^\alpha(\bar{\Delta})$ attached to a generic manifold M. Then f^* is a lift of f.*

Proof. Let $f^*(\zeta) = \bar{a}\zeta^{-1} + \bar{b} + \dots$. Then by Lemma 10.4 we have (10.3), which implies (10.4). By Theorem 10.5, there exists a lift $\tilde{f}^*$ of f such that $\tilde{f}^*(\zeta) = \bar{a}\zeta^{-1} + \bar{b} + \dots$. Then $\psi(\zeta) = \zeta^{-1}(\tilde{f}^*(\zeta) - f^*(\zeta))$ is holomorphic in Δ and $\operatorname{Re}\operatorname{Res}(\langle\psi, \dot{f}\rangle p) = 0$ for every infinitesimal perturbation $\dot{f}$ and real trigonometric polynomial p. We will show that this implies $\psi \equiv 0$, whence $f^* = \tilde{f}^*$ is a lift of f.

Take $p(\zeta) = c\zeta^m + \bar{c}\zeta^{-m}$, where $c \in \mathbf{C}$ and $m > 0$ is integer. Put $h = \langle \psi, \dot{f} \rangle$. Then $0 = \operatorname{Re}\operatorname{Res}(\langle \psi, \dot{f} \rangle p) = \operatorname{Re}\operatorname{Res}(hp) = \operatorname{Re}(\bar{c}\operatorname{Res}(h\zeta^{-m}))$ for all $c \in \mathbf{C}$. Then $\operatorname{Res}(h\zeta^{-m}) = 0$ for all integers $m > 0$. Hence $h \equiv 0$.

Now for every $\dot{f}$ we have $\langle \psi, \dot{f} \rangle = 0$. If ψ is not identically equal to zero, then $\psi(\zeta) = \lambda\zeta^m + O(\zeta^{m+1})$, for some integer $m \geq 0$ and $\lambda \neq 0$. Then $\langle \lambda, \dot{f}(0) \rangle = 0$ for all $\dot{f}$. Note the subspace $\{\dot{f}(0)\}$ spans $\mathbf{C}^N$ over $\mathbf{C}$ because the w- and y-components of $\dot{f}(0)$ are arbitrary. Hence $\lambda = 0$ and we come to a contradiction. The proof is complete.

References

[BER] M. S. Baouendi, P. Ebenfelt, and L. P. Rothschild, *Real Submanifolds in Complex Space and their Mappings*, Princeton Math. Series 47, Princeton Univ. Press, 1999.

[Bl] J. Bland, Contact geometry and CR structures on S^3, *Acta Math.* **172** (1994), 1–49.

[BD] J. Bland, T. Duchamp, Moduli for pointed convex domains, *Invent. Math.* **104** (1991), 61–112.

[Bo] A. Boggess, *CR manifolds and the tangential Cauchy-Riemann complex*, CRC Press, 1991.

[F1] F. Forstnerič, Mappings of strongly pseudoconvex Cauchy-Riemann manifolds, *Proc. Sump. Pure Math.* **52**, Part 1, 59–92, Amer. Math. Soc., Providence, 1991.

[F2] —————— , A reflection principle on strongly pseudoconvex domains with generic corners, *Math. Z.* **213** (1993), 49–64.

[F3] —————— , An elementary proof of Fefferman's theorem, *Exposition Math.* **10** (1992), 135–149.

[L1] L. Lempert, La métrique de Kobayashi et la représentation des domaines sur la boule, *Bull. Soc. Math. France* **109** (1981), 427–474.

[L2] —————— , Holomorphic invariants, normal forms, and the moduli space of convex domains, *Ann. of Math.* (2) **128** (1988), 43–78.

[PK] S. I. Pinchuk, S. V. Khasanov, Asymptotically holomorphic functions and their applications (Russian), *Mat. Sb. (N.S.)* **134(176)** (1987), 546–555.

[Sc] A. Scalari, Extremal discs and CR geometry, Ph.D. thesis, University of Illinois at Urbana-Champaign, 2001.

[Se] S. Semmes, A Generalization of Riemann Mappings and Geometric Structures on a Space of Domains in $\mathbf{C}^n$, *Mem. Am. Math. Soc.* **472**, Providence, 1992.

[Sl] Z. Slodkowski, Polynomial hulls with convex fibers and complex geodesics, *J. Funct. Anal.* **94** (1990), 156–176.

[T1] A. Tumanov, Analytic discs and the extendibility of CR functions. *Integral Geometry, Radon transforms, and Complex Analysis*, CIME Session, Venice, 1996 (*Lect. Notes in Math.* 1684, 123–141) Springer 1998.

[T2] —————— , Extremal discs and the regularity of CR mappings in higher codimension, *Amer. J. Math.* **123** (2001), 445–473.

[T3] —————— , On the regularity of CR mappings in higher codimension, Preprint, http://arxiv.org/abs/math.CV/0208103

List of Participants

1. Abate Marco, (lecturer)
   Università di Pisa, Italy
   abate@dm.unipi.it
2. Altomani Andrea
   Scuola Normale Superiore Pisa, Italy
   altomani@sns.it
3. Baracco Luca
   Università di Padova, Italy
   baracco@math.unipd.it
4. Barletta Elisabetta
   Università della Basilicata, Italy
   barletta@unibas.it
5. De Fabritiis Chiara
   Università di Ancona, Italy
   fabritii@dipmat.unian.it
6. Dini Gilberto
   Università di Firenze, Italy
   dini@math.unifi.it
7. Dragomir Sorin
   Università della Basilicata, Italy
   dragomir@unibas.it
8. Egorov Georgy
   Moscow State University, Russia
   egorovg@online.ru
9. Fornaess John Erik, (lecturer)
   University of Michigan, USA
   fornaess@umich.edu
10. Frosini Chiara
    Università di Firenze, Italy
    frosini@math.unifi.it
11. Geatti Laura
    Università di Roma 2, Italy
    geatti@mat.uniroma2.it
12. Huang Xiaojun, (lecturer)
    Rutgers University, USA
    huangx@math.rutgers.edu
13. Iannuzzi Andrea
    Università di Bologna, Italy
    iannuzzi@dm.unibo.it
14. Irgens Marius
    University of Michigan, USA
    irgens@umich.edu
15. Kazilo Aleksandra
    Moscow State University, Russia
    kazilo@mccme.ru
16. Kolar Martin
    Masartk University, Czech Rep.
    mkolar@math.muni.cz
17. Manjarin Monica
    Univ. Autonoma de Barcelona, Spain
    manjarin@mat.uab.es
18. Meneghini Claudio
    Università di Parma, Italy
    clamen@dimat.unipv.it
19. Minervini Giulio
    Università di Roma 1, Italy
    minervin@mat.uniroma1.it
20. Morando Giovanni
    Università di Padova, Italy
    gmorando@math.unipd.it
21. Morsli Nadia
    Univ. Sidi Bel Abbes, Algeria
    m_nadia_99@yahoo.fr
22. Munteanu Marian Ioan
    Univ. Al. I. Cuza Iasi, Romania
    munteanu2001@hotmail.com
23. Nordine Mir
    Univ. De Rouen, France
    Nordine.Mir@univ-rouen.fr
24. Parrini Carla
    Università di Firenze, Italy
    parrini@math.unifi.it
25. Pereldik Natalia
    Moscow State University, Russia
    pereldik@mccme.ru
26. Perotti Alessandro
    Università di Trento, Italy
    perotti@science.unitn.it

27. Peters Han
University of Michigan, USA
hanpet@umich.edu

28. Prelli Luca
Università di Padova, Italy
lprelli@libero.it

29. Prezelj Perman Jasna
Univ. Ljubljana, Slovenia
jasna.prezelj@fmf.uni-lj.si

30. Rosay Jean Pierre, (lecturer)
University of Wisconsin, USA
jrosay@math.wisc.edu

31. Sahraoui Fatiha
University of Sidi Bel Abbes, Algeria
douhy_fati@yahoo.fr

32. Scalari Alberto
Università di Padova, Italy
alberto.scalari@socgen.com

33. Selvaggi Primicerio Angela
Università di Firenze, Italy
asprimi@unifi.it

34. Siano Anna
Università di Padova, Italy
asiano@studenti.math.unipd.it

35. Tumanov Alexander, (lecturer)
University of Illinois
tumanov@uiuc.edu

36. Vlacci Fabio
Università di Firenze, Italy
fabio.vlacci@math.unifi.it

37. Zaitsev Dimitri, (editor)
University of Tuebingen, Germany
dimitri.zaitsev@uni-tuebingen.de

38. Zampieri Giuseppe, (editor)
Univesità di Padova, Italy
zampieri@math.unipd.it

39. Walker Ronald
University of Michigan, USA
rawalker@umich.edu

40. Wolf Christian
Insituto Sup. Tecnico, Germany
cwolf@math.ist.utl.pt

LIST OF C.I.M.E. SEMINARS

1996	126. Integral Geometry, Radon Transforms and Complex Analysis	(LNM 1684)	Springer-Verlag
	127. Calculus of Variations and Geometric Evolution Problems	(LNM 1713)	"
	128. Financial Mathematics	(LNM 1656)	"
1997	129. Mathematics Inspired by Biology	(LNM 1714)	"
	130. Advanced Numerical Approximation of Nonlinear Hyperbolic Equations	(LNM 1697)	"
	131. Arithmetic Theory of Elliptic Curves	(LNM 1716)	"
	132. Quantum Cohomology	(LNM 1776)	"
1998	133. Optimal Shape Design	(LNM 1740)	"
	134. Dynamical Systems and Small Divisors	(LNM 1784)	"
	135. Mathematical Problems in Semiconductor Physics	(LNM 1823)	"
	136. Stochastic PDE's and Kolmogorov Equations in Infinite Dimension	(LNM 1715)	"
	137. Filtration in Porous Media and Industrial Applications	(LNM 1734)	"
1999	138. Computational Mathematics driven by Industrial Applications	(LNM 1739)	"
	139. Iwahori-Hecke Algebras and Representation Theory	(LNM 1804)	"
	140. Theory and Applications of Hamiltonian Dynamics	to appear	"
	141. Global Theory of Minimal Surfaces in Flat Spaces	(LNM 1775)	"
	142. Direct and Inverse Methods in Solving Nonlinear Evolution Equations	(LNP 632)	"
2000	143. Dynamical Systems	(LNM 1822)	"
	144. Diophantine Approximation	(LNM 1819)	"
	145. Mathematical Aspects of Evolving Interfaces	(LNM 1812)	"
	146. Mathematical Methods for Protein Structure	(LNCS 2666)	"
	147. Noncommutative Geometry	(LNM 1831)	"
2001	148. Topological Fluid Mechanics	to appear	"
	149. Spatial Stochastic Processes	(LNM 1802)	"
	150. Optimal Transportation and Applications	(LNM 1813)	"
	151. Multiscale Problems and Methods in Numerical Simulations	(LNM 1825)	"
2002	152. Real Methods in Complex and CR Geometry	to appear	"
	153. Analytic Number Theory	to appear	"
	154. Imaging	to appear	"
2003	155. Stochastic Methods in Finance	to appear	
	156. Hyperbolic Systems of Balance Laws	to appear	
	157. Symplectic 4-Manifolds and Algebraic Surfaces	to appear	
	158. Mathematical Foundation of Turbulent Viscous Flows	to appear	
2004	159. Representation Theory and Complex Analysis	announced	
	160. Nonlinear and Optimal Control Theory	announced	
	161. Stochastic Geometry	announced	

Fondazione C.I.M.E.

Centro Internazionale Matematico Estivo
International Mathematical Summer Center
http://www.math.unifi.it/~cime
cime@math.unifi.it

2004 COURSES LIST

Representation Theory and Complex Analysis

June 10–17, Venezia

Course Directors:

Prof. Enrico Casadio Tarabusi (Università di Roma "La Sapienza")
Prof. Andrea D'Agnolo (Università di Padova)
Prof. Massimo A. Picardello (Università di Roma "Tor Vergata")

Nonlinear and Optimal Control Theory

June 21–29, Cetraro (Cosenza)

Course Directors:

Prof. Paolo Nistri (Università di Siena)
Prof. Gianna Stefani (Università di Firenze)

Stochastic Geometry

September 13–18, Martina Franca (Taranto)

Course Director:

Prof. W. Weil (Univ. of Karlsruhe, Karlsruhe, Germany)

Printing: Strauss GmbH, Mörlenbach
Binding: Schäffer, Grünstadt